SPRINGER HANDBOOK OF
AUDITORY RESEARCH

Series Editors: Richard R. Fay and Arthur N. Popper

Springer

New York
Berlin
Heidelberg
Barcelona
Hong Kong
London
Milan
Paris
Singapore
Tokyo

Springer Handbook of Auditory Research

Richard R. Fay
Arthur N. Popper
Editors

Comparative Hearing: Fish and Amphibians

With 116 Illustrations

Springer

Richard R. Fay
Parmly Hearing Institute and
Department of Psychology
Loyola University of Chicago
Chicago, IL 60626, USA

Arthur N. Popper
Department of Zoology
University of Maryland
College Park, MD 20742, USA

Series Editors: Richard R. Fay and Arthur N. Popper

Cover illustration: Evolutionary changes in the sarcopterygian ear.
The full figure appears on p. 29 of the text.

Library of Congress Cataloging in Publication Data
Comparative hearing. Fish and amphibians / [edited by] Richard R.
   Fay, Arthur N. Popper.
      p.   cm.—(Springer handbook of auditory research; v. 11)
   Includes bibliographical references (p.   ) and index.
   ISBN 0-387-98470-4 (hardcover : alk. paper)
      1. Hearing.   2. Fish—Physiology.   3. Amphibians—Physiology.
   I. Fay, Richard R.   II. Popper, Arthur N.   III. Series.
   QP461.C635   1998
   573.8'917—dc21   97-52161

Printed on acid-free paper.

Production managed by Terry Kornak; manufacturing supervised by Jacqui Ashri
Typeset by Best-set Typesetter Ltd., Hong Kong
Printed and bound by Maple-Vail Book Manufacturing Group, York, PA.
Printed in the United States of America.

9 8 7 6 5 4 3 2 1

ISBN 0-387-98470-4   Springer-Verlag New York Berlin Heidelberg   SPIN 10664783

The volume is dedicated to, and honors, the work of Robert R. Capranica and William N. Tavolga. Drs. Capranica and Tavolga are pioneers in leading to our current understanding of hearing and sound communication in fish and amphibians and they are the scientists who have ultimately been responsible for inspiring much of the work in this volume. In addition to being outstanding scientists and leaders in their fields, both Bill and Bob are exceptional teachers and role models. Most authors in this volume can proudly claim one or both of these men as valued friends, and either Bill or Bob as their academic parent or grandparent.

Robert R. Capranica

William N. Tavolga

Series Preface

The *Springer Handbook of Auditory Research* presents a series of comprehensive and synthetic reviews of the fundamental topics in modern auditory research. The volumes are aimed at all individuals with interests in hearing research, including advanced graduate students, postdoctoral researchers, and clinical investigators. The volumes are intended to introduce new investigators to important aspects of hearing science and to help established investigators to better understand the fundamental theories and data in fields of hearing that they may not normally follow closely.

Each volume is intended to present a particular topic comprehensively, and each chapter serves as a synthetic overview and guide to the literature. As such, the chapters present neither exhaustive data reviews nor original research that has not yet appeared in peer-reviewed journals. The volumes focus on topics that have developed a solid data and conceptual foundation, rather than on those for which a literature is only beginning to develop. New research areas will be covered on a timely basis in the series as they begin to mature.

Each volume in the series consists of five to eight substantial chapters on a particular topic. In some cases, the topics will be ones of traditional interest for which there is a substantial body of data and theory, such as auditory neuroanatomy (Vol. 1) and neurophysiology (Vol. 2). Other volumes in the series will deal with topics that have begun to mature more recently, such as development, plasticity, and computational models of neural processing. In many cases, the series editors will be joined by a co-editor having special expertise in the topic of the volume.

Richard R. Fay, Chicago, IL
Arthur N. Popper, College Park, MD

Preface

A major goal of the study of hearing is to explain how the human auditory system normally functions, and to help identify the causes of and treatments for hearing impairment. Experimental approaches to these questions make use of animal models, and the validity of these models will determine the success of this effort. Comparative hearing research establishes the context within which animal models can be developed, evaluated, validated, and successfully applied, and is therefore of fundamental importance to hearing research in general. For example, the observation that hair cells may regenerate in the ears of some anamniotes cannot be evaluated for its potential impact on human hearing without a comparative and evolutionary context within which observations on the animal model and humans can fit and be fully understood, and within which we can generalize observations from one species to another with any confidence. This confidence arises from comparative research that investigates the diverse structures, physiological functions, and hearing abilities of various vertebrate and invertebrate species in order to determine the fundamental principles by which structures are related to functions, and to help clarify the phyletic history of the auditory system among animals.

In light of the fundamental importance of comparative hearing research, we have decided to incorporate into the *Springer Handbook of Auditory Research* a series of books focusing specifically on hearing in nonhuman animals. The first in this series was *Comparative Hearing: Mammals* (editors Fay and Popper) and the companion volume *Hearing by Bats* (editors Popper and Fay). In addition to this current volume, recently published was *Comparative Hearing: Insects* (editors Hoy, Popper, and Fay), while soon to be forthcoming are *Comparative Hearing: Reptiles and Birds* (editors Dooling, Popper, and Fay), and *Comparative Hearing: Whales and Dolphins* (editors Au, Popper, and Fay). Our feeling is that in providing a comprehensive introduction to comparative hearing we will help all investigators in our field have a better appreciation for the diversity of animal models that could be incorporated into future research programs.

This volume deals with hearing by fishes and amphibians. In putting together these chapters we decided to take a new approach in which the authors combine material from these two groups in order to provide fresh views on comparative issues and themes. In Chapter 1, Fay and Popper provide an overview of the volume, and set out some of the dimensions on which the auditory systems of fishes and amphibians show similarities and differences. In Chapter 2, Fritzsch provides an overview of the main "cast of characters" among the fish and amphibians and puts their taxonomy into some perspective. He also provides a view into one of the most fascinating of all problems—how the ear changed when our sarcopterygian ancestors moved from living exclusively in water to living part-time, and then full-time, on land. Popper and Fay, in Chapter 3, discuss the peripheral structures and functions of the ear and auditory nerve in fishes, and Lewis and Narins do the same for amphibians in Chapter 4. In both chapters, issues of structure are combined with issues of function revealed primarily in the neural codes of the auditory nerve. We chose not to combine fishes and amphibians in these chapters since the literature is large and the ear structures are quite different. Fishes and amphibians are brought together in the remaining chapters. In Chapter 5, McCormick discusses the anatomy of the CNS, and its physiology is treated in Chapter 6 by Feng and Schellart. Psychophysics and complex sound source perception in fishes and amphibians, including the issues of sound source localization, are discussed in Chapter 7 by Fay and Megela-Simmons. The "other" part of the octavolateralis system, the lateral line, is considered in Chapter 8 by Coombs and Montgomery. It is now clear that the lateral line system and its functions are to a large degree independent of the auditory system. At the same time the two systems share many features including common sensory hair cells, a parallel central organization, the function of determining external sources, and perhaps evolutionary origins. Zelick, Mann, and Popper discuss the questions of acoustic communication in Chapter 9. Considering communication in fishes and amphibians together helps to balance the view, implicit in the experimental literature, that acoustic communication is more important for anurans than for fishes.

The chapters in this volume are closely related to chapters in other volumes of the *Springer Handbook of Auditory Research*. The chapters on structure and physiology of the auditory periphery in fish (Popper and Fay) and amphibians (Lewis and Narins) are closely related to the chapter on the mammalian ear by Slepecky in Vol. 8 of the series (*The Cochlea*), the chapter by Echteler, Fay, and Popper in Vol. 4 (*Comparative Hearing: Mammals*), the chapter on bat ears by Kössl and Vaster in Vol. 5 (*Hearing by Bats*), and the chapter on encoding in the auditory nerve by Ruggero in Vol. 2 of the series (*The Mammalian Auditory Pathway: Neurophysiology*). The material on CNS physiology by Feng and Schellart is complemented by many of the chapters in *The Mammalian Auditory Pathway: Neurophysiology*, while the chapter on CNS anatomy by McCormick has many parallels in the chapters of Vol. 1 (*The Mammalian Auditory Pathway: Neuroanatomy*). Finally, the chapter by Fay and Megela-Simmons in this volume closely parallels many of the chapters in Vol. 3 (*Human Psychophysics*).

Richard R. Fay, Chicago, Illinois
Arthur N. Popper, College Park, Maryland
September, 1997

Contents

Contributors

Sheryl Coombs
Parmly Hearing Institute, Loyola University of Chicago, Chicago, IL 60626, USA

Richard R. Fay
Parmly Hearing Institute and Department of Psychology, Loyola University of Chicago, Chicago, IL 60626, USA

Albert S. Feng
Department of Molecular and Integrative Physiology and Beckman Institute, University of Illinois, Urbana, IL 61801, USA

Bernd Fritzsch
Department of Biomedical Sciences, Creighton University, Omaha, NE 68178, USA

Edwin R. Lewis
Department of Electrical Engineering and Computer Science, University of California, Berkeley, CA 94720, USA

Catherine A. McCormick
Department of Biology, Oberlin College, Oberlin, OH 44074, USA

John C. Montgomery
School of Biological Science, University of Auckland, Private Bag 92019, Auckland, New Zealand

David A. Mann
Department of Biology, University of Maryland, College Park, MD 20742, USA

Peter M. Narins
Department of Physiological Science, University of California, Los Angeles, CA 90024, USA

Arthur N. Popper
Department of Biology, University of Maryland, College Park, MD 20742, USA

Nico A.M. Schellart
Laboratory of Medical Physics and Graduate School of Neurosciences, Amsterdam, University of Amsterdam, 1105 AZ Amsterdam, The Netherlands

Andrea Megela Simmons
Departments of Psychology and Neuroscience, Brown University, Providence, RI 02912, USA

Randy Zelick
Department of Biology, Portland State University, Portland, OR 97207, USA

1
Hearing in Fishes and Amphibians: An Introduction

Richard R. Fay and Arthur N. Popper

1. Introduction

As noted in the Preface to this volume, a major goal of hearing research is to explain how the human auditory system normally functions and to help identify the causes of, and treatments for, hearing impairment. Animal models are used extensively in this research, and valid generalizations from these models are required for progress to be made in understanding the human auditory system. In general, comparative hearing research establishes the biological and evolutionary context within which animal models can be developed, evaluated, validated, and successfully applied, and is therefore of fundamental importance to hearing research. The confidence that allows us to generalize some observations from one species to another arises from comparative research that investigates the structures, physiological functions, and hearing capabilities of various species in order to determine the fundamental principles by which structures determine functions in all vertebrate auditory systems.

This volume brings together our current understanding of the structures and functions of the auditory systems of two of the major vertebrate classes, fishes and amphibians. This comparison is rarely attempted in the literature, in part because of the breadth of knowledge and detail available for both groups, but more fundamentally because of traditional differences between the theoretical and experimental paradigms that underlie work on these groups. In most chapters of this volume, fishes and amphibians are treated together and compared in order to help us gain insight into broader comparative issues than would be possible if each group were treated separately. In chapters on the auditory periphery, however, the material to be integrated is so great that a single, combined chapter could not do justice to both groups.

2. Paradigms

Among anuran amphibians, the auditory system has tended to be seen as a component of a communication system because of the obvious importance of intraspecific vocal communication for the social and reproductive life of these animals (see Zelick, Mann, and Popper, Chapter 9). Thus, a great body of work on amphibian hearing has been in the tradition of neuroethology, with a focus on species-specific adaptations or specializations of the auditory system. One view arising from the literature on anurans is that the auditory system and sense of hearing of a given species may represent a set of special adaptations for the detection and processing of the species' own vocal sounds. The neuroethological paradigm is ideal for investigating species differences and special adaptations of the auditory system. However, the paradigm's focus on communication behaviors is less well suited for investigating those aspects of the sense of hearing that are primitive or shared among taxa. Thus, the dominant view of anuran auditory systems tends to emphasize those species-specific features that are revealed by the neuroethological paradigm as components of a communication system.

As discussed by Fay and Megela Simmons in Chapter 7, the traditions of the literature on hearing in fishes are quite different from those for anurans. Modern approaches to studying acoustic behavior began with the work in the early part of the 20th century of Parker, Bigelow, and Manning, and continued with the work of Karl von Frisch and his colleagues (these papers are all available in Tavolga 1976) using conditioning and psychophysical approaches (see Popper and Fay, Chapter 3). This early work led to the more recent psychophysical experiments by Tavolga and his colleagues (e.g., Tavolga and Wodinsky 1963), and to most of the recent behavioral work (see review by Fay 1988).

Conditioning paradigms often use simple and arbitrary sound stimuli that may not have any known communication value or other biological significance to the species studied. Moreover, psychophysical paradigms applied to fishes, as well as to many other vertebrates, tend to be derived from the mature literature on human hearing that takes an engineering approach to the auditory system as a generalized signal processor. Such studies have certainly revealed species differences, particularly in the frequency response of sound detection. However, the general view revealed by this paradigm is that of a set of shared or common principles of auditory reception and processing (see Popper and Fay 1997).

It has recently been concluded (Fay 1995) that this approach has so far failed to reveal fundamental qualitative differences between goldfish (*Carassius auratus*) and humans in auditory perception. Although the goldfish is the only fish species systematically studied with respect to complex sound perception, it appears likely that this fundamental similarity between the two species will be shared with other fishes, and with other vertebrates as well (Hulse et al. 1997). Since no clear species-specific specializations for speech perception have been revealed in psychoacoustical studies on humans, and since goldfish are not known to vocalize, we have concluded that the psychophysical approach to analyzing hearing capacities in these two species tends to reveal fundamental hearing processes that are shared and independent of vocal communication signals. Thus, the dominant view of fish auditory systems tends to emphasize those shared or primitive features that are revealed by the conditioning/psychophysical paradigm as components of a general signal processing system.

It is not likely that these two contrasting views are entirely adequate for the group to which they have been applied. Neither is it likely that either view is essentially incorrect or irrelevant. One of the motivations for combining discussions of fishes and amphibians in this volume is to suggest that both views tend to follow from the paradigms applied, and that both have value for understanding the dimensions on which fishes, anurans, and perhaps most vertebrate species show both differences and fundamental similarities in their sense of hearing.

3. What Do Fishes and Amphibians Listen To?

One of the potential difficulties in accepting the notion that vertebrates may share many fundamental features of a common sense of hearing is in formulating an acceptable answer to the question, What do fish (or any other group or species) listen to (Rogers 1986)? This question does not naturally arise for most anurans simply because their vocal and phonotactic behaviors are so well known and so obvious even to the most casual observer of a pond in early evening in the spring. In general, there is a tendency to accept vocal communication as a potential explanation for some of the characteristics of auditory systems (e.g., the matched filter hypothesis [Capranica and Moffat 1983]). The existence of vocal communication behaviors leaves no doubt that at least some sounds are biologically significant for a given species. However, there are many species that are not known to vocalize or otherwise communicate using sound (e.g., goldfish, salamanders), and we must confront the possibility that known communication signals cannot explain, in any satisfactory way, the fundamental characteristics of vertebrate auditory systems. So we are left with questions regarding the most general functions of auditory systems.

Rogers (1986) considered this question for goldfish and concluded that they listen to ambient noise. Rogers argued that any sound-scattering object in the nearby environment perturbs the ambient noise field and thereby potentially gives away its presence, and possibly its identity and location. This is an interesting hypothesis because it implies that continuous broadband noise may be as biologically significant as communication sounds, and that signals may consist of scattered noises. Thus, this hypothesis suggests that goldfish may receive information about the objects and events in their environment by listening to the local ambience, or to all detectable sounds. Although this suggestion may appear disappointingly inclusive, vague, and thus possibly meaningless, it is closely related to the recently proposed notion that the basic function of the human sense of hearing is to apprehend the "auditory scene" (Bregman 1990).

Experimental work on auditory scene analysis tends to focus anthropocentrically on speech and music perception (Handel 1989). However, the conceptual kernel of scene analysis is that the world of auditory objects and events is perceived as a sort of image. Although humans appear to be able to attend to only one object at a time, the overall effect is knowledge about the identity and location of the many sound-producing and sound-scattering objects that normally surround us. Stephen Hawking (1988) has reasoned that, "in any population of self-reproducing organisms, there will be variations in the genetic material and upbringing that different individuals have. These differences will mean that some individuals are better able than others to draw the right conclusions about the world around them, and to act accordingly. These individuals will be more likely to survive and reproduce and so their pattern of behavior and thought will come to dominate" (p. 12).

In the most general sense, "drawing the right conclusions about the world" would seem to be of primary survival value to individuals of all species, and the auditory system has probably had an important, general role to play in informing individuals about the world of sound sources and scatterers (Fay 1992). This notion is in accord with Rogers's (1986) speculation that goldfish listen to ambient noise, and may be a useful conception in considering the selective pressures that shaped the earliest auditory systems among fishes, and that have continued to maintain the auditory systems of extant species. Perhaps the shared characteristics of vertebrate auditory systems and sense of hearing have been successful adaptations to the physics of media and materials and the general requirements of auditory scene analysis. In this conception, we may not have to search for, or expect to find, specific biologically significant signals as an explanation for the structures and functions of most auditory systems.

4. What Is Shared and What Is Derived?

Comparisons of fishes and amphibians with respect to auditory structures and functions are not often made. To be sure, there are wide gulfs separating these groups, including a primarily aquatic versus terrestrial habitat, different sound conduction pathways to the ears, possibly nonhomologous auditory end organs, different paradigms used in behavioral studies, and somewhat different assumptions about what the two groups listen to and the relative importance of acoustic communication for behavior. Yet there are also many similarities, including the presence of a lateral line system in some species, the use of the saccule in acoustic motion detection, the existence of direct and indirect pathways of sound to the ears, inherent directionality of the ears, peripheral neural codes sharing features such as frequency selectivity and phase-locking, central auditory pathways that are generally similar in morphotype with several homologous nuclei, similar auditory capabilities revealed in conditioning and ethological studies, and the use of sound communication. Thus, while few peripheral structures are homologous, the functions of the auditory system, and the sense of hearing itself, are closely analogous and possibly homologous.

4.1 Sound Conduction Pathways

The mechanical pathways that transmit sound energy to the ears are discussed in this volume by Popper and Fay (Chapter 3) and by Lewis and Narins (Chapter 4) for fishes and amphibians, respectively. Among fishes, at least two major pathways for sound to get to the ear have been identified. The first, and most primitive, is the conduction of sound directly from the water medium to tissue and bone. In this case, the fish's body takes up the sound's acoustic particle motion, and hair cell stimulation occurs due to the difference in inertia (and resultant motion) between the hair cells and their overlying otoliths. The proximal stimulus is acoustic particle motion (displacement, velocity, or acceleration), but one could view this motion as resulting from pressure gradients. This mode of stimulation is a characteristic shared by all fishes, and is inherently directional because particle motion is a vector quantity.

The second, probably derived, sound pathway to the ears is indirect; the swim bladder or other gas bubble effectively near the ears expands and contracts in volume in response to sound pressure fluctuations, and this motion may be transmitted to the otoliths. This pathway is nondirectional in the sense that sound pressure is a scalar quantity, and because the indirect signal to the ears always arises from the gas bubble source, regardless of the location of the original sound source. Only some species of fish appear to be sound pressure sensitive via an effective indirect pathway to the ears (e.g., the otophysans [goldfish, catfish, and relatives], clupeids [herring, shad, anchovies], mormyrids [elephantfishes], and some others). These species have been termed the "hearing specialists," meaning only that their sound pressure sensitivity is high and that their upper frequency range of hearing is extended above those species that hear only via the direct pathway (the "hearing generalists," by default) (see Popper and Fay, Chapter 3).

Among amphibians, two major pathways of sound to the ears have also been identified: the columellar and opercular systems (see Lewis and Narins, Chapter 4). The columellar system, common among anurans and urodeles, conducts air- or water-borne sound to the periotic fluid system via the tympanum (in most species) and columellar middle ear. The lack of an effective tympanum and air-filled middle ear cavity in urodeles is thought to be a derived characteristic among amphibians. In species with a columellar system and a tympanum, at least two pathways of sound to the tympanum have been identified: a direct or usual route to the outer surface, and additional routes via the body cavity or mouth to the inner surface. This arrangement results in tympanic motion that is in proportion to the pressure difference, or gradient, between the two sides of the tympanum. This gradient can be dependent on the direction at which sound arrives at the animal, and thus confers on the ears an inherent directional selectivity (in some species and at some frequencies). Thus, both fishes and amphibians may have inherently directional ears.

The opercular system consists of a bony or cartilaginous plate attached to the oval window (along with the columella) and to the skeleton via muscle. It is thought that this system effectively couples the ears to the ground or substrate, and gives the amphibians sensitivity to low-frequency, substrate-borne energy. This system appears to be analogous to the direct pathway in fishes since in both groups, whole-body or skeletal motion is detected. It seems likely that the opercular system is the most primitive and widely shared among amphibians.

4.2 Auditory Organs

Most amphibians share all the organs of the fish ear: the three semicircular canals and their associated cristae, the saccule, lagena, and utricle, and the papilla (or macula) neglecta. In most fishes, the saccule is thought to be primarily an auditory organ, and the utricle primarily a vestibular organ. (Note, however, that the utricle appears to be adapted primarily as an auditory organ in clupeids, and possibly in some catfish.) In some amphibians, the saccule has been found to be highly sensitive to substrate-borne motion, and is generally thought to function as a sound- or vibration-sensitive organ rather than as a vestibular organ. This function was probably inherited from the fish-like ancestor of the modern amphibians. The lagena responds with great sensitivity to acoustic particle motion in goldfish, but its possible auditory and vestibular roles have not been systematically studied. The papilla neglecta is sensitive to vibration (Fay et al. 1974) and sound (Corwin 1981) in sharks, but has not been investigated in other fishes.

In addition, fishes, larval amphibians, and some postmetamorphic amphibians have a lateral line system. As discussed by Coombs and Montgomery (Chapter 8), the lateral line system is stimulated by low-frequency (generally below 50 Hz) relative movements between the organism and its water environment that can result from surface waves, the steep gradients of particle motion amplitude within the near field of a moving source, self-induced water currents, and flowing water. In lateral line canal organs, stimulation is caused by pressure gradients between adjacent pores. Questions concerning the extent to which the lateral line system can be considered a hearing organ, or even related to hearing as it is conventionally defined, is a matter of definition and opinion (see papers in Coombs et al. 1989). It is clear, however, that while the ears of fishes and amphibians can be stimulated by motion of the skeleton (the direct pathway for fishes and the opercular pathway for amphibians) or by sound pressure (the indirect pathway for fishes and the columellar pathway for amphibians), the lateral line system requires a relative motion between the skeleton and the surrounding medium. Thus, there are many stimuli (e.g., homogeneous sound pressure fields) that may stimulate the ears but not the lateral line, and other stimuli (e.g., laminar flow over the body surface) that may stimulate the lateral line but not the ears. In general, fishes and amphibians respond to and encode the directionality of pressure gradients using both the ears and lateral line.

In addition to these mechanosensory organs shared by most amphibians and fishes, anuran amphibians may also have two additional auditory organs: the amphibian and basilar papillae. The basilar papilla is considered the most primitive or widely shared of the two organs, and its absence in some amphibians is considered to be a derived characteristic (see Lewis and Narins, Chapter 4). Fritzsch (Chapter 2) argues that the basilar papilla arose among sarcopterygian fishes (as presently demonstrated in *Latimeria*) as a new organ associated with the lagenar recess, and that it is homologous among these fishes, amphibians, reptiles, birds, and mammals (the cochlea). This conclusion is presently controversial.

The amphibian papilla is an auditory organ thought to be derived de novo among amphibians, and thus is not found among fishes or other tetrapods. This organ appears to be the most sophisticated acoustic sensor of the amphibian ear. Recordings from the eighth nerve of the amphibian papilla show many functional similarities to those of reptiles, birds, and mammals, including similar sensitivity, tuning curve shapes, dynamic properties, phase-locking, tonotopy, two-tone suppression, efferent modulation, and distortion characteristics. In goldfish, catfish, and presumably other hearing specialists, saccular afferents show many of these same characteristics (see Popper and Fay, Chapter 3; Fay and Megela Simmons, Chapter 7; and Lewis and Narins, Chapter 4). Thus, primary auditory organs have developed independently several times among vertebrates: the saccule of most fishes, the utricle of clupeid fishes, the amphibian papilla of amphibians, the basilar papillae of sarcopterygian fishes and amphibians, and possibly the basilar papillae of reptiles, birds and mammals (i.e., the cochlea). Apparently, a shared sense of hearing does not necessarily arise from homologous acoustic organs. Rather, peripheral organs may have arisen or become adapted to serve the sense of hearing among vertebrates.

4.3 Hair Cells

As discussed by Popper and Fay (Chapter 3) and Lewis and Narins (Chapter 4), the hair cells of all auditory organs have many fundamental characteristics in common. Among all vertebrate groups, hair cells are mechanoreceptive transducers that communicate via chemical synapses with primary afferents. They have a bundle of stereocilia adjacent to an eccentrically placed cilium (kinocilium) or basal body. In all hair cells investigated, deflection of the stereovilli toward the kinocilium is depolarizing and excitatory, while deflection in the opposite direction is hyperpolarizing. The voltage response of hair cells is an approximate cosine function of the angle of ciliary deflection with respect to the axis from the center of the stereovillar bundle to the kinocilium. In fishes, amphibians, and all other vertebrate groups, these axes of best response direction are distributed in sometimes complex patterns over the surface of the sensory epithelia. These patterns of hair cell orientation may vary among organs, species, and families. In fishes (and for the amphibian saccule), patterns of hair cell orientation over the epithelia of otolith organs are the first step in encoding the axis of acoustic (and seismic) particle motion, and thus sound source direction. For the amphibian and basilar papillae, hair cell orientation patterns presumably arise to code the relative motions between the sensory epithelia and their overlying tectoria with greatest sensitivity. For saccular afferents of the bullfrog, goldfish, and toadfish, responses can be observed to whole-body displacements as small as 0.1 nanometer at frequencies between 50 and 140 Hz. It is not likely that hair cells of any vertebrate auditory organ are significantly more sensitive than this. In contrast to the case for the mammalian cochlea, hair cells of the lateral line system and auditory organs of fishes and amphibians proliferate throughout life and may regenerate after damage in adults.

4.4 Tuning Mechanisms

Some isolated hair cells of the sacculi of amphibians and fishes, and of the amphibian papilla, show electrochemical resonances. However, the extent to which these resonances play a role in the frequency selectivity of primary afferents is not clear. In the frog saccule, hair cell resonance appears not to play a direct role in neural tuning, but there is some evidence that in the toadfish saccule and the amphibian papilla, hair cell resonance may be an important first step in forming frequency selectivity and its diversity among primary afferents. Lewis and Narins (Chapter 4) point out that the steep band edges observed in the tuning curves of primary afferents cannot be accounted for by simple resonance, and suggest that tuning of high dynamic order arises from an effective coupling among hair cells that may take place through the tectorium, implying the presence of reverse transduction (electrochemical to mechanical). This suggestion is supported by the observations of otoacoustic emissions in some amphibians. In goldfish and toadfish, steep-edged tuning curves have also been observed in saccular afferents (see Popper and Fay, Chapter 3), suggesting an effective coupling among hair cells. The possible existence of otoacoustic emissions has not been investigated in fishes.

In goldfish and toadfish, there are at least two classes of tuned afferents with different characteristic frequency (CF). In fishes, frequency tuning and its diversity appear to arise from a combination of the electrochemical resonance of hair cells, micromechanical processes, and the combinations of inputs within the dendritic arbors of innervating eighth nerve cells. Although this sort of tuning is the simplest yet observed among vertebrates, its existence indicates that peripheral frequency analysis itself is a primitive character widely shared among vertebrates. At least some of the mechanisms underlying peripheral tuning and the central processing that follow it (see Feng and Schellart, Chapter 6) were probably first evolved among fishes.

4.5 Central Pathways and Processing

It is now clear that the information flow from the ears to the telencephalon follows the same general pattern among most vertebrates, including fishes and amphibians (see McCormick, Chapter 5). Thus, the auditory circuits of mammals and other amniotes most likely have developed from those of ancestral fishes and amphibians. These primitive circuits include first- and second-order medullary nuclei, a lateral lemniscus, at least one acoustic area of the midbrain (torus semicircularis), and multiple acoustic areas of the thalamus and forebrain.

Fishes and nonanuran amphibians share a complex of first-order nuclei in the ventral octaval column that receive input from otolith and semicircular canal organs, but apparently do not have the more dorsal nucleus that is exclusively auditory (or nearly so) in anurans and other amniotes. In general, these nuclei may have analogous functions to those of amniotes, but are probably not homologous among these groups. Similarly, second-order medullary nuclei of fishes and amphibians may be analogous to the superior olive of amniotes, but their possible homologies are not clear. Although the midbrain's torus semicircularis is more complexly organized in anurans than in fishes, it is apparently homologous among fishes, amphibians, and amniotes (the inferior colliculus of mammals). The relations among ascending pathways from the mesencephalon to the telencephalon in fishes and amphibians are not well understood, and it is not clear whether the various nuclei described may be primitive homologues of the lemniscal or extralemniscal auditory nuclei of mammals.

Relatively little is known about the response properties of central auditory neurons in fishes compared with anuran amphibians (see Feng and
Schellart, Chapter 6). For fishes, there are a few studies in several species
focusing on auditory nuclei of the medulla, midbrain (torus semicircularis),
thalamus (central posterior nucleus), and forebrain regions. For anurans,
there are far more central studies. In both groups, the torus semicircularis
is the most studied structure.

In fishes and amphibians, phase-locking to the sound waveform is generally robust in primary afferents and medullary nuclei, but tends to disappear
at midbrain levels and above. However, some cells recorded at the midbrain
level in fishes (possibly input neurons from lower centers) show more
precise phase-locking than those in the periphery. Frequency-threshold
(tuning) curves tend to become more complexly shaped and diverse in
CF progressing from the periphery to higher centers. At the midbrain
(and thalamus of anurans), tuning curves may have multiple CFs separated
by frequency regions where responses appear to be inhibited. In fishes,
the sharpness of tuning increases in the torus but widens again in the
thalamus. In general, inhibition appears to be an important process in
the torus of fishes and throughout the central pathway in anurans.
Tonotopic maps appear at least to the level of the torus in anurans, but their
existence in fishes has not been clearly demonstrated. In goldfish and some
anurans, two-tone rate-suppression and single-tone suppression have been
observed in a subset of low-CF primary afferents. Two-tone interactions
that may be inhibitory are found at least to the level of the torus in fishes,
and to the thalamus in anurans. Some neurons of the torus and thalamus
show a selectivity to temporal envelope patterns that is probably
synthesized using inhibition (see Crawford 1997 and Bodnar and Bass
1997 for recent reports on two species of vocal fishes). Temporal response patterns to tone bursts increase in type and complexity at least
to the level of the torus. Peristimulus time histograms from neurons of
the torus (and dorsal medullary nucleus of anurans) show patterns that
can be labeled as onset, onset with notch, pauser, chopper, buildup, and
primary-like. These patterns are qualitatively like those observed in central
auditory nuclei of amniotes. In fishes and anurans, cells of the torus
semicircularis show responsiveness that varies robustly with sound source
direction (in anurans) or with the axis of particle motion (in fishes). In many
cells of the goldfish torus, directionality in azimuth and elevation is sharpened, probably by inhibition from other directional inputs (Ma and Fay
1996).

Although the response properties of central auditory neurons in fishes
and amphibians are less extensively studied than in mammals and some
other amniotes, it is becoming clear that species of all vertebrate classes
share many of the same auditory processing mechanisms and strategies at
various levels of the brain. Since the torus semicircularis of fishes and
amphibians and the inferior colliculus of mammals are probably homolo-

gous structures, many of the processing strategies revealed at the midbrain level in mammals (see Fay and Popper 1992) may be primitive vertebrate characters that originated among fishes.

4.6 Psychophysics and Acoustic Behavior

Fay and Megela Simmons (Chapter 7) point out that auditory perception in fishes and anurans show many fundamental similarities that are shared with other vertebrates. In their usual environments, fishes and frogs solve problems of perceptual source segregation and localization of sound sources, and may use acoustic dimensions that control the perception of complex pitch and timbre for human listeners in similar ways. Fishes and frogs generally have a frequency range of hearing that is more restricted than those of other vertebrates, extending to 3 kHz in some fishes, and to 5 kHz in some anurans, although it has recently been demonstrated that at least one species of clupeid fish, the American shad, can detect sounds to over 180 kHz (Mann et al. 1997). However, best thresholds for the most sensitive species of both groups are comparable and fall within the range of all other vertebrate classes (-10 to 15 dB re: 20μPa). Level discrimination limens are comparable to those measured in many other vertebrates, and in fishes the effect of sound duration on level discrimination is similar to that shown in humans and other mammals. Frequency discrimination thresholds, psychophysical tuning curves, critical bandwidths, and critical masking ratios indicate a level of frequency analysis that falls only within the upper ends of the range of those measured in other vertebrates. This relatively poor frequency resolution appears to reflect a relatively crude peripheral frequency analysis (especially in fishes) and suggests a greater reliance on temporal processing than on spectral processing compared to birds and mammals. Both animal groups are able to localize sound sources in azimuth and elevation quite accurately ($10°$ to $20°$) in spite of small head widths, and for fishes an effective lack of interaural level and time-of-arrival differences. Fishes and amphibians share a reliance on pressure gradient processing for some aspects of directional hearing.

Some of the ethological data on anuran hearing suggest that anurans' sense of hearing might differ quantitatively in relative acuity and sensitivity from that measured in other vertebrates. These quantitative differences may be due to environmental and biological constraints that lead to an emphasis on one aspect of sound processing over another. However, these differences might also reflect the power of the behavioral techniques used. Much is known about the specific spectral and temporal acoustic features of communication sounds in anurans (see Zelick, Mann, and Popper, Chapter 9), but less about the acuity for the detection and discrimination of these features outside of the communication context. For fishes, in contrast, detection, discrimination, and perception of both simple and complex sounds are relatively well understood in a few species, but the perception of communication sounds is less well studied. New experiments and paradigms are needed to more precisely determine the limits of sound discrimination in anurans, and to better understand the perception of communication sounds in fishes. Such behavioral experiments will be useful in testing hypotheses about the peripheral and central determinants of auditory perception among all vertebrates.

It is self-evident to humans that the sense of hearing informs us about the physical characteristics of the sound producing and reflecting objects in the environment. These characteristics include source location, size, natural resonance frequencies, and vibration frequency. Since psychophysical experiments show that fishes and anuran amphibians appear to share many features of their sense of hearing with species of other vertebrate classes, including humans, we are led to suggest that the human sense of hearing revealed in classical psychoacoustical experiments is a primitive (i.e., shared) vertebrate character.

4.7 Communication

As already pointed out, we know a great deal more about acoustic communication in anuran amphibians than we do for fishes, and communications has been a driving force in establishing the extensive base of information on hearing that we have for anurans. The major limiting factor in learning more about fish acoustic communication results from the problems of observing fishes in their normal habitats. As pointed out by Zelick, Mann, and Popper (Chapter 9), there are extensive technical difficulties in observing and recording fish underwater. The very process of finding and then observing and recording fish is far more difficult underwater than working in air. Moreover, since humans cannot localize sounds very well underwater, determining which animal is producing a sound when there is more than one present is almost impossible. Thus, it becomes very difficult to correlate sound and behavior for fishes.

Because of the difficulties in obtaining data on fish communication, it appears (very possibly erroneously) that the number of fish species known to communicate acoustically is limited compared with anurans (see Zelick, Mann, and Popper, Chapter 9). We note that while data on acoustic communication are probably available for a fair proportion of anurans, limited data on sound production is probably available for fewer than 100 species of fish, or 0.4% of the known extant species. Since many of the species for which there are no data on vocalization have structures that suggest the possibility of sound production, as in the deep-sea myctophids (lantern fishes) (Marshall 1967), it may be that further observations will reveal many more species of fish that have complex acoustic repertoires, such as those that have been described in a number of mormyrids (e.g., Crawford et al. 1997).

5. Summary

In this chapter we have attempted to bring together some of the ideas on fishes and amphibians that are explored in this volume. In working with the various authors, and in reading the chapters as they were submitted, we have become more convinced that while there are differences in hearing and sound communication between fishes and amphibians, the study of each group would benefit from some of the insights we have gained from the other. Indeed, the power and usefulness of the comparative method is apparent throughout this volume.

References

Bodnar DA, Bass AH (1997) Temporal coding of concurrent acoustic signals in auditory midbrain. J Neurosci 17:7553–7564.

Bregman AS (1990) Auditory Scene Analysis: The Perceptual Organization of Sound. Cambridge, MA: MIT Press.

Capranica RR, Moffat AJM (1983) Neuroethological principles of acoustic communication in anurans. In: Ewert JP, Capranica RR, Ingle DJ (eds) Advances in Vertebrate Neuroethology. New York: Plenum Press, pp. 701–730.

Coombs SL, Görner P, Münz H (eds) (1989) The Mechanosensory Lateral Line: Neurobiology and Evolution. New York: Springer-Verlag.

Corwin JT (1981) Postembryonic production and aging in inner ear hair cells in sharks. J Comp Neurol 201:541–553.

Crawford JD (1997) Feature-detecting auditory neurons in the brain of a sound-producing fish. J Comp Physiol 180:439–450.

Crawford JD, Cook AP, Heberlein AS (1997) Bioacoustic behavior of African fishes (Mormyridae): Potential cues for species and individual recognition in *Pollimyrus*. J Acoust Soc Am 102:1–13.

Fay RR (1988) Hearing in Vertebrates: A Psychophysics Databook. Winnetka IL: Hill-Fay Associates.

Fay RR (1992) Structure and function in sound discrimination among vertebrates. In: Webster D, Fay RR, Popper AN (eds) The Evolutionary Biology of Hearing. New York: Springer-Verlag, pp. 229–263.

Fay RR (1995) Perception of spectrally and temporally complex sounds by the goldfish (*Carassius auratus*). Hear Res 89:146–154.

Fay RR, Popper AN (eds) (1992) The Auditory Pathway: Neurophysiology. New York: Springer-Verlag.

Fay RR, Kendall JI, Popper AN, Tester AL (1974) Vibration detection by the macula neglecta of sharks. Comp Biochem Physiol 47:1235–1240.

Handel SJ (1989) Listening: An Introduction to the Perception of Auditory Events. Cambridge, MA: MIT Press.

Hawking S (1988) A Brief History of Time. New York: Bantam.

Hulse SH, MacDougall-Shackelton SA, Wisniewski B (1997) Auditory scene analysis by songbirds: stream segregation of birdsong by European starlings (*Sturnus vulgaris*). J Comp Psychol 111:3–13.

Ma W-L, Fay RR (1996) Directional response properties of cells of the goldfish midbrain. J Acoust Soc Am 100:2786.

Mann DA, Lu Z, Popper AN (1997) Ultrasound detection by a teleost fish. Nature 389:341.

Marshall NB (1967) Sound-producing mechanisms and the biology of deep-sea fishes. In: Tavolga WN (ed) Marine Bio-Acoustics II. Oxford: Pergamon Press, pp. 123–133.

Popper AN, Fay RR (1997) Evolution of the ear and hearing: issues and questions. Brain Behav Evol 50:213–222.

Rogers P (1986) What do fish listen to? J Acoust Soc Am 79:S22.

Tavolga WN (ed) (1976) Sound Reception in Fishes—Benchmark Papers in Animal Behavior, Vol. 7. Stroudsburg, PA: Dowden, Hutchinson & Ross.

Tavolga WN, Wodinsky J (1963) Auditory capacities in fishes. Pure tone thresholds in nine species of marine teleosts. Bull Am Mus Nat Hist 126:177–240.

2
Hearing in Two Worlds: Theoretical and Actual Adaptive Changes of the Aquatic and Terrestrial Ear for Sound Reception

BERND FRITZSCH

1. Introduction

Sound in both water and air has two physical properties, the near-field particle motion generated by any moving object and the far-field pressure simultaneously generated by these objects (Kalmijn 1988). Based on the physical properties of the medium (the so-called characteristic impedance, which is about 3500 times larger for water than for air) and the steep loss of energy over distance (dipole source: $1/distance^3$; monopole source: $1/distance^2$) in the particle or direct sound in a frequency specific fashion (Schellart and Popper 1992), this component of sound carries enough energy to stimulate a receptor only over a very short range at low frequencies. In particular in air, this range is too short to play a role in terrestrial hearing in vertebrates and is used only by some insects (Michelsen 1992). While sound pressure reception opens up a wider range over which sound can be received as it falls off less steeply (1/distance), it cannot be extracted easily without specializations external to the inner ear. These adaptations to sound pressure reception, the way terrestrial and some aquatic vertebrates hear, essentially have to funnel the limited energy present in the far field with minimal loss to the appropriate receptor organs in the inner ear. The required series of changes in the otic region fall into three categories:

1. There have to be changes in the associated structures of the ear so that sound pressure changes move an unloaded structure, typically a membrane separating two or more distantly connected gas filled sacs (e.g., middle ear, outer ear canal) or a gas bubble (swim bladder, other gas bubbles), and transmit these sound pressure induced movements directly or indirectly to the inner ear. To achieve this, the ear has to provide a pathway of minimal resistance for sound pressure–induced lymphatic flow to minimize the energy loss. Otherwise almost all the energy would simply be reflected at the otic wall (terrestrial hearing) or pass through the ear with little energy transmitted to any given receptor organ (aquatic hearing). These pathways of low resistance are provided by the middle ear and perilymphatic

channels, which typically interconnect two or more foramina in the otic capsule.

2. Within the inner ear there have to be specific associated structures that transmit the periplymph (or endolymph) movements caused by the sound pressure fluctuations to inner ear end organs. These end organs (sensory epithelia) in turn have to be specialized for motion detection of these sound pressure–induced lymphatic movements. It would be important to know how these pathways channel sound in a frequency specific way through the ear, in particular in amphibians.

3. These specialized end organs of the ear have to be connected to an area in the brain that can process selectively the signals received via these sensory epithelia discretely from those received in the purely vestibular end organs of the ear. This will enable the animal to discriminate a sound from other signals perceived by other sensory epithelia of the inner ear (e.g., gravistatic, linear acceleration, near-field particle movement-generated linear acceleration, angular acceleration).

An additional factor that has been revealed in the last 20 years is a partial or complete segregation of projections of neurons from the central nervous system to the inner ear, the so-called efferents (Roberts and Meredith 1992). While those efferents projecting to the ear typically overlap in their distribution in vertebrates without specialized sound pressure reception, there is a progressive segregation of vestibular and auditory efferents among terrestrial vertebrates, in particular mammals.

In this overview I first discuss the systematic relationship of fishes and amphibians as revealed in most recent evolutionary analysis (Duellman and Trueb 1994; Janvier 1996; Stiassny et al. 1996) highlighting the positions of the few species studies with respect to sound pressure reception in some detail. Then I present the known and inferred evolutionary changes in the otic region of various vertebrates, followed by an overview of the known and presumed ontogenetic changes that underlie the adaptive reorganizations in the middle ear, the inner ear, the central nervous system connections of the inner ear, and the efferent system.

Throughout this presentation I follow the inside-out principle first highlighted by Lombard and Bolt (1988), that is, any change in the otic periphery will lead to the reception of a sound pressure signal only if the inner ear is able to receive that signal. This idea basically argues that changes in the inner ear have to happen first, followed by changes in the otic periphery. Alternative arguments assume explicitly or implicitly that changes in the otic periphery predate changes in the inner ear without providing any argument as to how these changes should be evolutionary stabilized in the absence of an apparent adaptive value, as they may not be received at all by an inner ear without specialized sound pressure receivers. It is unfortunate that fossil records are of limited help in elucidating most of these changes in the otic region as they either will not be preserved at all (neuronal changes, inner ear changes) or will be a source of various interpretations depending on the current status of recorded fossils (Bolt and Lombard 1992; Clack 1992; Ahlberg et al. 1996; Janvier 1996).

2. Taxonomy of Fishes and Amphibians

Vertebrates can be divided into two major taxa, those with and those without jaws. This fundamental division of vertebrates based on numerous derived characters is particularly obvious for the ear. All jawless fishes have at the most two semicircular canals, whereas all jawed fishes have three semicircular canals (Lewis et al. 1985). Clearly, fossil data indicate that the presence of two semicircular canals is primitive for all jawless vertebrates (Jarvik 1980; Janvier 1996). However, while the differences between various jawed fished can be dramatic with respect to their morphology (compare a bird and a shark, for example) the ear undergoes much less dramatic changes. Nevertheless, major taxa such as mammals and amphibians can be characterized by unique otic features, such as the mammalian cochlea or the amphibian papilla, respectively. In essence than, the otic characters reflect on a reduced scale the evolution of vertebrates in general.

Fishes have the most species of all vertebrate taxa (with over 25,000 species). Typically three major, separate evolutionary lines are identified among fishes (Fig. 2.1): the cartilaginous fishes (sharks, rays, and rat fishes), the ray-finned fishes (or actinopterygians, composed of polypterids, acipenserids, and neopterygii), and the lobe-finned fishes (or sarcopterygians, composed of coelacanths and lungfishes). Lobe-finned fishes also gave rise to terrestrial vertebrates with their two major lineages, amphibians and amniotes (Janvier 1996). While there is widespread agreement about these major divisions of vertebrates, the subdivisions within these lineages are still debated.

2.1 Cartilaginous Fishes

The inner ears of cartilaginous fishes have a single shared, derived character, a separate recess that contains the utricle. In all other jawed vertebrates the utricular sensory epithelium is contained within the enlarged horizontal common arm of the horizontal and anterior vertical canal. Interestingly, the only other vertebrate that shares this character is lungfish, which had already led Retzius (1881) to conclude that lungfish are cartilaginous fishes, a view still shared by some (Jarvik 1980). Most of the variation found in the ear of cartilaginous fishes concerns the separation of the lagenar sensory epithelium (macula) into a distinct recess. Another variation concerns the separation of the semicircular canals and the concomitant changes in the size of the papilla neglecta (Retzius 1881). The latter aspect is of interest here as it relates potentially to the known capacity of hearing in

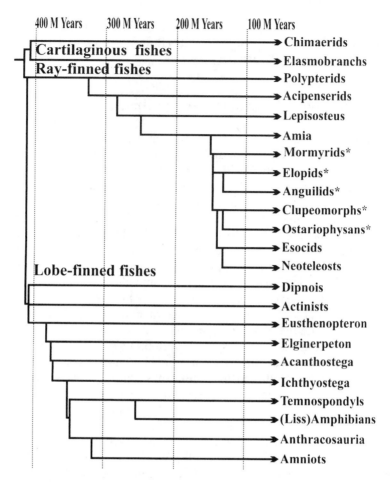

FIGURE 2.1. The likely interrelationships of aquatic jawed vertebrates is shown. Note that the divergence of the three major lineages of jawed vertebrates (cartilaginous fish, ray-finned fish, and lobe-finned fish) happened about 400 million years ago. While the radiation into the major lines of lobe-finned fishes, including the ancestors of tetrapods, was achieved more than 300 million years ago, the radiation of modern ray-finned fishes is a more recent event. Trifurcations indicate unresolved or disputed systematic affinities. Asterisk indicates ray-finned fishes with a known or suspected sound pressure connection between the ear and the swim bladder. (Modified from Bolt and Lombard 1992; Ahlberg et al. 1996; Janvier 1996; Stiassny et al. 1996; and Thulborn et al. 1996.)

elasmobranchs (Corwin 1981), although the mechanism by which this is achieved is still disputed (Kalmijn 1988). In the absence of any gas bubble in elasmobranchs, it appears that the physical properties for exquisite sound pressure detection do not exist in these vertebrates. Still, some sound pressure energy appears to be received (Schellart and Popper 1992) perhaps by using differential density of fluids around the ear.

2.2 Ray-Finned (Actinopterygian) Fishes

The ear of all nonteleostean actinopterygian fishes (Grande and Bemis 1996) does not differ much from that of ratfish (considered by most an ancient line of cartilaginous fish), with the notable exception of the position of the utricular macula. This organ is not in a separate recess, but in both sarcopterygian and actinopterygian fishes (except lungfishes) it is in the common crus of the horizontal and anterior vertical canal. Other features that show a progressive segregation in actinopterygian fishes is the degree of individualization of the lagenar macula in its own recess. In fact, the variation of size and position of the three sensory maculae (lagena, utricle, and saccule) is one of the main features of variation in the ear of neopterygian fishes, the other being variation in the papilla neglecta (Schellart and Popper 1992).

Another major variation concerns the association between the swim bladder and the inner ear. Apparently, the swim bladder evolved in the common ancestor of sarcopterygians and actinopterygians as an accessory breathing organ. For buoyancy, the swim bladder had to be close to the head in those animals with heavy armored heads (Jarvik 1980; Carrol 1988; Janvier 1996). This lung/swim bladder can either be in a direct contact with the ear (at least during development) or is connected via the famous Weberian ossicles to the ear. In fact, the interrelationship of the taxa in which this kind of swim bladder/ear connection for enhanced sound pressure reception exists (elopimorphs, mormyrids, herrings, ostariophysi) is still debated. Strong evidence exists for a sister taxon relationship between the clupeomorphs (herrings) and ostariophysi (otophysi consisting of e.g., goldfish, catfish; anotophysi consisting of e.g., chanidae; Fink and Fink 1996). Other taxa with a known swim bladder/ear connection such as notopterids, mormyrids, and albulids (Schellart and Popper 1992), or a suspected sound pressure sensitivity such as eels (Schellart and Popper 1992), could share as a primitive feature the ear/swim bladder connection (Lecointre and Nelson 1996).

However, numerous uncertainties exist in the systematic relationships of these taxa and the details of the swim bladder/ear connections. If indeed all basal neopterygians shared a swim bladder/ear connection, those apparently distinct taxa (Fig. 2.1) that show this connection today (notoperids, mormyrids, albulids, herrings, and otophysans such as catfish, goldfish, and knife fish) could have each independently modified an ancestral neoteleostean feature. It is important to note in this context that almost all sound pressure–hearing bony fish belonging to these taxa have an extended range into the far field and also an extended frequency range (Schellart and

Popper 1992). Indeed, the current lineage euteleostei comprises only taxa with no direct connection of the swim bladder to the ear (esociformes, salmoniformes, neoteleostei) and is thus in contrast to all other neoteleost consisting of otocephala (herrings, catfish, goldfish, etc.; Johnson and Patterson 1996), which do have such a connection between the swim bladder and the ear. Other connections of gas bubbles to the ear may exist in some euteleosts such as anabantids (Werner 1960) but they have not been as extensively tested for sound pressure reception as the more traditionally studied groups such as goldfish, catfish, and others with a swim bladder/ear connection (Schellart and Popper 1992; Popper and Fay, Chapter 3). In this context the sound pressure capacities of troglodytic catfish that have lost a swim bladder need to be investigated and characterized in comparison to sister taxa with a swim bladder and Weberian ossicle connection to the ear.

2.3 Sarcopterygians

Almost uniform agreement exists that sarcopterygians are monophyletic and comprise three lineages, the actinistia (represented by the coelacanth *Latimeria*), the dipnoi (represented by lungfish), and the ancestors of tetrapods (Fig. 2.1). The history of sarcopterygians is old and the fossil records of each of the three lineages with living representatives is filled with numerous uncertainties. This largely relates to the fact that three extinct groups each consists of highly specialized representatives of old groups that clearly are modified compared to the oldest fossils found for each group (Cloutier and Ahlberg 1996, Janvier 1996). For example, fossil data indicate for lungfish a transition from an open marine environment to apparently oxygen-poor coal swamps. Likewise, the fossils of many coelacanths indicate that they were living in estuaries and that the deep-sea environment of the living representative of the lineage, *Latimeria*, may be related to appropriate adaptations in this lineage of coelacanths. Thus the transition from an aquatic to a terrestrial environment known for fossil tetrapods (Cloutier and Ahlberg 1996) is in part matched by substantial changes in the environmental adaptation in the other two sarcopterygian lineages. Consequently, the relationship of the three taxa is not fully settled as many unique characters exist in each lineage in combination with primitive characters.

It is noteworthy in this context that the ear of lungfish and *Latimeria* represent this problem very well. Overall, the ear of lungfish resembles the ear of cartilaginous fishes more than that of tetrapods, whereas the ear of *Latimeria* shares unique characters with that of tetrapods (Fritzsch 1992). However, more recent investigations on the organization of hair cell orientation show that lungfish display characters that can be interpreted as primitive bony fish patterns (Platt and Popper 1996). This testifies to the long separate history of these three lineages, which have amassed a complex mix of primitive and derived characters with a potentially large set of character reversals.

2.4 Tetrapods

The monophyletic origin of tetrapods from a group of sarcopterygian fishes is well supported by paleontologic and neontologic data (Cloutier and Ahlberg 1996, Janvier 1996). However, the interrelationship among the two lineages of tetrapods—amphibians (consisting of caecilians, salamanders, and frogs) and amniots (consisting of reptiles, birds, and mammals)—is still debated. Interestingly, again the ear is rather revealing in that it provides a unique character: The amphibian papilla, that strongly supports the idea that amphibians are monophyletic and which likely is derived from the papilla neglecta (Fritzsch 1992), is present in all amphibians and only in amphibians.

The ear characteristics favor a systematic relationship among amphibians with the caecilians as the sister taxon for frogs and salamanders (Fritzsch and Wake 1988), also supported by other evidence (Duellman and Trueb 1994). However, this systematic relationship is not universally agreed upon (Hedges and Maxson 1993).

3. The Middle Ear: Sound Pressure Reception in Water and on Land Compared

As outlined above, any moving object produces both particle displacement and pressure waves. Pressure waves are the physical sources of hearing in air. However, there is a problem getting sound pressure into the ear, which is filled with fluid and surrounded by dense bone. Without special adaptations to funnel airborne sound energy to the inner ear, almost all energy would be reflected at the otic wall owing to the impedance mismatch between the media of different density (over 99%—van Bergijk 1966; Jaslow et al. 1988). In most terrestrial vertebrates this impedance matching is minimized by the middle ear consisting of a tympanum and one or more middle ear ossicles that conduct the vibrations of the tympanum to the inner ear. Understanding the evolution of terrestrial hearing thus requires an understanding of the evolution of the middle ear. Two issues are central here: the evolution of the middle ear ossicles and the evolution of the tympanic membrane. I will address only the evolution of the middle ear ossicles, as the evolution of the tympanic membrane is highly speculative in the absence of good fossil evidence.

On the basis of their comparative studies on the evolution of the terrestrial middle ear, Reichert (1837) and Gaupp (1898) suggested that a part of one gill arch, the hyoid arch, has been co-opted to serve a radically different function as a middle ear ossicle in tetrapods. Subsequent studies demonstrated that the hyomandibular bone, used by many anamniotic vertebrates as an additional support of the jaws, was independently freed three times (and likely two times separately among amniotes alone; Clack 1992) from that function. Apparently a fusion of the palatoquadrate bone with the skull to free the hyomandibular bone from its previous function to brace the lower jaw could occur only in those sarcopterygian lineages that had lost the intracranial joint, an ancestral feature of that lineage (Ahlberg et al. 1996). The fate of the hyomandibular bone was different in each lineage (Fig. 2.2):

1. The hyomandibular bone was used to support the first gill arch, its presumably even more primitive function in ratfish (Jarvik 1980; Carrol 1988; Clack 1992).

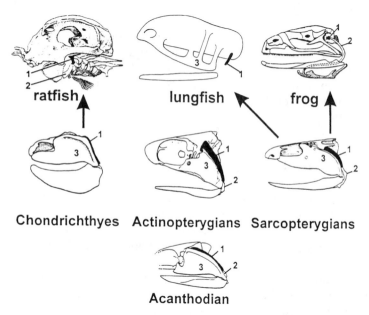

ratfish   lungfish   frog

Chondrichthyes Actinopterygians Sarcopterygians

Acanthodian

FIGURE 2.2. This scheme shows the fate of the hyomandibular bone in three verte-brate lineages with a fusion of the upper jaw with the skull, in ratfish, in lungfish, and in tetrapods (represented by a frog). Fossil data indicate that the condition in which the hyomandibular bone is an additional support for the jaws is primitive and exists in the earliest known fossils of jawed vertebrates, the acanthodians (430 million years ago). Likewise, fossils of all three jawed vertebrate radiations (middle row), suggest a similar association between the hyomandibular bone (epihyal; 1), the keratohyal (2), the palatoquadrate (3), and the brain case. Uncoupling from this function in lineages in which the maxilla is more or less fused with the neurocranium will make the hyomandibular bone obsolete in this function. This has led either to a function as a gill arch support in ratfish, to its almost complete regression in modern lungfish, or to its functional transformation into a middle ear ossicle in tetrapods (frogs). (Modified from Starck 1979 and Janvier 1996).

2. In most modern lungfish, the hyomandibular bone (epihyal) apparently is reduced to a small appendix of unknown functional significance (Jarvik 1980; Carrol 1988). The ancestors of modern lungfish had a short hyomandibula and the palate was already autostylic (Campbell and Barwick 1986), that is, the palatoquadrate bone was fused with the skull after they had lost the intracranial joint (Ahlberg et al. 1996). The overall skull morphology of these lungfish resembled *Acanthostega* and primitive actinopterygian fishes (Ahlberg et al. 1996). In contrast, *Latimeria* and other coelacanths have long hyomandibular bones, like other basal osteichtyans, and an intracranial joint (Ahlberg et al. 1996).

3. In the tetrapod lineage the hyomandibular bone eventually became situated between a tympanic membrane (where this is present) and the oval window, after the intracranial joint was eliminated (Ahlberg et al. 1996) and after being freed from its previous function. Instead of bracing the lower jaw it now served a novel function—to transmit vibrations of the tympanic membrane to the inner ear. It has been suggested that this happened at least twice independently in terrestrial vertebrates (Bolt and Lombard 1992; Clark 1992) (Fig. 2.2).

The hyoid arch as well as all or parts of the stapes of salamanders, chicken, and mice form during development from the neural crest (Toerien 1963; Noden 1987). It is not the adult hyomandibular bone but its embryonic anlage that was apparently remodeled and inserted either into the oval window of the otic capsule or fused with the otic capsule (Frizsch 1992). This ontogenetic transformation (Northcutt 1992) occurred near the middle ear/spiracular pouch and in conjunction with the need for a structure that can transmit sound-induced vibrations of the tympanum to the ear to minimize impedance mismatch in terrestrial hearing. All these temporal and spatial coincidences led only in the tetrapod lineage to the functional transformation of the hyomandibular bone anlage into a middle ear ossicle.

Obviously, this adaptation was not achieved in a single step and the fossil data at hand point to a large diversity of intermediate changes of the hyomandibular bone into a true stapes (Bolt and Lombard 1992; Clack 1992). Unfortunately, evolution of the perilymphatic system that connects the oval to the round window and thus allows the stapes footplate to drive fluid movements is not studied in significant depth at all to warrant a discussion.

The stapes of basic tetrapod-like vertebrates tends to be massive (Bolt and Lombard 1992; Clack 1992). This by itself limits sound transmission to low frequencies but does not preclude a role in sound transmission. The ear of salamanders shows that sound transmission can be performed even in the absence of a tympanic membrane (Jaslow et al. 1988). This does not preclude that the hyomandibular bone, once freed from its previous function, can assume yet a different function before it functions as the stapes (Clack 1992). At the moment it is unknown how many genes are involved in sculpting either the stapes or the hyomandibular bone. A comparative study of messenger RNA (mRNA) expression in developing stapes and hyomandibular bone is needed to estimate the complexity of the reorganization in terms of the genes involved. Each of the developmental changes caused by the genetic rearrangement should have led to some improvement of the sound pressure impedance matching of the stapes, even if this appears insignificant as viewed from the contemporary high efficiency of sound transmission in the middle ear. The current paleontological data as well as these theoretical arguments suggest that this adaptive process was a checkered path rather than a straightforward, streamlined, goal directed process. One important but largely unresolved problem is the association of the stapes with the tympanum, a membrane that is closing during development the spiracular pouch. It is highly speculative how often this association occurred. Arguments in the literature range from the tympanum being monophyletic (i.e., was formed only once) to the tympanum being polyphyletic and arose independently in each major terrestrial radiation (Clack 1992; Janvier 1996).

Irrespective to these unsolved issues of how the hyomandibular bone was transformed to serve a novel function to minimize the impedance mismatch between the airborne sound and the inner ear (Bolt and Lombard 1992; Clack 1992), the concept of functional transformation of a homologous structure, now more than 100 years old, has withstood the test of time. In this context it is important that transformations comparable in their magnitude have been proposed for the gills of arthropods, which supposedly have become wings (Averof and Cohen 1997). However, some related ideas for the evolution of sound pressure receiving parts of the central nervous system did not fare as well (see below).

4. Reorganization of the Ear and the Formation of the Basilar Papilla

As outlined above, hearing is the reception of either of two particle motions created by any oscillating body: near-field particle motion and pressure waves (or sound; Webster 1992). The near field will fall off with 1/distance^3 (distance), whereas the pressure waves reach into the far field and fall off with 1/distance (Kalmijn 1988). Near-field particle motion can be detected with either an otolithic ear or the lateral line (Coombs and Montgomery, Chapter 8). The lateral line perceives the relative movement of water with respect to the body surface (Coombs et al. 1992; Coombs 1994). In contrast, the otolithic ear will use the inertia of the denser otolith to create a shearing force on hair cells by moving inner ear sensory epithelia with respect to the lagging otolith (Schellart and Popper 1992; Popper and Fay, Chapter 3). While not tested experimentally, it is possible that all animals with an otolith potentially may be able to detect this kind of near-field motion (or displacement; Schellart and Popper 1992; Canfield and Rose 1996) for hearing in water (Fig. 2.3). While pressure waves reach into the far field and fall off with 1/distance (Kalmijn 1988), only animals with a compressible, gas-filled chamber in their body may be able to detect the far-field pressure component of sound in water (Fig. 2.3).

Such a gas-filled chamber evolved in bony fish and the sarcopterygian lineage with a swim bladder/lung. Apparently, sound pressure perception with the swim bladder has been perfected independently several times among bony fish (Schellart and Popper 1992). In addition, gas bubbles in the buccal cavity of anabantid fish may be used in a similar way (Schellart and Popper 1992). In many cases, the sensory epithelium (which is either the saccule or the utricle) overlies a perilymphatic space (Fig. 2.3) that is in direct or indirect contact with a gas bubble. A pressure release window may also exist, which may lead to a lateral line canal (Bleckmann et al. 1991). Any near-field particle motion, being from the swim bladder or the sound source, induces a relative movement of the sensory epithelium with respect to the covering structure, typically a modified otolith (Werner 1960). This creates a complex and hard-to-interpret pattern of motion in the vicinity of the sound source (Fritzsch 1992) that varies with frequency, distance, and direction from the source (Fig. 2.3).

In principle it would be possible to compute the indirect and the direct sound independently (Buwalda et al. 1983; Schellart and Popper 1992). However, this would require further information processing in the central nervous system with a concomitant small decrease in reaction time, a factor that may play a key role in predator avoidance (Canfield and Rose 1996). I propose here another solution that uses a separate, non-otolithic sensory epithelium for sound pressure reception and an exclusive restriction of the otolithic epithelium for near-field particle motion detection. This requires creation of two receivers, each one specialized for a particular component of sound—otolithic organ(s) for near-field particle motion detection and non-otolithic organ(s) for far-field pressure detection—thus avoiding the need for a computational analysis altogether (Fig. 2.3).

Comparable changes in the "hardware" for stimulus acquisition are known for the semicircular canals in which the primitive pattern with only two vertical canals in jawless vertebrates has been changed into a derived pattern with a horizontal canal being added for motion detection exclusively in the horizontal plane in jawed vertebrates (Lewis et al. 1985; Janvier 1996). The only apparent advantages of this change are (1) the possibly improved speed of extracting horizontal angular movement that would otherwise be computed from the residual horizontal vector present in the tilted vertical canals of the two canals present in jawless vertebrates, and (2) a less ambiguous movement orientation in three-dimensional space. It would be important to measure the speed of horizontal nystagmus in lampreys (which have only two vertical canals) and a jawed vertebrate and the accuracy of orientation in space to have some evidence for these

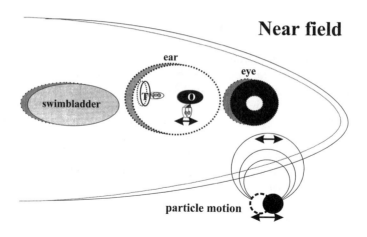

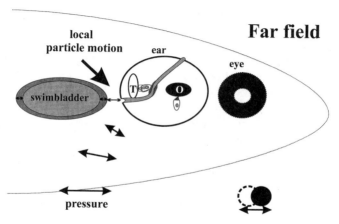

FIGURE 2.3. The relative movement of the otolith (o) with respect to a sensory epithelium for the near field (top) and the local particle motion that is generated near the swim bladder by the pressure changes in the far field (bottom) are shown. The entire fish will move as one if near-field particle motion is generated by an oscillating sphere (right). Thus the fish and its sensory epithelia in the ear will move with respect to the otolith, which will lag owing to its higher density. In the far field, pressure changes on the swim bladder will generate secondary near-field particle motions that impact on a nearby non-otolithic receptor (T) through a perilymphatic system that interconnects two or more foramina in the otic capsule. Note that the near field has a different orientation in space than the far field and will be perceived predominantly by an otolithic organ (top). In contrast the far field will be perceived by any organ within the secondary particle motion induced near a gas bubble such as the swim bladder (bottom). Segregation of the two physical components of sound (near-field particle motion and far-field pressure) to distinct receptors in the ear as indicated here could help minimize confusion of signals generated by both components of sound in a single receptor as will happen in the near-field.

proposition. Clearly, at the moment the advantage of having this distinct horizontal canal is not too obvious.

The coelacanth *Latimeria* has a sensory epithelia covered by a non-calcareous tectorial membrane beside the otolith-covered maculae of the saccule, lagena, and/or utricle in the ear (Fritzsch 1987). The function of the basilar papilla of *Latimeria* has never been tested experimentally. However, its structural features make it a likely candidate to perceive minute movements induced by pressure differences between the perilymphatic space underneath it and the endolymph filled ear (Fritzsch 1992).

These considerations of how to extract different aspects of sound in water only relate to sound pressure acting in water. How sound pressure can be generated to direct formation of a middle ear on land without the presence of an already existing sound pressure receptor in the inner ear is more difficult to understand. In the absence of such a receiver in the inner ear there would not be much of a selective advantage to early adaptive changes in the middle ear. Thus, a major issue to be discussed next is the evolution of the basilar papilla, the major sound pressure receiver in all amniotes that also plays a role in many amphibians. Consequently, understanding the evolution of the basilar papilla of tetrapods will elucidate a crucial step in the evolution of tetrapod hearing in general. The following reasoning assumes that this organ arose only once in evolution and did so in water (Fritzsch 1992). Should fossil data be found that indicate otherwise, most of the arguments raised below would be falsified.

To form a basilar papilla, otolith-bearing organs must be transformed in the following ways to eliminate any detection of near-field particle motion, which could hardly be avoided otherwise in water:

1. The basilar papilla has to be generated by either an additional segregation from the dispersing sensory primordia of the developing ear (Fritzsch et al. 1998) or by a novel proliferation (Fekete 1996).
2. The new epithelium must develop a tectorial membrane instead of an otolith/otoconia covering to avoid direct sound.
3. The epithelium must be positioned selectively near the border of an endolymphatic and a perilymphatic space to be excited by indirect sound.

I propose that the changes needed to achieve this transformation were likely associated with a rather unrelated event: the formation of a caudal extension of the ear to create a separate lagenar recess. Formation of a lagenar recess and formation of a lagenar sensory epithelium are independent events, because presence or absence of either or both can occur among sarcopterygians (Fritzsch and Wake 1988). Caudal extensions of the ear can be induced by treatment with certain teratogens such as lithium (Gutknecht and Fritzsch 1990), which are known to influence expression of developmental regulatory genes (Christian and Moon 1993). This indicates that perhaps only minor adjustments of the developmental program are necessary for this event. Clearly, *Latimeria* has a lagenar recess, whereas lungfish have none (Fritzsch 1992). A lagenar recess apparently evolved at least three times independently: (1) in the elasmobranch lineage, (2) in the teleost lineage, and (3) in the sarcopterygian lineage (Fritzsch 1992). I propose that following the formation of a new lagenar recess, the saccular/lagenar sensory anlage had to extend into this recess to provide it with the lagenar sensory epithelium (Fig. 2.4).

It is conceivable that the basilar papilla of *Latimeria* formed as a consequence of such a process. I assume that this segregation of a novel sensory epithelium happened in evolution only once, thus providing the sarcopterygian lineage with a unique opportunity. It is unclear how the other properties of the basilar papilla, formation of a tectorial membrane and association with the perilymphatic space, were achieved. Nevertheless, the topological position in the caudomedial aspect of the ear brought this epithelium into the closest possible proximity to the sound pressure receiving swimbladder in the belly. On the basis of this proximity to the swimbladder and the potential adaptive need for a sensory organ that can detect sound pressure motion separately, the newly formed epithelium may have been transformed from an otolithic organ precursor into a sound pressure receiving organ. Thus the fortuitous formation of a functionally redundant and rather uncommitted sensory epithelium in the sarcopterygian lineage alone may have been the base on which selection has worked and has ultimately led to the major sound pressure receiver of the inner ear of terrestrial vertebrates, the basilar papilla.

In summary, I propose that the formation of the basilar papilla is correlated with the de novo formation of a lagenar recess among sarcopterygians. Creation of a new, functionally redundant and uncommitted sensory epithelium in the general proximity of the sound pressure–receiving swim bladder may have been a fortuitous coincidence that was exploited by

---

FIGURE 2.4. The right ears of several sarcopterygian vertebrates are shown as viewed from the cranium; anterior is to the left, dorsal is up. *Latimeria* has a lagenar recess with the lagenar sensory epithelium (L), which is within the saccular recess (S) in lungfish. It is speculated that the formation of a separate lagenar recess in the lobe-finned fish lineage may have led to the formation of a redundant and functionally uncommitted sensory epithelium that eventually was transformed into a tectorial membrane-bearing perilymphatic organ, the basilar papilla (BP). This organ may have been inherited in tetrapods. However, amphibians also evolved a second sound pressure receiver, the amphibian papilla (AP), which likely represents a transformed papilla neglecta (PN) found in most other vertebrates. Note that the relationship of lungfish and *Latimeria* as well as the three amphibian taxa (salamanders, caecilians, frogs) are disputed and thus treated as unresolved trifurcation.

Salamanders   **Caecilians**   **Frogs**

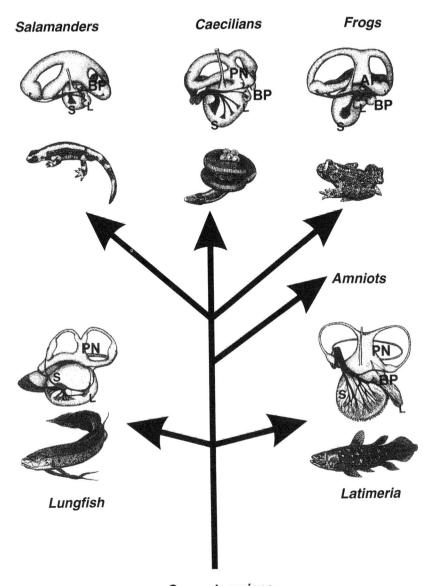

Sarcopterygians

evolution. Eventually, this exploitation may have resulted in the stabilization of structural transformations of this end organ that turned it into a tectorial membrane-bearing, perilymphatic end organ exclusively committed to a novel function: sound pressure perception. While this all started in water, the middle ear evolved as an adaptation to minimize the unavoidable impedance mismatch once the ancestral tetrapods crawled on land.

5. Evolution of Auditory Projections and Nuclei

While the evolutionary and developmental appearance of sound pressure conducting pathways (middle ear) as well as sound pressure receiving end organs in the inner ear are complex reorganizations in the developmental program of these structures, the formation of sound pressure receiving nuclei in the brain and a selective projection to these neurons requires reorganizations in an apparently developmentally unrelated system, the central nervous system. However, recent molecular data strongly reinforce the interaction between brain stem and ear development (McKay et al. 1996; Fritzsch et al. 1998). Thus it is not impossible that a single mutation effects both the hindbrain and also leads to the formation of auditory nuclei and the formation of the basilar papilla in the ear. Such regulatory genes are now characterized in ascidians, where a single gene is apparently responsible for the formation of the entire tail (Swalla and Jeffrey 1996). In the following discussion I use the terms *auditory nuclei* and *auditory projections* only for structures related to epithelia known or presumed to be involved in sound pressure perception. Answers to two questions are paramount for an understanding of the evolution of auditory nuclei and projections:

1. What makes sound pressure receiving end organs project to distinctly different terminal areas in the hindbrain?
2. What are the embryonic cell sources of the auditory nuclei?

In analogy to the middle ear transformation, a transformation of ancestral lateral line nuclei into auditory nuclei was proposed (Larsell 1967). However, recent data have shown that the lateral line system of amphibians consists of two distinct sensory systems, the electroreceptive ampullary organs and the mechanosensory lateral line (Fritzsch 1992) and many amphibians lose either of these organ types alone or both together at various developmental stages (Fritzsch 1989). Only anurans, which never develop electroreceptive ampullary organs, form a distinct auditory nucleus. The embryological source(s) of these neurons, called the dorsolateral (auditory) nucleus (Will and Fritzsch 1988), is not as yet fully determined.

For a scenario of potential functional transformation of central neurons, it is important that loss of input through an ontogenetic transformation of the periphery will not immediately cause degeneration of all second-order auditory and vestibular neurons in frogs (Fritzsch 1990), mammals (Moore 1992), and birds (Peusner 1992) and likely in the lateral line and electroreceptive nuclei as well. As in the case of the oculomotor system (Fritzsch et al. 1995) or the extension of inner ear afferents along a partially denervated cochlea (Fritzsch et al. 1997), these second-order neurons, once deprived of their afferents, may attract other efferent fibers. On the basis of the coincidence of the loss of electroreception and the gain of an auditory nucleus in frogs, there is a chance that some neurons, which would have contributed in the anuran ancestors to the dorsal electroreceptive nucleus, may now be functionally respecified and contribute instead to the auditory nucleus (Fritzsch 1992). In this context it is important that the rostrocaudal extent of the electroceptive nucleus matches that of the auditory nucleus (from trigeminal to glossopharyngeal root) and may indicate at least the contribution from the same rhombomeres 2–6 (Fritzsch 1996; Lumsden and Krumlauf 1996). Any relevance of this anuran scenario for amniotes is highly speculative, and other cell sources may be possible, such as duplication and functional respecification of vestibular nuclei (Fritzsch 1992; McCormick 1992).

In summary, functional transformation of parts of the central nervous system, while possible in cases of experimental manipulation (Metin and Frost 1991; Pallas 1991) and perhaps evolution (Bronchti et al. 1989; but see Cooper et al. 1993 for a different view) of the thalamus, have not been convincingly demonstrated for the reorganization of the hindbrain. It is unclear whether the coincidence of the loss of ampullary electroreceptors and the gain of a sound pressure–receiving auditory nuclei in anuran ancestors were causally related (Fritzsch 1992). In any case, a major issue for either scenario is what directs the auditory projection to end specifically on these (new or respecified) cells, which appears to happen rather readily as it is also found in all sound pressure–receiving bony fishes that have been thoroughly investigated (Bleckmann et al. 1991; McCormick, Chapter 5).

The three kinds of organs of the octavolateral system (ampullary organs, mechanosensory lateral line, inner ear) project in a segregated, non-overlapping fashion into the alar plate, the dorsolateral area of sensory fiber termination in the hindbrain (Fig. 2.5). This segregation may occur because of three factors recognized in the development of other sensory projections, alone or combined: (1) timing of arrival, (2) biochemical specificity, and (3) activity.

Evidenc of the possible importance of time of arrival can be found during development; typically the ear develops first and the ampullary organs last (Northcutt 1992). Our data suggest that indeed the inner ear projection extends first into the hindbrain alar plate followed by the mechanosensory lateral line and, last, by the electroreceptive ampullary projection (Gregory, Fritzsch, and Dulka, in preparation). However, a critical test would be to provoke a simultaneous regeneration of ampullary organ afferents, lateral line afferents, and inner ear afferents. This has never been performed and

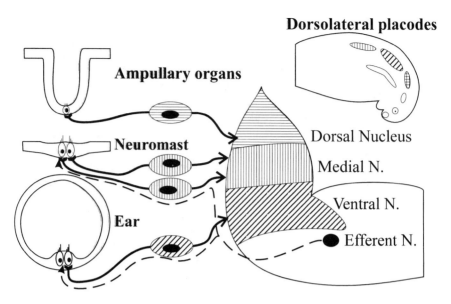

FIGURE 2.5. This scheme of a larval salamander shows (from left to right, top to bottom) the three octavolateral organs: the ampullary electroreceptors, the mechanosensory lateral line neuromast, and the vestibular part of the ear. These organs are connected through their specific afferent ganglia to their respective nuclei in the alar plate of the hindbrain: the dorsal nucleus for the ampullary organs (horizontal lines), the medial nucleus for the neuromasts (vertical lines), and the ventral nucleus for the ear (oblique lines). The hemisected transverse scheme of the hindbrain shows the position of efferent cells in the area of the facial nucleus and their branching fibers, which run to neuromasts and the ear. Insert on top right shows the distribution of the dorsolateral placodes that produce the octavolateral system as well as the lens placode and olfactory placode.

could provide also a critical assessment of the possible importance of the time of arrival for the observed segregation of afferents.

Evidence for the existence of biochemical gradients are plentiful in the visual system (reviewed in Tessier-Lavigne and Goodman 1996). However, the principle of this system (mapping one surface onto another one) is very different from the lateral line, ampullary electrorceptors, vestibular and auditory system in which no obvious topography between peripheral distribution or organs and their central projection exists, except for the cochleotopic organization of the cochlear projection. And even the cochleotopic projection could reflect simply a developmental gradient of differentiation of spiral ganglion neurons. It would be important to know how the specification of the polarity of the otic vesicle is related molecularly to the functional specification of otic ganglia, that is, through which process a ganglion cell is determined to project to a specific organ. Choice tests should be conducted in which vestibular neurons from different areas of the

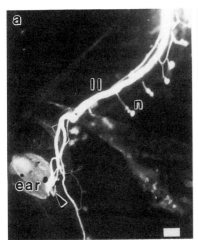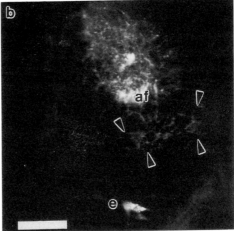

FIGURE 2.6. An ear was transplanted in a stage-30 *Xenopus* embryo into the dorsal fin. The pattern of lateral line fibers on the same side was revealed in their peripheral (a) and central distribution (b) at stage 50 employing Texas red dextran amine as a tracer; flat mounted skin (a) and 15-µm-thick frozen section (b). Note that some of the lateral line (ll) fibers enter the inner ear (arrowhead) as well as neuromasts (n). Note that centrally some fibers extend beyond the lateral line afferents (af) and project into the ventral, vestibular nucleus (arrowheads). An efferent cell (e) is also retrogradely filled. Bar indicates 100 µm.

vestibular ganglion known to project to different sensory epithelia are provided a choice between their ususal target and an unrelated epithelium. Beyond the known attraction of neurites to sensory epithelia (Fritzsch et al. 1997), this could provide answers for the specificity of connections. However, clear indications are present in the hindbrain with respect to longitudinal columns of expression of certain genes (Myat et al. 1996) and dorsoventral gradients (Tanabe and Jessell 1996). Thus it is at the moment unclear what role biochemical gradients may play in the hindbrain other than specifying the alar plate by mechanisms that reject the growth of afferents toward the floor plate (Tessier-Lavigne and Goodman 1996).

Evidence of the importance of activity may be obtained by either changing the development of the end organs (i.e., provoking differentiation of ampullary organs into mechanosensory lateral line organs) or connecting one type of afferent fiber to a different type of endorgan. To test the latter possibility, I have transplanted *Xenopus* ears into the dorsal fin. These ears were innervated by the posterior lateral line nerve (Fig. 2.6). Labeling of these lateral line nerves with tracers revealed a segregation of some lateral line afferents into the projection area of the ear, something lateral line fibers will nerve do even in the absence of inner ear afferents (Fritzsch

1990). Two reasons for this may be that (1) the lateral line afferents receive a stimulus from the ear that is different from the lateral line and cause a segregation comparable to that of retinal projections from two eyes in one tectum (Wilm and Fritzsch 1992) or of ocular dominance columns (Katz and Shatz 1996), or (2) the lateral line ganglia innervating the ear may be respecified into "otic ganglia." In this context the existence of the neurotrophin brain drived nuratrophic factor (BDNF) in the vestibular and cochlear system (Fritzsch, et al. 1997) as well as the correlation of activity with upregulated expression of neurotrophins (Katz and Shatz 1996) may be relevant. For example, BDNF has been implicated in the formation of ocular dominance columns, which has previously been shown to be activity dependent (Katz and Schatz 1996). Either change can explain why these fibers project differently from other lateral line fibers.

What is the relevance of these data for out problem of segregation of afferents derived from sound pressure–receiving perilymphatic end organs? Obviously, even when different organs would project centrally at the same time and in an overlapping manner, they could segregate. In fact, segregation will always be achieved as soon as there are two (or more) functionally different peripheral organs, either because they are activated by a different stimulus or because ganglion cells may be respecified in their central projection according to their peripheral connection or their development. Thus, projections from sound pressure–receiving organs could be forced to project to different areas within a nucleus or, if present, to other nearby areas like the newly forming auditory nuclei.

In summary, I suggest that the potential to segregate fibers from sound pressure–receiving end organs will be realized even if no new central target is available. Viewed from this perspective, knowledge about the ontogenetic origin of the cell source of the auditory nuclei would be the less important of the two events.

6. Evidence that Otic Efferents are Rerouted Facial Motoneurons

Inner ears of virtually all vertebrates receive an efferent innervation. Comparative data in many vertebrates show that these efferents are often colocalized with the facial motoneurons (Fig. 2.7) and thus suggest that facial motoneurons may have been rerouted to be efferent to the ear (Roberts and Meredith 1992). Recent developmental evidence from chickens and mice indeed snuggest that efferent cells to the ear are in fact ontogenetically rerouted facial motoneurons (Fritzsch 1996).

In larval and adult *Xenopus* and lampreys, the facial motoneurons and the efferent cells to the ear overlap completely (Fig. 2.7). At the earliest developmental stages in which facial motoneurons and octaval efferent

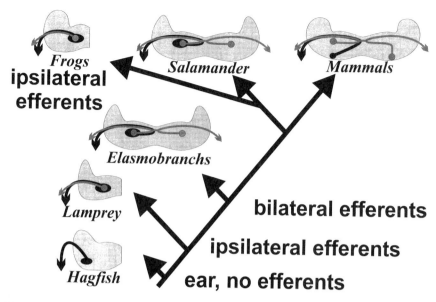

FIGURE 2.7. This scheme shows the evolution of efferents in craniate vertebrates. Hagfish have a facial motor nucleus (black) but are the only craniates without efferents to the ear. Lampreys have efferents to the ear (gray), the cell bodies of which are coextensive with facial motoneurons (black). In elasmobranchs there is some segregation and a bilateral distribution and projection of efferents. This bilateral distribution of efferents and their fibers is likely primitive for jawed vertebrates. Frogs and bony fish (not shown) have ipsilateral efferents coextensive with facial motoneurons, whereas salamanders show bilateral distribution of efferents. In mammals, vestibular and cochlear efferents are segregated from each other and from the facial motor nucleus. It appears that absence of efferents is primitive for craniates, unilateral efferents to the ear alone are primitive for lampreys, and bilateral efferents that are also common to the inner ear and the lateral line are primitive for jawed vertebrates. Phylogenetic and ontogenetic evidence supports the idea that efferents to the ear are redirected facial motoneurons.

cells can be labeled from the periphery in mice and in chickens they show an overlapping distribution of their perikarya along the floor plate (Fritzsch 1996). It is only in later development that facial motoneurons and efferent cells migrate differentially into the distinctly different adult positions (Bruce et al. 1997).

These data show that efferents to the ear are in fact best regarded as rerouted facial motoneurons that do not innervate mesodermally derived muscle fibers but instead placodally derived hair cells or ganglia (Fritzsch 1996). Obviously, they must therefore have acquired a rather different function than causing contraction of muscle cells. It has been claimed that the somite underlying the developing ear is crushed by the developing ear in lampreys (Jefferies 1986). If this can be proven with more detailed embryological studies, these facial motoneurons, destined to innervate this crushed somite, were then faced with two alternatives: reroute to the replacement of their previous target (the inner ear) or degenerate. Clearly, experimental evidence suggests that rerouting of motoneurons to a different target is possible (Landmesser 1972; Fritzsch et al. 1995), thus giving some credibility to the scenario of ontogenetic reorganization outlined above. We have recently removed the developing ears in chicken and found that the contralateral migrating efferents (Fritzsch 1996) do develop and send processes with facial motoneurons to the branchial arches, thus indicating that, if experimentally challenged, these cells can still reveal their motoneuron nature.

In summary, it appears possible that a regressive change (loss of one somite) caused by an unrelated progressive change (gain of the ear) are causally linked to the reorganization of the trajectories of facial motoneurons that now form the efferents to the ear in lampreys and gnathostomes instead of branchial motoneurons.

7. Conclusion: From Molecules to Morphological Evolution

In recent years the role of developmental regulatory genes has been extensively demonstrated through experimental means and the evolution of these genes is currently studied to unravel the molecular machinery underlying morphological evolution. Although we are still far away from a detailed understanding of the molecular mechanisms that regulate ear morphogenesis, some rather crude effects (using generic engineering of certain genes or differential expression of genes with certain teratogens/morphogens) indicate the potential of this research direction for the problems outlined above. For example, it has been shown that single genes like the *int-2* (FGF-3) gene (Mansour et al. 1993) or the Kreisler gene (McKay et al. 1996) can seriously disrupt normal development of the ear and, in addition, the adjacent area of the hindbrain.

Moreover, some of these genes may be regulated to some extent by retinoic acid, a potential morphogen. Clearly retinoic acid plays a major role in anteroposterior axis specification in developing vertebrate embryos, in particular in the hindbrain region (Lumsden and Krumlauf 1996). Application of retinoic acid causes, for example, anterior ectopic expression of Hox genes (Lumsden and Krumlauf 1996) and anterior expression of certain hindbrain neurons such as Mauthner cells (Manns and Fritzsch 1992). Interestingly, lithium chloride, which is also known to be able to transform the ear (Gutknecht and Fritzsch 1990) has recently also been implicated in affecting the expression of developmental regulatory genes (Christian and

Moon 1993). Based on these data it seems reasonable to speculate that perhaps some of these bewildering reorganizations of the otic region are governed by spatial, temporal, or intensity changes in the expression of a few genes such as genes encoding transcription factors (e.g., dlx-3), secreted factors (e.g., FGF-3), and others (Fekete 1996; Fritzsch et al. 1998), some of which may respond to retinoic acid. In this context it is important that a tentative connection between the transformation of limbs and the ear was recently noticed in fossils, too (Ahlberg et al. 1996), another system in which retinoic acid plays a major role in pattern formation (Fritzsch et al. 1998). Clearly, more details are needed before the evolutionary reorganizations in the is ear with their adaptive importance for sound pressure reception can be understood in terms of the underlying modifications of gene expression patterns.

8. Summary

The evolution of sound pressure receivers and all its components such as the middle ear, the inner ear, the central nervous system nuclei, and the efferent system were reviewed, highlighting both the strengths and the limits of fossil data as well as the importance of a theoretical model to search for the adaptations imposed on the various subsystems. While the middle ear represents a prime example of co-option of a system evolved in a different context (supporting the lower jaw) to a new function (conducting airborne sound from the tympanic membrane to the inner ear), this example is of limited importance for the inner ear and the central nervous system. Nevertheless, the evolution of the basilar papilla and the cochlear nuclei may be related to formation of novel, uncommitted organs and cells that could have been differentially specified owing to their relaxed commitment. How the formation of novel structure or the retention and differential specification of existing structures is achieved through molecular modifications of the developmental pathways is still unclear. Nevertheless, gathering data on genes involved in early pattern formation of both the ear and the brain stem may eventually help to solve this issue. A prime example of the apparent involvement of chance and necessity seems to be provided by the efferent system to the inner ear, which appears to be a developmentally and evolutionary respecified motor system that can still be experimentally rerouted to innervate branchial muscle fibers.

Acknowledgments. This work was supported in part by a National Institutes of Health (NIH) grant (50 DC00215-09). I thank Dr. T. J. Neary for linguistic advice, Dr. J. A. Clack and Dr. A.N. Popper for helpful suggestions on an earlier version of this manuscript, and Dr. M.-D. Crapon de Caprona for the drawings shown in Figure 2.3.

References

Ahlberg PE, Clack JA, Luksevics E (1996) Rapid braincase evolution between Panderichthyes and the earliest tetrapods. Nature 381:61–63.

Averof M, Cohen SM (1997) Evolutionary origin of insect wings from ancestral gills. Nature 385:627–630.

Bleckmann H, Niemann U, Fritzsch B (1991) Peripheral and central aspects of the acoustic and lateral line system of a bottom dwelling catfish, *Ancistrus* sp. J Comp Neurol 314:452–466.

Bolt JR, Lombard ER (1992) Nature and quality of the fossil evidence of otic evolution in early tetrapods. In: Webster DB, Popper AN, Fay RR (eds) The Evolutionary Biology of Hearing. New York: Springer-Verlag, pp. 377–403.

Bronchti G, Heil P, Scheich H, Wollberg Z (1989) Auditory pathway and auditory activation of primary visual targets in the blind mole rat, (*Spalax ehrenbergi*): 1,2-deoxyglucose study of subcortical centers. J Comp Neurol 284:253–274.

Bruce LL, Kingsley J, Nichols DH, Fritzsch B (1997) The development of vestibulocochlear efferents and cochlear afferents in mice. Int J Dev Neurosci 15:567–584.

Buwalda RJA, Schuijf A, Hawkins AD (1983) Discrimination by the cod of sound from opposing directions. J Comp Physiol 150:175–184.

Campbell KSW, Barwick RE (1986) Paleozoic lungfishes—a review. J Morphol Suppl 1:93–131.

Canfield JC, Rose GJ (1996) Hierarchical sensory guidance of Mauthner-mediated escape response in goldfish (*Carassius auratus*) and cichlids (*Haplochromis burtoni*). Brain Behav Evol 48:137–156.

Carrol RL (1988) Vertebrate Palaeontology and Evolution. New York: Freeman.

Christian JL, Moon RT (1993) Interactions between *Xwnt-8* and Spemann organizer signaling pathways generate dorsoventral pattern in the embryonic mesoderm of *Xenopus*. Genes Dev 7:13–28.

Clack JA (1992) The stapes of *Acanthostega gunnari* and the role of the stapes in early tetrapods. In: Webster DB, Popper AN, Fay RR (eds) The Evolutionary Biology of Hearing. New York: Springer-Verlag, pp. 405–419.

Cloutier R, Ahlberg PE (1996) Morphology, characters, and the interrelationships of basal sarcopterygians. In: Stiassny MLJ, Parenti LR, Johnson GD (eds) Interrelationship of Fishes. San Diego: Academic Press, pp. 445–480.

Coombs S (1994) Nearfield detection of dipole sources by the goldfish (*Carassius auratus*) and the mottled sculpin (*Cottus bairdi*). J Exp Biol 190:109–129.

Coombs S, Jansen J, Montgomery J (1992) Functional and evolutionary implications of peripheral diversity in lateral line systems. In: Webster DB, Popper AN, Fay RR (eds) The Evolutionary Biology of Hearing. New York: Springer-Verlag, pp. 267–293.

Cooper HM, Herbin M, Nevo E (1993) Visual system of a naturally microphthalmic mammal: the blind mole rat *Spalax ehrenbergi*. J Comp Neurol 328:313–350.

Corwin JY (1981) Audition in elasmobranchs. In: Tavolga WN, Popper AN, Fay RR (eds) Hearing and Sound Communication in Fishes. New York: Springer-Verlag, pp. 81–105.

Duellman WE, Trueb L (1994) Biology of Amphibians. Baltimore: The Johns Hopkins University Press.

Fekete DM (1996) Cell fate specification in the inner ear. Curr Opin Neurobiol 6:533–541.

Fink SA, Fink WL (1996) Interrelationship of ostariophysin fishes (teleostei). In: Stiassny MLJ, Parenti LR, Johnson GD (eds) Interrelationship of Fishes. San Diego: Academic Press, pp. 209–250.

Fritzsch B (1987) The inner ear of the coelacanth fish *Latimeria* has tetrapod affinities. Nature 327:153–154.

Fritzsch B (1989) Diversity and regression in the amphibian lateral line system. In: Coombs S, Görner P, Münz H (eds) The Mechanosensory Lateral Line Neurobiology and Evolution. New York: Springer-Verlag, pp. 99–115.

Fritzsch B (1990) Experimental reorganization in the alar plate of the clawed toad *Xenopus laevis*. I. Quantitative and qualitative effects of embryonic otocyst extirpation . Dev Brain Res 51:113–122.

Fritzsch B (1992) The water-to-land transition: evolution of the tetrapod basilar papilla middle ear and auditory nuclei. In: Webster DB, Popper AN, Fay RR (eds) The Evolutionary Biology of Hearing. New York: Springer-Verlag; pp. 351–375.

Fritzsch B (1996) Development of the labyrinthine efferent system. Ann NY Acad Sci 781:21–33.

Fritzsch B, Wake MH (1988) The inner ear of gymnophione amphibians and its nerve supply: a comparative study of regressive events in a complex sensory system. Zoomorphology 108:210–217.

Fritzsch B, Nichols DH, Echelard Y, McMahon AP (1995) The development of midbrain and anterior hindbrain ocular motoneurons in normal and in Wnt-1 knockout mice. J Neurobiol 27:457–469.

Fritzsch B, Barald K, Lomax M (1998) Early embryology of the vertebrate ear. In: Rubel EW, Popper AN, Fay RR (eds) Development of the Auditory System. New York: Springer-Verlag, 80–145.

Fritzsch B, Silos-Santiago I, Bianchi L, Farinas I (1997) The role of neurotrophic factors in regulating inner ear innervation. TINS 20:159–164.

Gaupp E (1898) Ontogenese und Phylogenese des schalleitenden Apparates bei den Wirbeltieren. Erg Anat Entwicklungsgesch 8:990–1149.

Grande L, Bemis WE (1996) Interrelationships of acipenseriformes, with comments on "Chondrostei." In: Stiassny MLJ, Parenti LR, Johnson GD (eds) Interrelationship of Fishes. San Diego: Academic Press, pp. 85–116.

Gutknecht D, Fritzsch B (1990) Lithium induces multiple ear vesicles in *Xenopus laevis* embryos. Naturwissenschaften 77:235–237.

Hedges SB, Maxson LR (1993) A molecular perspective on lissamphibian phylogeny. Herpetol Monogr 7:27–42.

Janvier P (1996) Early Vertebrates. Oxford, England: Oxford University Press.

Jarvik E (1980) Basic Structure and Evolution of Vertebrates, Vol. 1. London: Academic Press.

Jaslow AP, Hetherington TE, Lombard RE (1988) Structure and function of the amphibian middle ear. In: Fritzsch B, Ryan M, Wilczynski W, Hetherington T, Walkowiak W (eds) The Evolution of the Amphibian Auditory System. New York: Wiley, pp. 69–91.

Jefferies RPS (1986) The Ancestry of the Vertebrates. London: British Museum (Natural History).

Johnson GD, Patterson C (1996) Relationships of lower euteleostean fishes. In: Stiassny MLJ, Parenti LR, Johnson GD (eds) Interrelationship of Fishes. San Diego: Academic Press, pp. 251–332.

Kalmijn AJ (1988) Hydrodynamic and acoustic field detection. In: Atema J, Fay RR, Popper AN, Tavolga WN (eds) Sensory Biology of Aquatic Animals. New York: Springer-Verlag, pp. 84–130.

Katz LC, Shatz CJ (1996) Synaptic activity and the construction of cortical circuits. Science 274:1133–1138.

Landmesser L (1972) Pharmacological properties cholinesterase activity and anatomy of nerve-muscle junctions in vagus-innervated frog sartorius. J Physiol 220:243–256.

Larsell O (1967) In: Jansen J (ed) The Comparative Anatomy and Histology of the Cerebellum from Myxinoids through Birds. Minneapolis; University of Minnesota Press, pp. 163–178.

Lecointre G, Nelson G (1996) Clupeomorpha, sister-group of ostariophysi. In: Stiassny MLJ, Parenti LR, Johnson GD (eds) Interrelationship of Fishes. San Diego: Academic Press, pp. 193–208.

Lewis ER, Leverenz EL, Bialek WS (1985) The vertebrate Inner Ear. Boca Raton, FL: CRC Press.

Lombard RE, Bolt JR (1988) Evolution of the stapes in paleozoic tetrapods. In: Fritzsch B, Ryan M, Wilczynski W, Hetherington T, Walkowiak W (eds) The Evolution of the Amphibian Auditory System. New York: Wiley, pp. 37–67.

Lumsden A, Krumlauf R (1996) Patterning the vertebrate axis. Science 274:1109–1115.

Manns M, Fritzsch B (1992) Retinoic acid affects the organization of reticulospinal neurons in developing Xenopus. Neurosci Lett 139:253–256.

Mansour SL, Goddard JM, Capecchi MR (1993) Mice homozygous for a targeted disruption of the proto-oncogene int-2 have developmental defects in the tail and inner ear. Development 117:13–28.

McCormick CA (1992) Evolution of central auditory pathways in anamniotes. In: Webster DB, Popper AN, Fay RR (eds) The Evolutionary Biology of Hearing. New York: Springer-Verlag; pp. 323–349.

McKay IJ, Lewis J, Lumsden A (1996) The role of FGF-3 in early inner ear development: an analysis in normal and kreisler mutant mice. Dev Biol 174:370–378.

Metin C, Frost DO (1991) Visual responses on neurons in somatosensory cortex of hamster with experimentally induced retinal projections to somatosensory thalamus. In: Finlay BL, Innocenti G, Scheich H (eds) The Neocortex: Ontogeny and Phylogeny. London: Plenum Press, pp. 219–228.

Michelsen A (1992) Hearing and sound communciation in small animals: evolutionary adaptation to the law of physics. In: Webster DB, Popper AN, Fay RR (eds) The Evolutionary Biology of Hearing. New York: Springer-Verlag, pp. 61–78.

Moore DR (1992) Developmental plasticity of the brainstem and midbrain auditory nuclei. In: Romand R (ed) Development of Auditory and Vestibular Systems 2. Amsterdam: Elsevier, pp. 297–320.

Myat A, Henrique D, Irsh-Horowicz D, Lewis J (1996) A chick homologue of serrate and its relationship with notch and delta homologues during central neurogenesis. Dev Biol 174:233–247.

Noden DM (1987) Interactions between cephalic neural crest and mesodermal populations. In: Maderson PFA (ed) Developmental and Evolutionary Aspects of the Neural Crest. New York: Wiley, pp. 89–120.

Northcutt RG (1992) The phylogeny of octavolateralis ontogenies: a reaffirmation of Garstang's phylogenetic hypothesis. In: Webster DB, Popper AN, Fay RR (eds) The Evolutionary Biology of Hearing. New York: Springer-Verlag, pp. 21–47.

Pallas SL (1991) Cross-modal plasticity in sensory cortex. In: Finlay BL, Innocenti G, Scheich H (eds) The Neocortex: Ontogeny and Phylogeny. London: Plenum Press, pp. 205–218.

Peusner KD (1992) Development of vestibular nuclei. In: Romand R (ed) Development of Auditory and Vestibular Systems 2. Amsterdam: Elsevier, pp. 489–518.

Platt C, Popper AN (1996) Sensory hair cell arrays in lungfish inner ear suggest retention of the primitive patterns for bony fishes. Soc Neurosci Abstr 22:1819.

Reichert C (1837) Über die Visceralbögen der Wirbeltiere im allgemeinen und deren Metamorphose bei den Vögeln und Säugetieren. Arch Anat Physiol 120–222.

Retzius G (1881) Das Gehörorgan der Wirbeltiere: I. Das Gehörorgan der Fische und Amphibien. Stockholm: Samson and Wallin.

Roberts BL, Meredith GE (1992) The efferent innervation of the ear: variations on an enigma In: Webster DB, Popper AN, Fay RR (eds) The Evolutionary Biology of Hearing. New York: Springer-Verlag, pp. 182–210.

Schellart NAM, Popper AN (1992) Functional aspects of the evolution of the auditory system of actinopterygian fish. In: Webster DB, Popper AN, Fay RR (eds) The Evolutionary Biology of Hearing. New York: Springer-Verlag, pp. 295–321.

Starck D (1979) Vergleichende Anatomie der Wirbeltiere. Berlin: Springer-Verlag.

Stiassny MLJ, Parenti LR, Johnson GD (1996) Interrelationship of Fishes. San Diego: Academic Press.

Swalla BJ, Jeffrey WR (1996) Requirement of the Manx gene for expression of chordate features in a tailless ascidian larva. Science 274:1205–1208.

Tanabe Y, Jessell TM (1996) Diversity and pattern in the developing spinal cord. Science 274:1115–1123.

Tessier-Lavigne M, Goodman CS (1996) The molecular biology of axon guidance. Science 274:1123–1133.

Thulborn T, Warren A, Turner S, Hamley T (1996) Early carboniferous tetrapods in Australia. Nature 381:777–780.

Toerien MJ (1963) Experimental studies on the origin of the cartilage of the auditory capsule and columella in *Ambystoma*. J Embryol Exp Morphol 11:459–473.

Van Bergijk WA (1966) Evolution of the sense of hearing in vertebrates. Am Zool 6:371–377.

Webster DB (1992) Epilogue to the conference on the evolutionary biology of hearing. In: Webster DB, Popper AN, Fay RR (eds) The Evolutionary Biology of Hearing. New York: Springer-Verlag, pp. 787–793.

Werner G (1960) Das Labyrinth der Wirbeltiere. Jena: Fischer Verlag.

Will U, Fritzsch B (1988) The octavus nerve of amphibians: patterns of afferents and efferents. In: Fritzsch B, Ryan M, Wilczynski W, Hetherington T, Walkowiak W (eds) The Evolution of the Amphibian Auditory System. New York: Wiley, pp. 159–184.

Wilm C, Fritzsch B (1992) Ipsilateral retinal projections into the tectum during regeneration of the optic nerve in the cichlid fish *Haplochromis burtoni*. J Neurobiol 23:692–707.

3
The Auditory Periphery in Fishes

Arthur N. Popper and Richard R. Fay

1. Introduction

Fossil evidence shows that an inner ear is found in the most primitive of jawless vertebrates (e.g., Stensiö 1927; Jarvick 1980; Long 1995). It may never be known whether these vertebrates actually were able to "hear" or whether the earliest ear may have been only a vestibular organ for the detection of angular and linear accelerations of the head. However, it is not hard to imagine that such a system could have ultimately evolved into a system for detection of somewhat higher frequency sounds during early vertebrate evolution (van Bergeijk 1967; Popper and Fay 1997). Although some might argue that sound detection would not have evolved until fish, or predators, started to make sounds, this may not be a valid argument. In fact, it is very likely that the earliest role, and still the most general role, for sound detection is to enable a fish to gain information about its environment from the environment's acoustic signature (e.g., Popper and Fay 1993). Such a signature results from the ways a sound field produced by sources such as surface waves, wind, rain, and moving animals is scattered by things like the water surface, bottom, and other objects.

It was suggested late in the 1800s that the ear arose as an invagination of the lateral line and that the initial function of the ear was as a gravity receptor (reviewed in van Bergeijk 1967; Popper et al. 1992). A number of arguments were made for the common origin of the ear and lateral line, including a supposed commonality of embryonic origin and innervation as well as function. This view led to the acousticolateralis hypothesis (reviewed by Wever 1974; Popper et al. 1992). Although it is now clear that the acousticolateralis hypothesis was incorrect in many ways, including the notions of the common embryonic origin and innervation of the ear and lateral line (Wever 1974; Northcutt 1980; Popper et al. 1992), there are certainly many structural and functional similarities between these two systems.

Thus the use of the anatomic term "octavolateralis" has come into favor and applies to the mechano- and electrosensory systems utilizing hair cells as receptors and utilizing branches of the eighth (octaval) and the lateral line nerves (McCormick 1978, 1982; Northcutt 1980; Popper et al. 1992). The term octavolateralis does not imply common acoustic function or common evolutionary or ontogenetic origins to these systems, but it does acknowledge basic morphological and functional similarity among the various mechanosensory end organs (see Coombs and Montgomery, Chapter 8, for a discussion of the lateral line system).

Although it is not certain that hearing is as old as the most primitive fishes, it is likely that the auditory role of the ears and associated peripheral structures has been evolving in fishes for hundreds of millions of years. Thus it is not surprising that we find extraordinary diversity in ear structures among fishes (e.g., Weber 1820; Retzius 1881; Corwin 1981a; Platt and Popper 1981; Popper and Coombs 1982). We do not yet know whether the diversity in structure is also reflected in a similar diversity of function (see Fay 1988), but behavioral and neurophysiological data on a few species suggest that basic hearing functions are more conservative than the structures that underlie them (Popper and Fay 1997).

1.1 Cast of Characters

This chapter explores the diversity of structure and function of the auditory system peripheral to the brain. Most of the discussion concentrates on the bony fishes (the class Actinopterygii) because it is for these species that most data are available. Of the bony fishes studied, the vast majority are members of the subclass Neopterygii and the division Teleostei (see Nelson 1994 for a discussion of the fish phylogeny that is used in this chapter; also see Long 1995 for a discussion of primitive fishes). However, some data are available for nonteleost bony fishes, the jawless fishes (the superclass Agnatha), and the cartilaginous fishes (the class Chondrichthyes), although these will not be gone into in great detail in this chapter.

For the sake of clarity, the word "fish" refers to actinopterygians and Chondrichthyes. The term "bony fishes" refers to members of the actinopterygians, and the term "cartilaginous fishes" refers only to the Chondrichthyes. We will also refer to gnathostomes, or jawed vertebrates, in contrast to the jawless agnathans, which include the lampreys and hagfish. The chapter deals primarily with the inner ear and auditory nerve, but when considering bony fishes, there will also be a discussion of the structures and functions of the swim bladder and other ancillary structures that are thought to enhance hearing capabilities.

Finally, among the fishes, we frequently refer to hearing "generalists" and "specialists." The hearing specialists are species having special peripheral structures that enhance hearing, whereas hearing generalists (or "nonspecialists") are fishes without such specializations. The most widely known specialists, and the ones that we refer to most often, are members of the superorder Ostariophysi, in which all members of one major group, the series Otophysi (catfish, carp, and relatives), have a particular specialization, the Weberian ossicles, for hearing. Other hearing specialists are found in the subdivision Clupeomorpha (herrings and relatives) and scattered among most of the other major bony fish groups.

2. Underwater Acoustics

Sound is propagated in all media as longitudinal waves carried by pressure fluctuations and particle motions. Particle motions have been described as those occurring in the "near-field" and the "far-field" of sound sources (e.g., van Bergeijk 1967; Kalmijn 1988a,b, 1989). Far-field particle motions accompanying propagated sound in a free field can be predicted with pressure measurements and the acoustic impedance of the medium. Near-field particle motions are hydrodynamic flows that occur near vibrating sources and attenuate very rapidly with distance from the source. The distance over which hydrodynamic motion exceeds particle motion associated with the propagating sound wave is defined in wavelength units and is thus frequency dependent. For a sphere fluctuating symmetrically in volume (monopole), the distance at which hydrodynamic and acoustic particle motions are equal in amplitude occurs at $\frac{1}{2\pi}$ wavelength (approximately one-sixth of a wavelength) from the source. For more complex (and realistic) sources such as the vibrating sphere (dipole), this critical distance is greater by a factor of ~1.4 (reviewed in Kalmijn 1988a,b, 1989; Rogers and Cox 1988; see also Coombs and Montgomery, Chapter 8). The region from the source to this point has been traditionally called the acoustic near-field, whereas the region beyond this point has been called the acoustic far-field (van Bergeijk 1967). In water, the near-field of a source extends large distances compared with air, in part because the speed of sound (and hence wavelength) is correspondingly greater in water.

An important characteristic of the near-field region is that particle motion amplitude attenuates rapidly with distance from the source, producing rather steep spatial gradients. Acoustic pressure and particle motion attenuate as $1/r$ for a propagated spherical wave in the far-field, where r is proportional to distance from the source. By contrast, the amplitude of hydrodynamic flow attenuates as $1/r^2$ for a monopole source and as $1/r^3$ for a dipole source. It is important to understand, however, that the boundary as defined above is somewhat arbitrary and that pressure fluctuations and particle motions occur within both the near- and far-fields.

The propagation of sound underwater depends on frequency and depth (Rogers and Cox 1988). In general, the deeper the water, the lower the frequencies that can be propagated. For example, the lowest frequency that will be propagated in water 1 m deep over a rock bottom is ~300 Hz, whereas in water 10 m deep, the lowest frequency would be ~30 Hz. For "softer" bottoms (e.g., silt, sand), the cutoff frequency is higher. In general, fishes moving into shallow-water environments will be at a disadvantage in detecting distant sources. Rogers and Cox (1988) point out that many shallow-water species exhibit special adaptations for high-frequency hearing (see below). Thus high-frequency hearing among fishes, such as in many otophysans, may be an adaptation for detecting sources at greater distances in shallow water.

2.1 Sound Detection Mechanisms

Near-field hydrodynamic flow may be detected by lateral line organs (e.g., Denton and Gray 1983, 1989; Coombs et al. 1996; see also Coombs and Montgomery, Chapter 8) and by the otolith organs of the ear by somewhat different processes. Stimulation of the lateral line can occur when there is a relative movement between the body and the surrounding medium caused by rather steep spatial gradients of pressure or particle motion amplitude (Coombs and Montgomery, Chapter 8). Otolith organs may respond directly to motion of the body as inertial detectors (de Vries 1950; Fay 1984; Fay et al. 1994; Lu et al. 1996; Fay and Edds-Walton 1997a,b). As the fish moves with the particle motion of the surrounding medium, the otoliths are thought to move at a different amplitude and phase due to their greater density. In this way, a relative displacement between the otolith and underlying hair cells occurs that is proportional to acoustic particle motion (i.e., displacement, velocity, or acceleration, depending on the mode of attachment between the otolith and hair cells). It is likely that all otolith organs in all species tend to respond to sound-induced motions of the fish's body in both the near- and far-fields.

In many fish species, the otoliths may also receive a displacement input from the swim bladder or other gas-filled chamber near the ears (see Section 4.4). Motions of the walls of these chambers are created by changes in their volume as sound pressure fluctuates. These displacements may be efficiently transmitted to one or more of the otolith organs by a variety of specialized pathways. Thus the ears of many fishes may respond in proportion to acoustic pressure as well as particle motion. Species having a particularly efficient mechanical coupling between the gas-filled chamber and the otolith organs (i.e. the hearing specialists; see below) tend to have high sensitivity to sound pressure and may hear in a relatively wide frequency range (up to, or above, 2 kHz).

As is discussed by Fay and Megela Simmons (Chapter 7), this dual sensitivity to pressure and particle motion may provide the animal with valuable information about sound source characteristics, including distance and location (Buwalda 1981; Schuijf and Hawkins 1983; Fay 1984; Schellart and de Munck 1987; Popper et al. 1988; Rogers et al. 1988). However, sensitivity to both sound pressure and particle motion has made the study of hearing in fish rather difficult and at times confusing. For example, specifying a sound detection threshold in a behavioral or physiological experiment requires a determination of whether pressure or particle motion is the effective stimulus. The answer may depend on species, type of source, frequency, distance from the source, and the acoustic characteristics of the environment.

3. Cells of the Octavolateralis System

Before considering the detailed structure and physiology of the ear, it is important to have some understanding of the transduction and support cells of the octavolateralis system (Fig. 3.1). The sensory hair cells are the transducing elements of the octavolateralis system, whereas the support cells appear to provide the proper environment for hair cell function.

3.1 Supporting Cells

As seen in Figure 3.1, support cells extend from the surface of the sensory epithelium to the basal membrane. The supporting cells surround each hair cell and, as in tetrapod ears, are likely to provide an interlocking surface to the epithelium to maintain fluids of different ionic concentration (i.e., perilymph vs. endolymph) across the tops and bottoms of the hair cells. The apical surfaces of the supporting cells are generally covered with short microvilli (Figs. 3.1 and 3.2), which presumably anchor the otolithic membrane and overlying otolith (or cupula) in place. The supporting cells may also be involved in the generation of new sensory hair cells (Corwin and Warchol 1991; Lanford et al. 1996; Presson et al. 1996).

3.2 Sensory Hair Cells

The sensory hair cells found in the octavolateralis system of all fishes are similar to the sensory hair cells of tetrapods. Hair cells are elongate epithelial cells with an apically located ciliary bundle (Figs. 3.1 and 3.2). Each ciliary bundle contains a large number of microvillus-like stereocilia (also called stereovilli) and a single, eccentrically placed kinocilium. The kinocilium is a true cilium with a 9 + 2 filament pattern (Fig. 3.1C). The stereocilia are often graded in size, with the longest lying closest to the kinocilium.

The lengths of the ciliary bundles vary depending on the location of the hair cell on the epithelium. Typically, the hair cells closest to the edges of the otolithic epithelia have long kinocilia and short stereocilia. In contrast, hair cells in other epithelial regions may have short or long ciliary bundles. The functional significance of the different bundle lengths has not been experimentally verified in fish. In general, however, hair cells with the longest bundles respond to the lowest frequencies (e.g., those of the semicircular canal cristae), whereas hair cells with the shortest bundles (e.g.,

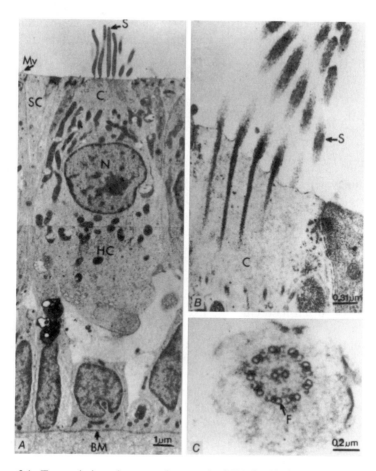

FIGURE 3.1. Transmission electron micrographs (TEM) of hair cells of the saccular macula of *Gymnothorax* sp., a moray eel. A: a single hair cell (HC) surrounded by a number of support cells (SC). The cell has a number of stereocilia (S) that are embedded in a cuticular plate (C). B: higher magnification of a different cell showing the stereocilia embedded in the cuticular plate. C: cross section through a single kinocilium showing the typical 9 + 2 filament pattern that is found in motile cilia. BM, basement membrane; Mv, microvilli on supporting cell apical surface; N, nucleus of hair cell. (From Popper 1983, with permission of University of Michigan Press.)

those of the rostral end of the saccule of otophysan fishes) respond best to higher frequencies (Popper and Platt 1983; Platt and Popper 1984; Sugihara and Furukawa 1989).

The only significant difference in hair cell ultrastructure from the described pattern has been reported for the hagfish *Myxine glutinosa* (Lowenstein and Thornhill 1970). Most notably, the kinocilia in *Myxine*

FIGURE 3.2. Scanning electron micrograph of the saccular macula of the Hawaiian lizardfish *Saurida gracilis* illustrating a shift in the orientation of the ciliary bundles. These hair cells are divided into two groups on opposite sides of the dashed line. All of the ciliary bundles in each group are oriented in the same direction as indicated by the arrows. Ciliary bundle orientation is defined by the side of the bundle on which the kinocilium is located. In each case, the kinocilium is the longest cilium in a bundle. (From Popper 1983, with permission of University of Michigan Press.)

lack the central pair of microfilaments and so have a 9 + 0 pattern as opposed to the 9 + 2 pattern in all other vertebrates. In addition, Lowenstein and Thornhill occasionally encountered kinocilia in the center of the bundle of stereocilia rather than at the edge, and other cells that had two kinocilia close to one another. The functional and/or evolutionary significance of these observations is unknown, although it is possible that the differences in the ciliary structure from that found in all other vertebrates is a derived characteristic and does not represent ancestral ciliary patterns.

3.3 Physiology of Hair Cells

Much of what we understand about the physiology of hair cells in fishes comes from studies from the lateral line (e.g., Flock 1965, 1971), the saccule of frogs (e.g., Hudspeth and Corey 1977; Hudspeth 1983), and the auditory papillae of terrestrial animals (e.g., Weiss et al. 1976; reviewed in Corwin and Warchol 1991). It is generally believed that all vertebrate hair cells function according to similar principles. A displacement or pivoting of the stiff stereocilia in the direction of the kinocilium results in a depolarization of the hair cell membrane potential, whereas displacement in the opposite direction causes hyperpolarization. The potential change is asymmetrical, with larger saturated depolarizations than hyperpolarizations. This asymmetry leads to an effective DC component of the voltage response to sinusoidal deflection and to harmonic distortion. The depolarized membrane potential generally saturates for displacements >100 nm (e.g., Dallos 1996; Kros 1996).

Sensory hair cells are directional transducers in the sense that the cell's response is determined only by the amplitude of stereociliary deflection along a single axis (Flock 1965; Hudspeth and Corey 1977) defined by a line from the center of the hair bundle to the kinocilium, parallel to the apical surface of the cell. Membrane potential change is approximately a cosine function of the deflection angle with respect to this axis. Because the voltage function of deflection magnitude shows a compressive nolinearity, the cosine function of deflection angle may be distorted at high stimulus levels.

Electrophysiological studies on isolated saccular hair cells of the goldfish (*Carassius auratus*) and toadfish (*Opsanus tau*) have identified at least two different cell types with respect to basolateral membrane properties. Resonating and spiking hair cells have also been reported (Sugihara and Furukawa 1989) in the goldfish saccule. Resonance (40 to over 200 Hz) in response to a current step was observed in relatively short cells from the rostral region of the saccular epithelium. These cells exhibited a calcium-activated potassium current and an "A"-type current. Taller hair cells from the caudal regions of the saccule produced spikes and subsequent voltage plateaus in response to current steps. These cells exhibited sodium and calcium conductances in addition to several potassium conductances (Sugihara and Furukawa 1989), and they also have ultrastructural differences that may be correlated with differences in conductance (Lanford and Popper 1994; Saidel et al. 1995).

In *Opsanus* (Steinacker and Romero 1992), cells with calcium-activated potassium conductances were shown to exhibit a "ringing" or resonant response to current commands with center frequencies averaging 142 Hz. A second class of cells having calcium-activated, voltage-controlled, and A-type potassium currents produced spikes but not resonant behavior. The functional significance of resonant hair cells is that they may begin to accomplish a peripheral frequency analysis, possibly in concert with micromechanical mechanisms (Fay 1997; Fay and Edds-Walton 1997b). The functional significance of spiking hair cells is not clear.

Ultrastructural results also show differences between rostral and caudal saccular hair cells. Hair cells from the rostral portion of the epithelium tend to have short cell bodies (Sento and Furukawa 1987; Saidel et al. 1995) and short stereocilia (Platt and Popper 1984), are reactive to the calcium-binding protein antibody S-100 (Saidel et al. 1995), have numerous small synaptic bodies (Lanford and Popper 1994), and tend to oscillate in response to depolarizing current clamps as summarized above. In contrast, caudal hair cells have long stereocilia, do not react to S-100, and have larger synaptic bodies. In goldfish, large-diameter saccular afferents tend to innervate the rostral saccular epithelium, whereas the more numerous smaller diameter fibers tend to terminate caudally (Furukawa and Ishii 1967; also see Saidel et al. 1995). The physiological response properties of afferents innervating these different saccular regions are discussed in Section 5.

3.4 Hair Cell Heterogeneity

Before discussing sensory hair cells in fishes, it is important to note that, historically, it was believed that fishes only had a single type of sensory hair cell. This belief resulted from early electron-microscopic data, which showed that amniotes had two types of vestibular hair cells (called types I and II), whereas anamniotes only had a single hair cell type that closely resembled the amniote type II cell. These cells differed in shape and inner-vation, with the type I cell being longer and surrounded by a nerve calyx, whereas the type II cell was shorter and rounder and had more individual synapses on its basal surface. The type II cell was assumed to be more primitive (e.g., Wersäll et al. 1965). With more recent studies, we now know that amniotes not only have two types of vestibular hair cells but that birds and mammals have at least two additional types of hair cells in their audi-tory end organs. Birds have tall and short hair cells in the basilar papilla, whereas mammals have inner and outer hair cells in the organ of Corti (Manley and Gleich 1992; Slepecky 1996).

3.4.1 Hair Cell Heterogeneity in Fishes

Recent studies of the ultrastructure of the hair cell bodies have demon-strated that at least some species of fish, such as amniotes, have several types of sensory hair cell in their otolithic end organs (Fig. 3.3). These findings have led to the suggestion that in fishes, as in amniotes, physiologi-cal properties of hair cells may be correlated with ultrastructure (e.g., Saidel et al. 1995), as discussed in Section 3.3.

Detailed analysis with a variety of techniques shows that there is signifi-cant hair cell heterogeneity within the ears of fishes (Wegner 1982; Chang et al. 1992; Popper et al. 1993; Saidel et al. 1995; Lanford and Popper 1996). Hair cell heterogeneity can be seen in illustrations in publications about several elasmobranch species (Lowenstein et al. 1964; Corwin 1977). Work by Hoshino (1975) also demonstrated the presence of at least two types of hair cells in the ears of the lamprey *Entosephenus japonicus*, although

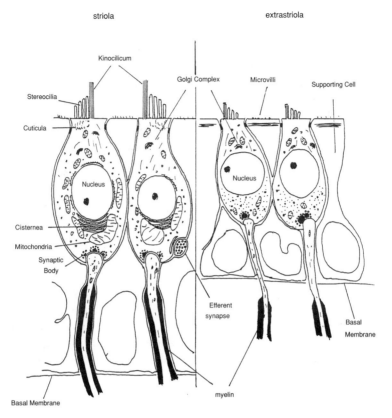

FIGURE 3.3. Schematic diagram of sensory hair cells from the oscar *Astronotus ocellatus*. The cells at the left, now called type I-like hair cells, are found in the striola region of the utricle and in the central part of the saccular epithelium, whereas the cells at the right, type II hair cells, are found outside the utricular striolar region and at the margins of the saccule. Similar types of hair cells are found in the ears of other species as well as in the lateral line. Each fish sensory hair cell is surrounded by supporting cells that reach from the apical membrane of the epithelium to the basal membrane. The hair cells extend only part way to the basal membrane. The apical end of each hair cell has a cuticular plate into which is embedded a series of microvillus-like stereocilia. At one end of the group of stereocilia is a single true cilium, or kinocilium. This ciliary bundle projects into the lumen of the end organ. Type I-like cells receive both afferent and efferent innervation, whereas type II cells often only have afferent innervation. (From Chang et al. 1992, © John Wiley and Sons, Inc. 1992.)

these do not have the same ultrastructural characteristics as those that define hair cell types in bony and cartilaginous fishes. Finally, we now have evidence that heterogeneity also occurs in the teleost lateral line (Song et al. 1996).

Initially, two types of hair cells were identified in each of the otolith end organs of the cichlid *Astronotus ocellatus* (the oscar) (Chang et al. 1992; Popper et al. 1993) (Fig. 3.3). One of these cells was first identified in the striolar region of the utricle of *Astronotus*, whereas the other cell type was found in the extrastriolar region. The striolar region of the epithelium is a zone characterized by hair cells having relatively short ciliary bundles, with adjacent hair cells having opposed directional orientations. The striolar hair cells were found to have extensive rough endoplasmic reticulum (ER) just below the nucleus, very large mitochondria, and very small synaptic bodies associated with the synaptic regions. In contrast, the synaptic bodies of the extrastriolar hair cells are large with no subnuclear ER and relatively small mitochondria. Several structural and biochemical studies on the utricle of *Astronotus* revealed differences between striolar and extrastriolar hair cells. For example, striolar cells are reactive with an antibody to the calcium-binding protein S-100 and express cytochrome oxidase, whereas the extrastriolar hair cells do not show these reactions (Saidel et al. 1990; Saidel and Crowder 1997).

Morphologically, extrastriolar hair cells closely resemble amniote type II hair cells and have been given that name. The striolar hair cells bear striking resemblance to mammalian type I hair cells. However, the definition of amniote type I hair cells includes the presence of a unique "chalice" or calyx innervation in which an afferent fiber envelops most of the hair cell body. Because this type of innervation has not been observed in the otolithic end organs of fishes, striolar cells have been called type I-like hair cells (Chang et al. 1992). At the same time, calyxlike endings have been found associated with hair cells in the semicircular canal cristae of *Carassius auratus* (Lanford and Popper 1996).

3.4.2 Evolution of Hair Cell Heterogeneity

Several arguments support the idea that two hair cell types evolved very early in the evolution of the octavolateralis system. First, the two types are found in diverse teleost taxa (Wegner 1982; Chang et al. 1992; Popper et al. 1993; Saidel et al. 1995; Lanford and Popper 1996). Second, there is evidence that the same types of cells are also found in the lateral line of *Astronotus* (Song et al. 1996), suggesting that they were present very early in the evolution of sensory hair cells. Third, examination of published figures from studies on elasmobranchs (Lowenstein et al. 1964; Corwin 1977) shows structures in some hair cells that resemble teleost type I-like hair cells.

3.5 *Hair Cell and Nerve Fiber Addition*

Unlike most other vertebrates, fishes continue to add large numbers of sensory hair cells in the otic end organs for much (if not all) of an animal's life (Corwin 1981a,b, 1983; Popper and Hoxter 1984; Lombarte and Popper 1994). This is in contrast to amphibians (Corwin 1985) and birds (Jørgensen and Mathiesen 1988), which only add a small number of hair cells postembryonically, and mammals, which may not add hair cells at all (e.g., Forge et al. 1995; Rubel et al. 1995).

Analysis of several end organs of the elasmobranch ear demonstrates that hair cell addition continues for a number of years (Corwin 1981b, 1983). Evidence for hair cell addition is also found among diverse bony fishes including goldfish (Platt 1977) and *Astronotus* (Popper and Hoxter 1984). The most comprehensive example of addition comes from a study of the saccule, lagena, and utricle of the hake *Merluccius merluccius*, a commercially important fish whose life history is well known (Lombarte and Popper 1994). Figure 3.4 shows that a 6-month-old hake has ~5000 saccular hair cells, whereas a 9-year-old hake will have over one million. Interestingly, hair cell addition is greatest in the saccule of this species and particularly in the region of the saccule closest to the swim bladder (Lombarte and Popper 1994).

Is the addition of hair cells accompanied by an increase in the number of afferent neurons? This question has not been studied extensively, and the results are contradictory. No increase in the number of eighth neurons was observed in the ray (*Raja clavata*) as hair cells were added, suggesting that the number of hair cells innervated by each neuron increased with the age of the animal (Corwin 1983). In contrast, there was a small but significant

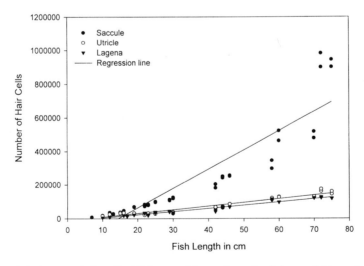

FIGURE 3.4. Sensory hair cell addition vs. fish age in each of the otolithic end organs of the hake (*Merluccius merluccius*) for fish from ~7-cm total length (6 months of age) to 75 cm (9 years old). Note that the greatest amount of hair cell addition occurred in the saccule, whereas hair cell addition was about comparable in the lagena and utricle. (Data from Lombarte and Popper 1994.)

increase in the number of nerve fibers to the saccule of *Astronotus* as the number of hair cells increased (Popper and Hoxter 1984), and this increase has been corroborated in developmental studies (Presson and Popper 1990). However, although the number of hair cells in *Astronotus* increased by at least 100-fold, the number of neurons only increased ~4.8 times. As in the ray, this results in a substantial increase in the number of hair cells innervated by each neuron as *Astronotus* grows, with the result being a ratio of hair cells to neurons of 30:1 in small fish to over 300:1 in the largest fish (Popper and Hoxter 1984).

3.5.1 Significance of Hair Cell Addition

The functional significance of the postembryonic addition of hair cells is not really understood. In the macula neglecta of *Raja clavata*, Corwin (1983) found evidence of increased neural sensitivity to sound as the number of hair cells innervated by each neuron increased. In contrast, Rogers and colleagues (Popper et al. 1988; Rogers et al. 1988) noted that as fishes grow, changes occur in the relative sizes and positions of the different structures associated with hearing. They developed a model that predicts that growth in the number of hair cells tends to maintain sound detection and processing capabilities in the fact of this structural change. At this point, there are no other data to directly test these competing hypotheses. The only indirect data come from a study on goldfish 45 and 120 mm in standard length, showing that hearing sensitivity and frequency range do not vary with size (Popper 1971). Yet it is possible that this size difference was not large enough to show differences in hearing sensitivity.

4. The Auditory Periphery

The auditory periphery in bony fishes includes the inner ear and, in many species, ancillary structures that may enhance hearing capabilities (both sensitivity and bandwidth). No ancillary structures have been described for elasmobranchs and agnathans.

4.1 *The Ear*

The inner ear in fishes and elasmobranchs includes three semicircular canals and associated sensory regions (cristae ampullaris) and three otolithic end organs, the saccule, utricle, and lagena (Figs. 3.5, 3.6, and 3.7). Some species of fish and all elasmobranch (Fig. 3.7) have a seventh end organ, the macula neglecta (Retzius 1881; Corwin 1977, 1981a). The ear in agnathans (Fig. 3.8) is strikingly different from that in jawed fishes (Retzius 1881; Lowenstein et al. 1968; Lowenstein and Thornhill 1970) in that various species have one or two canals as opposed to three canals found in all other vertebrates. Although these species have otolithic organs, they are

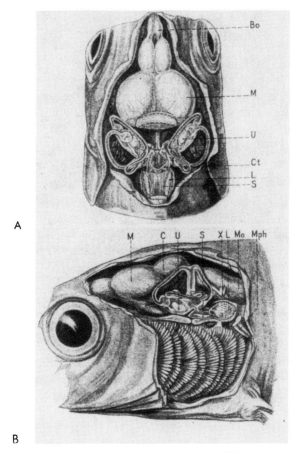

FIGURE 3.5. Head of an otophysan fish, the minnow (*Phoxinus laevis*), showing the ear and brain. This ear is very similar to that of the goldfish, a closely related otophysan. A: dorsal view. B: lateral view. Aa, Ae, Ap, ampullae of semicircular canals; Bo, olfactory bulb; C, cerebellum; Ca, Ch, Cp, semicircular canals; Ct, transverse canal; L, lagena; M, midbrain; Mo, medulla; Mph, pharyngeal muscles; Raa, Rl, Rs, rami of eighth cranial nerve innervating ear; S, saccule; Ss, common crus; U, utricle; X, cranial nerve X. (From von Frisch and Stetter 1932). See Figure 3.5A for details of the structure of the ear of *Phoxinus*.

not likely to be serially homologous to the comparably named otolithic end organs of fishes (Popper and Hoxter 1987).

4.2 General Structure of the Ear in Fishes

4.2.1 Teleosts

Within teleosts, there is considerable variation in the structure of the otolithic end organs, as seen in Figure 3.6. The greatest variability is found

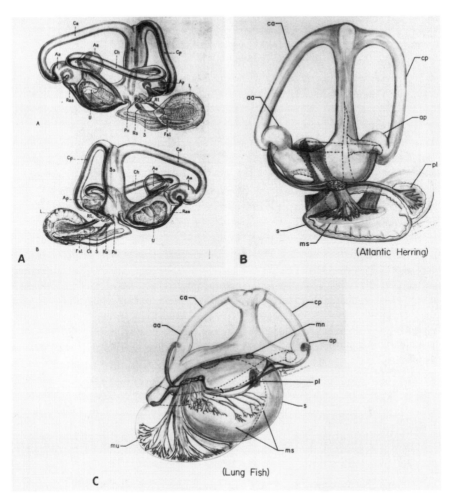

FIGURE 3.6. Drawings of the ears of three bony fishes to show the variation in structure encountered. A: minnow *Phoxinus laevis*, an otophysan (from Wohlfahrr 1933). Top lateral view, bottom medial view. B: Atlantic herring *Clupea harengus* (redrawn from Retzius 1881). C: lungfish *Protopterus* (redrawn from Retzius 1881). Aa, Ae, Ap, cristae of semicircular canals; Ca, Ch, Cp, semicircular canals; Ct, transverse canal (connecting left and right saccules); L, lagena; Fsl, foramen between saccule and lagena; mn, Pn, papilla (macula) neglecta; ms, saccular macula; mu, utricular macula; pl, lagenar macula; Raa, Rl, Rs, branches of eighth nerve; S, saccule; Ss, common canal; U, utricle.

in the end organs associated with hearing—the saccule in most species and the utricle in clupeids (Blaxter et al. 1981; Platt and Popper 1981). This variation includes the gross shapes and sizes of the end organs, as well as the shapes and sizes of the otoliths, the extent of sensory epithelium

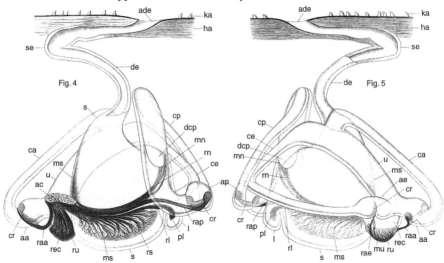

FIGURE 3.7. Drawing of the ear of the angle shark *Squantia angelus* (from Retzius 1881). Left is medial view and right is lateral view. Selected structures: aa, ae, ap, ampullae of semicircular canals; ac, eighth nerve; ade, opening to endolymphatic duct; ca, ch, cp, semicircular canals; cr, cristae of semicircular canals; de, endolymphatic duct; l, lagena; mn, macula neglecta; ms, saccular macula; mu, utricular macula; pl, lagenar macula; raa, rl, rs, ru, branches of eighth nerve; s, saccule; u, utricle.

covered by the otoliths, the shapes and sizes of the sensory epithelia, the hair cell orientation patterns on the epithelia, the lengths of the ciliary bundles in different epithelial regions, and the distribution of different hair cell types.

Perhaps the most pronounced differences in gross organ structure are the relative sizes of the saccule and lagena in otophysan fishes (including goldfish, carps, and catfishes) (Fig. 3.6A) compared with all other teleosts (often informally called "nonotophysans") (Fig. 3.6B). In the nonotophysan fishes, the saccule is generally much larger than the lagena (Fig. 3.6B). In contrast, the otophysan lagena has approximately the same epithelial area as the saccule (Fig. 3.6A).

The functional significance of this difference has not been shown experimentally, but there is reason to suggest that the saccule and lagena may play somewhat different roles in hearing in otophysan and nonotophysan fishes (Schellart and Popper 1992; Popper and Fay 1993). In otophysans, the saccule receives nondirectional, pressure-dependent input from the swim bladder, and the lagena responds primarily to particle motion (Fay 1984) (see Section 5.7). In some nonotophysans, the saccule may respond to both of these sound components. The function of the diminutive nonotophysan lagena is not known.

In general, the utricle does not vary substantially in gross structure among fishes or among most vertebrates other than the agnathans (Platt and Popper 1981). The only exceptions are found in fishes that use the utricle in sound detection. These include the clupeid (herringlike) fishes where the swim bladder actually projects to an air bubble that is intimate to the utricle (e.g., Blaxter et al. 1981) and where the utricle may be involved in detection ultrasonic sounds for predator avoidance (Mann et al. 1997). In addition, at least some of the Ariid catfishes (e.g., *Arius felis*) have an enlarged utricle compared with other fishes, and it has been suggested that this end organ is adapted for detection of low-frequency sounds used by these species in communication and in a form of echolocation (Popper and Tavolga 1981).

4.2.2 Elasmobranchs

The structural plan of the ear in elasmobranchs is basically the same as in other vertebrates (Fig. 3.7), with three semicircular canals, three otolithic end organs, and a macula neglecta (e.g., Tester et al. 1972; Corwin 1977, 1981a; Popper and Fay 1977). However, unlike most other vertebrates, the macula neglecta is often quite large (Corwin 1977, 1978, 1981a, 1989). It is located in the posterior canal duct (Fig. 3.7, de-endolymphatic duct) and consists of paired sensory epithelia overlaid by a single gelatinous cupula. The elasmobranch ear is embedded in cartilaginous otic capsules that lie just below a tiny pair of endolymphatic pores that open to the dorsal surface of the animal. These pores mark the position of the endolymphatic (or parietal) fossa, which is a "dished-out" area in the chondrocranium covered by a taut layer of skin (Corwin 1977, 1981a, 1989; Popper and Fay 1977).

4.2.3 Agnathans

The gross structure of the agnathan ear (Fig. 3.8) differs from that in other vertebrates. The lampreys (Petromyzoniformes) have two semicircular canals and a single, elongate, sensory epithelium that has been called the macula communis (Lowenstein et al. 1968) (Fig. 3.8B). The ear also has enlarged chambers that open to the macula communis and contain numerous multiciliated epithelial cells lining its walls. These ciliated cells, which are not found in any other vertebrate ears, appear to cause fluid movements within the chamber (Lowenstein et al. 1968; Popper and Hoxter 1987). The functional significance of this movement is not known.

The ear of the hagfish *Myxine glutinosa* was first described by Retzius (1881) and most recently by Lowenstein and Thornhill (1970) (Fig. 3.8A). If anything, the hagfish ear is even simpler than that in the lamprey. The labyrinth is toroidal in shape and is, for all practical purposes, a single canal that contains two semicircular canallike cristae that Lowenstein and Thornhill called anterior and posterior. Lying between the cristae,

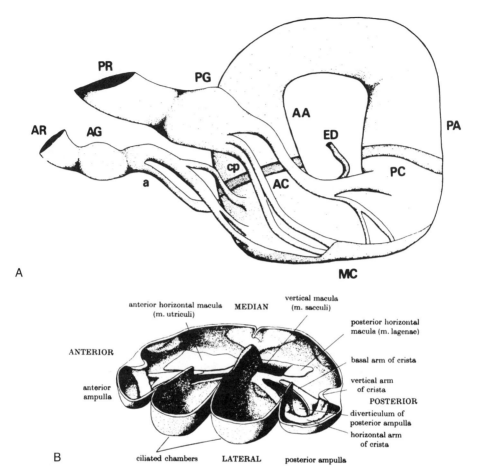

FIGURE 3.8. Schematic drawings of the ears of two jawless fishes. A: hagfish *Myxine glutinosa* (from Lowenstein and Thornhill 1970). a, Anterior nerve branches; AA, anterior ampulla; AC, anterior canal crista; AG, anterior ganglion (eighth nerve); AR, anterior ramus of eighth nerve; cp, central and posterior nerve branches; ED, endolymphatic duct (proximal part); MC, macula communis; PA, posterior ampulla; PC, posterior crista; PG, posterior ganglion (eighth nerve); PR, posterior ramus of eighth nerve. B: lamprey *Lampetra fluviatilis*. This schematic drawing shows the lower half of the left labyrinth. Note the locations of the ampullae of the two semicircular canals and their cristae as well as the position of the macula communis, which is made up of the m. utriculi, m. sacculi, and m. lagenae. The large ciliated chambers contain numerous cells, each of which has multiple kinocilia on its apical surface (Popper and Hoxter 1987). (From Lowenstein et al. 1968.)

essentially at the base of the toroid, is a macula communis. This macula is elongate and extends up into both the anterior and posterior chambers of the toroid, which results in sensory hair cells of the macula being located on several different planes of the hagfish.

4.3 Otolithic End Organs

Each of the gnathostome otolithic end organs, the saccule, utricle, and lagena, is a sac that includes two major structures (Fig. 3.6). Lining a portion of the wall of the sac is a sensory epithelium (or macula) that contains hair cells and supporting cells (Figs. 3.1 and 3.2). A calcium carbonate mass fills a portion of the sac, lying very close to the sensory epithelium (Fig. 3.6). In nonteleost bony fishes and in cartilaginous fishes and agnathans (as well as in all tetrapods), the mass is composed of elongate or fusiform crystals embedded in a gelatinous matrix (Carlström 1963; Gauldie 1996). In most teleosts, however, the crystals are fused into a solid mass, the otolith (Fig. 3.9). The functional difference between an otoconial mass and an otolith has not been investigated.

Both the otolith and otoconial mass are far denser than the rest of the fish's body. They are kept in position near the epithelium via a thin gelatinous otolithic "membrane" (Dunkelberger et al. 1980), which appears to connect to the surface of the epithelium and to the otolith (Popper 1977). In other fishes (and tetrapods), the otoconia are actually embedded in the otolithic membrane that is attached to the surface of the sensory epithelium.

The shape of the otoliths in teleost fishes, particularly the saccular otolith, is highly species specific (Fig. 3.9), and has been used to classify fossil fishes (see Gauldie 1996). The functional significance of this shape, if any, to audition is not known.

4.4 Ultrastructure of the Sensory Epithelium

4.4.1 Teleost Fishes

The sensory epithelium in each of the otolithic end organs contains hair cells and supporting cells (see Sections 3.1 and 3.2). The hair cells in each end organ are divided into "orientation groups," each of which has cells oriented in approximately the same direction as defined by having their kinocilia on the same side of the ciliary bundle (Fig. 3.2). The orientation directions may shift abruptly, creating groups of oppositely oriented hair cells separated by an uneven dividing line, often called the "striola." There is considerable interspecific variation in the hair cell orientation patterns in different fish species (Fig. 3.10).

The orientation pattern in the lagena and utricle tend to be relatively conservative among fishes, and the usual fish utricular pattern is also similar to the general tetrapod pattern (Fig. 3.10A). The only exception to this rule is in species where the utricle is adapted for sound detection, such as in the clupeids and the marine catfish (*Arius felis*) (see Section 4.2.1).

The saccular hair cell orientation pattern tends to be similar in species that do not have specializations thought to enhance hearing. These fishes generally only respond to sounds up to 300–500 Hz with relatively poor sensitivity (see Fay and Megela Simmons, Chapter 7). In contrast, hearing

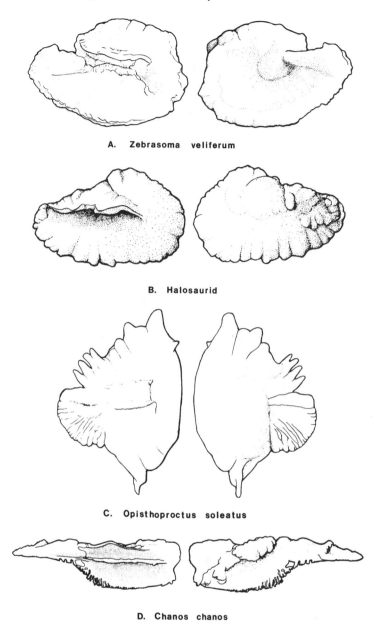

A. Zebrasoma veliferum

B. Halosaurid

C. Opisthoproctus soleatus

D. Chanos chanos

FIGURE 3.9. Saccular otoliths from four teleost species. In each case, the medial side of the otolith is on the left and the lateral side on the right. A: *Zebrasoma veliferum*. B: a halosaurid. C: *Opisthoproctus soleatus*. D: *Chanos chanos*. *Zebrasoma* and the halosaurid are typical of many nonotophysan species in having relatively ellipsoid otoliths with a deep groove (sulcus) on the medial side in which sits the sensory epithelium. *Opisthoproctus* and *Chanos* (an ancestor to otophysans) demonstrate some extremes in the shapes of saccular otoliths. (From Platt and Popper 1981.)

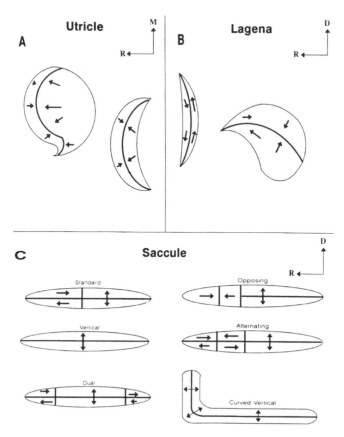

FIGURE 3.10. Schematic drawings of the sensory hair cell orientation patterns on otolithic organs from different fish species. These patterns represent the vast majority of fish species that have been studied and are drawn to demonstrate that the same basic patterns are found among a wide range of fishes. The arrows approximate the orientation of the major portion of the hair cells in each epithelial region (see Fig. 3.2). These regions are separated by darker lines. A: utricular patterns. The pattern on the left is the most commonly found among teleosts, whereas that on the right is less common. B: lagenar pattern on the left is the most common among fishes, but the one on the right has been found among all otophysan fishes. C: six different saccular patterns have been identified (see Popper and Coombs 1982). See text for discussion. D, dorsal; M, mediolateral; R, rostral. (Reprinted from Popper and Platt 1993. Copyright CRC Press, Boca Raton, Florida, © 1993.)

specialists, species that have wider hearing ranges and greater sensitivity than generalists, tend to have not only specializations that enhance sensitivity peripheral to the ear but also specializations within the otic end organ most closely linked to these specializations (Popper and Coombs 1982).

Analysis of the hair cells among bony fishes has demonstrated the presence of six basic hair cell orientation patterns (Fig. 3.10C) (Popper and Coombs 1982; Popper and Platt 1993). These are found across taxonomic groups, and it has been suggested that several patterns arose multiple times during the evolution of fishes in response to similarities in adaptive pressures for certain kinds of sound detection or processing (Popper and Coombs 1982).

The most common, or "standard," pattern is found in all teleost fishes that are thought to be hearing generalists based on their having a narrow bandwidth of hearing (e.g., Fay 1988; Fay and Megela Simmons, Chapter 7) or because they lack morphological structures usually associated with hearing specialization. Four patterns (opposing, crossing, vertical, and dual) are found in fishes that, based on behavioral data or morphological specializations, appear to be hearing specialists that use the saccule for audition (Popper and Coombs 1982).

For example, the vertical pattern is found in all otophysan fishes (Popper and Platt 1983) as well as in the unrelated mormyrids (Popper 1981). Otophysans have Weberian ossicles providing acoustic coupling between the swim bladder and the saccular chamber, whereas the mormyrids have an auxiliary air bubble attached to the saccule (Stipetić 1939). Similarly, the opposing pattern is found among a number of species in which the swim bladder makes close contact with the saccule. The relationship between peripheral specializations and specializations in the saccule is strikingly seen in two squirrelfishes (Holocentridae). One species, *Myripristis kuntee*, has the swim bladder contacting the saccule and an opposing saccular orientation pattern, whereas the standard pattern is found in *Adioryx xantherythrus*, a related species that does not have a connection between the swim bladder and saccule (Popper 1977). Based on behavioral data, *Adioryx* is considered a hearing generalist, whereas *Myripristis* is a specialist (Coombs and Popper 1979).

Although most of the interspecific variation found in otolithic end organs is associated with the saccule, there are at least two groups of teleosts in which the utricle is associated with specializations for hearing, and it is that end organ that has specializations in hair cell orientation patterns. One such species is the marine catfish, *Arius felis*, an otophysan species that has particularly acute hearing at ~200 Hz (Popper and Tavolga 1981), whereas other otophysans have their best hearing from 500 to 1,000 Hz. *Arius* has an enlarged utricle with hair cells only along the striolar region as opposed to all other otophysans, which have the normal pattern of extensive extrastriolar hair cells as well as striolar hair cells.

Another specialization in the utricle is found in the Clupeomorph fishes. These species have an air-filled auditory bulla associated with the utricle (Blaxter et al. 1981). Unlike other fishes, the utricle in clupeids is divided into three parts, and the combined hair cell orientation patterns in these regions are similar to the specialized patterns found in the saccules of hearing specialists (Popper 1977; Popper and Platt 1979; Popper and Coombs 1982). In contrast, the saccule in clupeids has the standard pattern found in species that do not have saccular specializations for hearing. At least some clupeids can detect ultrasound up to 180 kHz, and it has been suggested that the specialized utricle and associated structures may be involved in this capability (Mann et al. 1997).

4.4.2 Nonteleost Bony Fishes

The basic anatomy of the ear does not differ between teleost and nonteleost bony fishes. Species that have been studied include members of the Actinopterygian subclass Chondrostei, the bichir *Polypterus bichir* (Polypteriformes) and the shovel-nose sturgeon *Scaphirhynchus platorynchus* (Acipenseriformes) (Popper 1978), and two nonteleost members of the subclass Neopterygii, including the gar *Lepisosteus osseus* (Semionotiformes) (Mathiesen and Popper 1987) and the bowfin *Amia calva* (Amiiformes) (Popper and Northcutt 1983). Data are also available for representatives of each of the three lungfish groups (Platt and Popper 1996, unpublished data) and for the lagena of the coelacanth *Latimeria chulumnae* (Platt 1994).

The basic pattern of the lagena and utricle of the nonteleost bony fishes is similar to that of teleosts, although there are some small differences that may ultimately prove to be functionally significant if and when there are studies of hearing and inner ear function in these species. The saccule differs from teleosts in several ways. Most importantly, the hair cell orientation pattern for all the nonteleosts most closely resembled the vertical pattern (Fig. 3.10C, curved vertical pattern). However, although the vertical pattern appears to be derived in teleosts from a four-quadrant pattern (e.g., Popper and Northcutt 1983), in nonteleosts, the vertical pattern appears to be the primitive form of orientation. At the same time, hair cells in the saccule of nonteleosts are oriented in four directions. But rather than having hair cells in four distinct groups as in the teleosts, the shift in orientation from vertical to horizontal, when it occurs in nonteleosts, is related to developmental changes in the curvature of the epithelium and the resultant shift in the direction of hair cells (Fig. 3.10C) (Popper 1978; Popper and Northcutt 1983; Mathiesen and Popper 1987; Platt and Popper 1996, unpublished data).

It is also of interest that the nonteleost saccular hair cell pattern is reminiscent of that found in both the elasmobranch (see Section 4.1.3) and the tetrapod saccule (Spoendlin 1964), supporting the argument that these fishes may be more closely ancestral to tetrapods than to teleosts. Second, it is clear that the vertical pattern in these fishes is not related to that found in the otophysans or the mormyrids. It has been suggested that the vertical pattern found in teleosts is secondarily derived from fishes having a standard teleost pattern (Popper and Platt 1983).

4.4.3 Chondrichthyes

The ears of the sharks and rays are basically similar to the ears of other fishes (Fig. 3.7), with the exception that the chondrichthyes often have a large macula neglecta (e.g., Retzius 1881; Corwin 1977) that may be a primary sound detector (Corwin 1978). The ear is located just ventral to the skull, and elongate endolymphatic ducts project to an endolymphatic fossa in the dorsal chondrocranium.

The sensory epithelia in the ears of the few elasmobranchs that have been studied have orientation patterns that are generally similar to patterns in the primitive bony fishes (Section 4.1.2) (e.g., Lowenstein et al. 1964; Corwin 1977, 1978; Barber and Emerson 1980). The utricular orientation patterns are basically similar to those found in other fishes and tetrapods (e.g., Barber and Emerson 1980), although Lowenstein et al. (1964), using transmission electron microscopy (EM) rather than scanning EM (SEM), reported that the orientation pattern in the utricle of *Raja clavata* is random rather than organized as in other fishes.

The saccular pattern in the skate (*Raja ocellata*), studied using SEM, is close to a vertical pattern, although the cells at the rostral end of the epithelium are rotated somewhat rostrally and caudally in a pattern that resembles the teleost standard pattern (Barber and Emerson 1980). However, because there is no detailed analysis of orientation patterns of the saccule in this species, it is very possible that the pattern more closely resembles the curved vertical pattern of primitive teleosts (Fig. 3.10C). In contrast, Lowenstein et al. (1964) reported a totally uniform vertical pattern in the saccule of *R. clavata*.

The lagena in both species of *Raja* that have been studied have hair cells oriented dorsally and ventrally, but cells with opposing orientations are intermixed as opposed to being in separate groups as in other fishes and tetrapods. Finally, the macula neglecta is divided into two separate epithelia in elasmobranchs. All of the hair cells on each epithelium are oriented in generally the same direction, and the orientations on the two epithelia are opposite to one another (Corwin 1977, 1978). Corwin (1978) has demonstrated that there is substantial interspecific differences in the size of the macula neglecta in six different elasmobranchs, and he suggested a correlation between size and use of the macula neglecta in food finding.

4.4.4 Agnathans

The macula communis of the lamprey lies along the medial floor of the ear and is divided into three contiguous regions that have been called the anterior, middle, and posterior maculae (Lowenstein et al. 1968). These regions have, at times, been called the saccule, lagena, and utricle, with the assumption that these are homologous to like-named end organs of other vertebrates (Lowenstein et al. 1968). However, there is no evidence to demonstrate such homology (Popper and Hoxter 1987). The whole macula communis is covered by a thick otoconial layer, and the ciliary bundles on the hair cells of the macula are similar to those found in other fishes. There are no data to indicate whether lampreys are capable of detecting sound, although Lowenstein (1970) showed a sensitive response to vibration from the eighth nerve of isolated ears of *Lampetra*.

The macula communis of *Myxine* (the only hagfish for which there are data) is a single structure that lies at the base of the single toroidal chamber and extends into both the anterior and posterior vertical arms of the toroid (Lowenstein and Thornhill 1970). The macula is covered by a mass of otoconia and contains sensory hair cells that have a wide range of orientations. In general, those at either end of the macula appear to be generally directed upward into the arms and toward the cristae, whereas the hair cells on the part of the macula lying at the base appear to be oriented in a wide range of directions, with a tendency to be oriented perpendicular to the cells in the arms (Lowenstein and Thornhill 1970).

4.5 Functional Significance of Ear Structure in Bony Fishes

The functional significance of diversity in ear structure has not been explored systematically beyond the observations that one or the other of the otolith organs of the otophysan and nonotophysan hearing specialists appear to be adapted for receiving input from the swim bladder or another nearby gas bubble. Questions remain about the functional significance of differences in size and shape of the end organs and their otoliths, the extent of the sensory epithelium covered by the otolith (Popper 1978), and the degree to which end organs are segregated in separate sacs.

Although still speculative, there are three aspects of ear ultrastructure that may be correlated with function and that may ultimately provide insight not only into how individual ears function but also into the broader issue of why ears of various species are morphologically distinct. These three areas are (1) hair cell ultrastructure, (2) ciliary bundle length, and (3) hair cell orientation patterns. The significance of hair cell ultrastructure was considered in Sections 3.3 and 3.4 and ciliary bundle length in Section 3.2.

As discussed in Section 3.3, sensory hair cells are morphologically and physiologically polarized so that their response to hair bundle deflection is proportional to a cosine function of the direction of stimulation relative to the most sensitive axis of the cell. This axis is defined as a line from the center of the hair bundle through the eccentrically located kinocilium. The very fact that hair cells on any epithelium are divided into orientation groups leads to the suggestion that the patterns on the epithelium may function in directional hearing.

One might imagine, using a very simple model of the standard pattern (Fig. 3.10), that motion of the otolith relative to the sensory epithelium along the rostrocaudal axis would maximally excite cells in the rostral end of the epithelium, whereas there would be minimum stimulation to cells that are oriented orthogonally to this direction (the vertically oriented cells.) Conversely, motion on the vertical axis would maximally stimulate caudal cells and minimally stimulate the horizontally oriented cells. Finally, stimulation along an axis that is at a 45° angle would produce the same level of stimulation in both the horizontal and vertical cell groups.

If the assumption is then made that otolithic afferents are "labeled" according to the orientation of the hair cells that they innervate, at least to the first level of the auditory central nervous system (e.g., Fay 1984; Fay and Edds-Walton 1997a; Lu et al. 1998), then it could be possible to devise a model that computes the direction of a signal source by comparing across-neuron patterns of activity (e.g., Buwalda 1981; Schellart and de Munck 1987; Popper et al. 1988; Rogers et al. 1988). For example, many sensory epithelia have a complex curvature, thus increasing the range of stimulus directions represented. In addition, each end organ in an ear is oriented on nonparallel planes, and corresponding organs of the two ears are oriented along nonparallel axes. Thus a fish can potentially obtain a good deal of directional information by comparing separate inputs from hair cells from both ears and all end organs. It is not clear, however, whether inputs to the brain from differently oriented hair cells are spatially mapped (however, see Wubbles et al. 1993) or functionally labeled in any way. This is an important issue for most models of directional hearing (e.g., Schuijf 1975; Buwalda et al. 1983; Saidel and Popper 1983a,b; Rogers et al. 1988).

Although this model suggests a way that the axis of particle motion could be resolved, it still leaves open the question of determining the direction of wave propagation. In other words, which end of the axis points to the sound source? This has been called the "180° ambiguity" by Schuijf (1975). In an attempt to solve this problem, Schuijf and colleagues (Schuijf 1975; Schuijf and Buwalda 1980; Buwalda et al. 1983) proposed that fishes not only gain directional information from the orientation patterns of the sensory epithelia but compare phase information from organs that differ in their sensitivity to sound pressure and particle motion. This is discussed in greater detail by Fay and Megela Simmons (Chapter 7). It is worthwhile to point out here that the few behavioral tests of this method to resolve ambiguity support its existence; however, the Schuijf model would only work in fishes that can detect both sound pressure and particle motion (generally, hearing specialists).

Also supportive of the model suggesting that hair cell orientation patterns give rise to directional information are results showing that many otolithic afferents in several unrelated species, the goldfish, *Opsanus*, and a goby (*Dormiator latifrons*), exhibit directional sensitivity that would be expected if the afferent innervated one hair cell or a group of hair cells having the same directional orientation (Hawkins and Horner 1981; Fay 1984; Fay et al. 1994; Fay and Edds-Walton 1997a; Lu and Popper 1997; Lu et al. 1998). Moreover, the axis of particle motion may be spatially mapped in the midbrain of a trout, *Oncorhynchus mykiss* (see Feng and Schellart, Chapter 6), and there is evidence for binaural interactions in the brain of the cod *Gadus morhua* (Horner et al. 1980), an important prerequisite for most models of directional hearing (also see McCormick, Chapter 5, and Fay and Megela Simmons, Chapter 7).

The final question to be asked concerns the functional significance of the different patterns of hair cell orientation in both the saccule and lagena. Are the different patterns all achieving the same final goals for signal detection and processing, or do they represent different mechanisms for directional detection and processing? Interestingly, as mentioned earlier, the same basic patterns reappear in taxonomically unrelated species, and the standard pattern is nerve encountered in hearing specialists. Although there are not sufficient data to directly address these issues, the fact that certain patterns have reevolved suggests that there are only a limited number of detection and processing schemes possible among fishes.

4.6 Innervation of the Ear

As discussed in Section 4.2, several models have been proposed for hearing in fishes that assume an anatomic or functional segregation in the brain of information from discrete regions of the sensory epithelium (e.g., Buwalda 1981; Schellart and de Munck 1987; Popper et al. 1988; Rogers et al. 1988). However, little is actually known about the regional innervation of the otolith organs or of topographic projections to brain stem nuclei.

The only systematic data on innervation of the saccular epithelium are for *Astronotus* (Saidel and Popper 1983a,b; Popper and Saidel 1990; Presson et al. 1992), *Carassius* (Furukawa 1981; Sento and Furukawa 1987), and *Opsanus tau* (Edds-Walton et al. 1996). In these species, many afferent neurons divide rather widely to innervate large regions of the epithelium (Sento and Furukawa 1987; Presson et al. 1992). However, there is some evidence that most neurons are restricted enough in their field of innervation to stay within one hair cell orientation group or region (Furukawa and Ishii 1967; Furukawa 1981; Saidel and Popper 1983a; Edds-Walton et al. 1996), and there appear to be separate neurons innervating the marginal and central regions of the epithelium in *Astronotus* (Presson et al. 1992). Furukawa and Ishii (1967) reported a few neurons that responded as if they innervated hair cells from both orientations, but it is not known whether similar types of neurons are found in other species.

The size of the dendritic arbor of each neuron varies (Presson et al. 1992), and we do not know how many hair cells are innervated by any individual neuron, although Sento and Furukawa (1987) report at least 7–10 terminals per afferent in the goldfish saccule, whereas Edds et al. (1989) found up to 50 terminals per arbor in the same species. Edds-Walton et al. (1996) also report up to 100 terminals for some arbors in the saccule of *Opsanus tau*. It is likely that each neuron must innervate large numbers of hair cells because there are far more hair cells than afferent fibers (Popper and Hoxter 1984; Mathiesen and Popper 1987). It is also likely that each neuron makes more than one synapse per hair cell (Popper and Saidel 1990). At the same time, in both the saccule of *Astronotus* and the lagena of the anabantid *Colisa labiosa*, there is considerable inter-hair cell variation in the number and distribution of both afferent and efferent synapses (Wegner 1982; Popper and Saidel 1990).

Little is known about efferent innervation of the ear and the relationship between efferent and afferent innervation. It is clear, however, that there are far fewer efferent neurons than hair cells (e.g., Sans and Highstein 1984; Roberts and Meredith 1992) and the majority of hair cells, at least in the saccule of *Astronotus*, receive efferent innervation (Popper and Saidel 1990). Although efferent neurons are known to modulate the responses of sensory hair cells in the lateral line of some elasmobranchs and *Opsanus* (Roberts and Russell 1972; Tricas and Highstein 1990, 1991) and in the ears of at least some tetrapods (Guinan 1996), there are no data to address this issue in the fish ear.

Several questions need to be asked with regard to innervation of the ear. Some of the most interesting are: (1) Is the interspecific variation found in numbers of neurons and synapses on individual hair cells meaningful or only a consequence of the small sample size (both inter- and intraspecific)? (2) What is the functional significance of a variable number of afferent or efferent synapses per cell and in having efferent synapses on some populations of hair cells and not others? (3) What is the area of the epithelium innervated by a single efferent neuron? The answer could have important consequences for the function of the efferent system because an efferent neuron that innervates a broad expanse of sensory epithelium would have much coarser effects than a neuron that innervates only a small epithelial region. (4) Is the innervation found in *Astronotus* typical of all fishes or are there differences in innervation among different species, and particularly between hearing specialists and nonspecialists? (5) Most importantly, what are the effects of efferent activation on the encoding of sound by the ear in fishes (Tricas and Highstein 1990, 1991)?

4.7 Ancillary Structures

It has been clearly demonstrated that the hearing specialists have ancillary structures that function to help increase signal level and bandwidth at the ear. The most widely found auxiliary structure is one or more gas-filled bladders that are mechanically coupled to the inner ear. It should be noted, however, that the presence of the abdominal swim bladder itself does not mean that a fish is a hearing specialist, and it is only when there is some means of mechanical coupling to the ear that specialization occurs. In addition to the swim bladder, some species of bony fish have secondary air bubbles located in the head near the ears (e.g., mormyrids, Stipetić 1939;

anabantids, Wegner 1979; Saidel and Popper 1983a,b; clupeids, Denton and Gray 1980) that appear to enhance hearing sensitivity and bandwidth in these species.

Other than the swim bladder, the best-known specializations are found in the otophysan fishes where a series of bones, the Weberian ossicles (Weber 1820), physically couple the swim bladder to the ear, thereby providing a direct path for motion of the swim bladder walls to be transmitted to the fluids of the ear (Alexander 1962). The mechanical transmission and filtering characteristics of the Weberian ossicles have not been systematically investigated.

4.7.1 Swim Bladder and Other Gas Bubbles

The swim bladder (Fig. 3.5) serves a variety of roles for bony fishes, including helping the fish maintain its position in the water column and produce and detect sound (Blaxter 1981). Due to the low density and high compressibility of the swim bladder gas, the bladder walls tend to move at a relatively high amplitude as its volume fluctuates in a sound pressure field (Rogers and Cox 1988). The motions of the wall, in effect, reradiate the energy in the sound signal, and the swim bladder becomes a secondary near-field source that can potentially stimulate the otolith organs.

The distance between the rostral end of the swim bladder and the ear appears to be of significance for the role of the swim bladder in audition. The swim bladder is a monopole source because it fluctuates in volume and its nearfield attenuates steeply with distance. As a consequence, the signal level at the ear is a function of the distance between the swim bladder and the ear and of the characteristics of intervening structures such as bones, muscles, and other tissues. However, there are very few data on the actual function of these systems. Most of what can be said is speculative.

There is generally some distance between the rostral end of the swim bladder and the inner ear in hearing generalists. In contrast, a number of species have evolved specializations that in some way bring the swim bladder closer to the auditory regions of the inner ear. The most common adaptation is a rostral extension of the swim bladder toward the auditory bulla as found in the hearing specialist *Myripristis*, in the gadids (cods and relatives), and in many other species in widely divergent taxonomic groups. In other cases, the hearing adaptation may be a small bubble of air in the head region. In some cases, as in the anabantid fishes (bubble-nest builders), the bubble is totally separate from the swim bladder (Saidel and Popper 1987), whereas in others, such as in the clupeids, there is a small tube connecting the inner ear bubble to the swim bladder (Blaxter et al. 1981). It appears that the air bubbles in the head region enhance hearing sensitivity, and in those cases where behavioral comparisons were made between related species that have and do not have the bubbles, the species with the bubbles always showed a wider bandwidth and greater sensitivity than the species without the bubbles (anabantids, Saidel and Popper 1987; squirrelfish, Coombs and Popper 1979).

5. Response of Otolithic Afferents to Sound

The responses of otolithic organs to sound are encoded in the response patterns of innervating eighth nerve neurons. Studies of these activity patterns help reveal the acoustic response properties of the otolithic organs and their ancillary structures, the functional characteristics of hair cells and their synapses on primary afferents, and the dimensions of neural activity that represent acoustic features of sound sources such as level, frequency, and location. In addition, afferent activity patterns contain all the information that is used by the brain in computing the characteristics of sound sources (see Feng and Schellart, Chapter 6). Quantitative analyses of peripheral neural codes and acoustic behaviors help reveal what dimensions of neural activity are used by the brain in computing sound source characteristics (see Fay and Megela Simmons, Chapter 7). In a comparative context, these analyses help to specify which structural features of the ears have functional significance for the sense of hearing and which do not. Finally, comparisons between fishes and other taxa with regard to peripheral representations of sound features help to suggest which neural representations are primitive and which are derived among the vertebrates.

The responses of primary otolith afferents in response to sound and head motion have been systematically studied in only a few fish species. These species include goldfish (*Carassius auratus*, reviewed in detail below), catfish (*Ictalurus punctatus*, Moeng and Popper 1984), bullhead (*Cottus scorpius*, Enger 1963), cod (*Gadus morhua*, Horner et al. 1981), tench (*Tinca tinca*, Grozinger 1967), and the oyster toadfish (*Opsanus tau*, Fay and Edds-Walton 1997a,b). The following discussion focuses on the goldfish, with treatment of other species investigated where appropriate.

5.1 Postsynaptic Potentials and the Hair Cell Synapse

Excitatory postsynaptic potentials (EPSPs) in saccular afferents of the goldfish have been extensively studied by Furukawa and his colleagues (Ishii et al. 1971; Furukawa et al. 1972; Furukawa et al. 1978; Furukawa et al. 1982; Furukawa 1981; Suzue et al. 1987). This work has led to a multiple-release-site model of the hair cell synapse (Furukawa 1986). The fundamental observations are that EPSPs are graded in amplitude and decline with time during a stimulus tone, thus suggesting the hair cell synapse as the site of adaptation (Furukawa et al. 1978). A statistical analysis of EPSP amplitudes showed that the adaptive rundown of EPSPs was due to a reduction of the number of transmitter quanta available at the presynaptic sites, given the stimulus level, and not to the probability of a given quantum being released.

These and other studies provided evidence that (1) there are numerous presynaptic release sites, (2) each release site has a different threshold and is activated only if the stimulus level reaches that threshold, (3) a single synaptic vesicle is allocated to each release site, and (4) once a vesicle is released, the site remains empty until replenished from a larger store. Among other things, this model explains why an increment in stimulus level results in a robust spike response from highly adapted afferents (the existence of release sites having higher thresholds) and why a small stimulus level decrement may result in a transient loss of all spikes (only empty, low-threshold sites are activated). Furukawa et al. (1982) also obtained evidence that vacant release sites are replenished in an order from high threshold to low and that sites with thresholds below the stimulus level are not replenished as long as the stimulus level remains above their threshold.

Among other effects, the multiple-release-site model predicts very high sensitivity to both increments and decrements (amplitude modulations) in the level of an ongoing sound. Behavioral studies (Fay 1980, 1985) indicate that goldfish are able to detect amplitude increments and decrements as small as 0.1 dB. No other vertebrate, including humans, has been shown to be as sensitive to amplitude fluctuations (Fay 1988). Because responses to stimulus decrements are less pronounced in mammalian cochlear afferents than in goldfish saccular afferents (Fay 1980), it appears that some aspects of the multiple-release-site model may not apply as well to mammals.

5.2 Spontaneous Activity

As in all vertebrate auditory systems investigated, primary afferents of several fish species investigated show varying degrees and patterns of spontaneous activity. In goldfish (Fay 1978a,b), the catfish *Ictalurus nebulosus* (Moeng and Popper 1984), *Gadus morhua* (Horner et al. 1981), and *Opsanus* (Fay and Edds-Walton 1997a), saccular afferents generally fall into four groups with respect to spontaneous activity patterns: (1) those without spontaneous firing, (2) those with approximately random interspike-interval distributions, (3) those that show random bursts of spikes giving bimodal distributions of interspike intervals, and (4) small proportion of saccular neurons that show highly regular spontaneous patterns. These fibers are very insensitive to sound and may serve vestibular rather than auditory functions or may be efferent neurons. Spontaneous rates can range up to 250 spikes/sec. As is the case for mammals (Ruggero 1992), afferents showing no spontaneous activity are among the least sensitive and afferents with low, irregular spontaneous activity are the most sensitive.

The functional significance of spontaneous activity and its variation among afferents is not clear. However, it has been noted in goldfish (Fay et al. 1996b) and *Opsanus* (Fay and Edds-Walton 1997a) that the encoding of a sound's waveform differs somewhat for highly spontaneous and low or zero spontaneous afferents. In highly spontaneous afferents, spikes tend to occur at the same times or phase with respect to the stimulating waveform regardless of sound level within an afferent's dynamic range. In this case, the sound waveform simply modulates spike probability in a linear manner. In zero and low spontaneous afferents, spikes advance in phase over a 90° range as sound level increases above threshold. Here, afferents respond at the time at which the stimulating waveform reaches threshold; this threshold crossing occurs earlier in time as the stimulus level is raised. These latter afferents thus represent excitation level in terms of response time or latency as well as spike rate or probability. This effect has implications for sound source localization, as discussed in Section 5.7.

5.3 Frequency Selectivity of Auditory Afferents

The question of the degree to which auditory afferents of fishes are frequency selective is important with respect to understanding the micromechanics and biophysics of hair cells and with respect to understanding the computational strategies of the brain that help in the detection and determination of sound sources based on their spectral characteristics. As in all vertebrates, fishes encode sound signals through phase locking in the time domain (e.g., Fay 1978a; see Section 5.6) and through filtering in the frequency domain (e.g., Furukawa and Ishii 1967) (see Figs. 3.11 and 3.13). The relative importance of these two codes for mammals has long been a matter of debate (Wever 1949), and the debate has now extended to all vertebrate groups including fishes.

Long ago, it was pointed out that a mechanical frequency-to-place transformation (von Békésy 1960) was unlikely to occur in otolith organs (von Frisch 1938) and thus that any frequency analysis that did occur probably depended on time-domain computations based on phase-locked inputs to the brain. More recently, however, it has become clear that hair cell resonance (Crawford and Fettiplace 1981) and local, micromechanical factors (reviewed by Patuzzi 1996) could achieve frequency selectivity and tonotopy in the absence of a macromechanical traveling wave (Holton and Weiss 1983).

Quantitative data on frequency selectivity of primary auditory afferents exist only for a few fish species including goldfish (e.g., Furukawa and Ishii 1967; Fay and Ream 1986; Fay 1977), *Ictalurus* (Moeng and Popper 1984), *Gadus* (Horner et al. 1981), and *Opsanus* (Fay and Edds-Walton 1997b). For these species, it is clear that saccular afferents differ with respect to the center frequency and bandwidth of response. However, it has proven difficult to characterize the frequency response or tuning characteristics of afferents in a single, simple way. This difficulty apparently arises from nonlinearities in the afferent response, including frequency after saturation (see Section 5.6), frequency-dependent adaptation (see Section 5.1), single-tone suppression (see Section 5.5), and the operational definitions used to characterize frequency selectivity.

Furukawa and Ishii (1967) first described saccular afferents of the goldfish with respect to frequency selectivity and categorized them into two groups: S1 and S2. Briefly, afferents classified as S1 respond best at high frequencies (>500 Hz) and have larger fiber diameters and no spontaneous activity. S1 afferents innervate primarily the rostral region of the saccule. Hair cells of the rostral saccule tend to have short cell bodies and short stereocilia (Platt and Popper 1984) and tend to exhibit a damped oscillation of the membrane potential in response to depolarizing current steps (Sugihara and Furukawa 1989). Afferents classified as S2 respond best at low frequencies (<300 Hz), are smaller in diameter, and are the major projections from the caudal region of the saccule where hair cells are tall, with stereocilia generally longer than those of the rostral hair cells. Tall hair cells do not resonate but produce a "spike-plateau" response when depolarized. The S1–S2 classification scheme focuses on a suite of afferent characteristics but does not include quantitative descriptions of frequency selectivity.

The frequency-response properties of goldfish saccular afferents have been quantitatively described using several methods including frequency-threshold (tuning) curves based on phase-locking (Fay 1978b) and spike rate (Fay and Ream 1986) criteria, frequency-by-level response areas based on spike rate criteria (Fay 1990, 1991; Lu and Fay 1996), and the reverse correlation (revcor) method (Fay 1997) in response to wide-band, flat-spectrum noise. The revcor method was first used as a way to characterize the filtering functions of auditory nerve fibers by de Boer and de Jongh (1978). Briefly, averaging hydrophone recordings of the acoustic noise triggered by spike times produces an impulse response (revcor) that estimates the response of the linear filtering that precedes spike generation. The spectrum of the revcor estimates an afferent's filter shape. In general, the picture that emerges is that each of these measures provides a somewhat different view of frequency selectivity. The most revealing measures are the frequency-by-level response areas and the revcor filter functions.

Figure 3.11 compares these two measures for four representative saccular afferents of the goldfish (Fay 1997). In this figure, the continuous lines are the revcor filter shapes at three overall noise levels, and the functions composed of straight line segments plot spike rate as a function of frequency for tone burst stimuli, with overall level as the parameter. These latter functions will be referred to as response areas (RAs).

The RAs for the two low-frequency afferents (bottom panels) show characteristic frequencies (CFs; the frequency at which threshold is lowest) in the region of 200 Hz. For afferent C15 (lower left), the best frequency

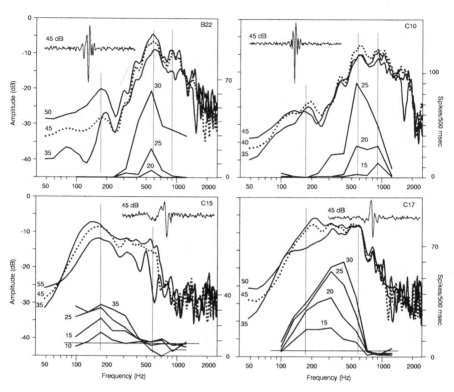

FIGURE 3.11. Frequency-response data for four representative saccular afferents of the goldfish *Carassius auratus*. The revcor function (40 msec in duration) is shown as an inset. The filter shapes (smooth curves) are the spectra of the revcors after subtracting the averaged, smoothed acoustic spectrum and then smoothing with a 5-point moving average (9.77 Hz/point). The left ordinate applies to the revcor filter shapes (in dB with an arbitrary reference). The dotted and dashed lines serve to visually separate the functions obtained at different stimulus levels (spectrum levels in dB: re: 1 dyne cochlear microphonics^{-2}). Isolevel spike rate functions of tone-burst frequency (response areas, RAs) are shown as lines connecting data points referred to the right ordinate. Numbers are sound pressure levels in dB re: 1 dyne cochlear microphonics^{-2}. Spontaneous rates are: 0 for B22 and C10, 16 spikes/sec for C15, and 4.4 spikes/sec for C17. (Modified from Fay 1997.)

(BF; the frequency producing the most spikes) remains at CF as level increases. In contrast, afferent C17 (lower right) exhibits an upward shift in BF, reaching 500 Hz as the level increases. This level-dependent "BF shift" is common among low-CF saccular afferents (Lu and Fay 1993). At frequencies above BF, both afferents show a relatively steep decline in tone-evoked activity that converges at the spontaneous rate between 500 (C15) and 700 Hz (C17) and falls below the spontaneous rate at higher frequencies. This single-tone suppression is common among low-CF saccular afferents in goldfish (Fay 1990; also see Section 5.5).

Revcor filter functions for C15 and C17 show relatively broad tuning with corners at ~150–200 and 600 Hz (thin vertical lines in the bottom panels). Although the revcor features near 200 Hz tend to correspond with CF determined from the RAs, the corner features at 600 Hz do not. For C15, 600-Hz tones are suppressive rather than excitatory, and for C17, the RA functions decline steeply above 600 Hz. Thus revcor filter functions present a different view of frequency selectivity than do the RAs.

Afferent B22 (upper left) has a CF near 600 Hz and a BF that remains at this frequency throughout the dynamic range. The RA for afferent C10 (upper right) shows a CF at ~900 Hz. Note that the BF for this afferent shifts downward to ~600 Hz as level is raised. This is common among high-frequency saccular afferents (Lu and Fay 1993).

The revcor filter functions for these high-frequency afferents show a peak at ~600 Hz, with roll-offs toward lower and higher frequencies. Thin vertical lines approximately locate features (peaks or corners) of the filter functions. In general, these features include a major peak at 600 Hz, and minor peaks at ~170 and 900 Hz. Note that for afferent C10, the 900-Hz peak in the filter functions corresponds to the CF determined from the RA. For B22, the 600-Hz peak in the filter functions corresponds to CF.

These results on the frequency selectivity of goldfish saccular afferents are rather complex, illustrating that different experimental paradigms may produce contradictory results and may lead to different conclusions. In general, RAs are more informative than frequency-threshold curves, in part because suprathreshold phenomena such as the BF shift are not observable in frequency-threshold curves. Interestingly, the revcor filter functions present the simplest view of frequency selectivity, indicating that saccular afferents can be placed in two major categories: those with a major peak at ~200 Hz and a plateau extending to 600 Hz and those with prominent peaks at ~600 and 900 Hz. Perhaps the simplicity of the revcor results reflects the more natural, wide-band stimuli used to obtain them. In any case, there is no doubt that the saccule of the goldfish parses the sound spectrum into at least two frequency regions and encodes energy in these regions somewhat independently. Thus frequency selectivity exists for the goldfish saccule as it does for the auditory receptor organs investigated among all vertebrate classes (e.g., Sachs and Kiang 1968; Köppl and Manley 1992; Lewis 1992; Manley and Gleich 1992).

There are limited data on frequency selectivity of saccular afferents in other fish species, including another hearing specialist (*Ictalurus*, Moeng and Popper 1984) and two hearing generalists, *Gadus* (Horner et al. 1981) and *Opsanus* (e.g., Fay and Edds-Walton 1997b; see Fig. 3.12). In general, these data are consistent with those for the goldfish in indicating a small number of differently tuned peripheral channels. *Opsanus* and *Gadus* hear only up to several hundred hertz and their tuned channels are therefore restricted to the very low frequencies. Recent data for *Opsanus* using the revcor method (Fay and Edds-Walton 1997b) reveal two filters with peaks at 74 and 140 Hz and a third population that apparently combines these two filters to varying degrees.

Fishes differ from other vertebrates in having a small number of differently tuned channels, and questions arise about the utility of such a simple system compared with the continuously variable tuning observed in all other vertebrate classes. We point out here that many cells of the auditory midbrain (central nucleus of the torus semicircularis) in goldfish show sharper tuning and a more continuous distribution of CFs than those observed in primary afferents (Lu and Fay 1993; see also Feng and Schellart, Chapter 6). Sharpening and the dispersion in CF are created by inhibitory interactions observable at the level of the midbrain (Lu and Fay 1996). For the goldfish, then, a set of auditory filters with continuously distributed CFs is synthesized in the brain using only two differently tuned peripheral inputs. We suggest that similar processing occurs in other fishes as well. Indirect evidence that such processing occurs in both hearing specialists and generalists comes from psychophysical experiments (see Fay and Megela Simmons, Chapter 7) demonstrating sharply tuned and continuously distributed auditory filters (e.g., Hawkins and Chapman 1975; Fay et al. 1978).

5.4 Origins of Frequency Selectivity

It is very unlikely that frequency selectivity observed in saccular afferents arises from the sort of macromechanical tuning characteristic of the mammalian cochlea and the bird basilar papilla. More likely explanations include hair cell resonance and micromechanical mechanisms that are local to hair cells and their ciliary attachments to the otoliths. Hair cell resonance has been studied in isolated saccular hair cells of the goldfish (Sugihara and Furukawa 1989) and *Opsanus* (Steinacker and Romero 1992). In both species, two general classes of hair cells have been found: those that produce a damped oscillation (resonance) to a current step and those that produce a spike. In goldfish, the resonance frequency of hair cells from the rostral saccular region varies between 40 and 200 Hz. Because these cells likely provide input to the high-frequency saccular afferents that respond best in the 600- to 900-Hz range (Fig. 3.11), it is unclear how the 200-Hz resonance observed in vitro could contribute to the frequency response of these afferents.

In *Opsanus*, the resonance frequency of saccular hair cells averages 142 Hz (Steinacker and Romero 1992), a value equal to the peak frequency of the high-frequency afferents (Fig. 3.12). This suggests that the tuning of these afferents is caused, at least in part, by hair cell resonance. At the same time, however, the simple second-order resonance typical of hair cells alone cannot entirely explain the filter shapes that have been observed (Figs. 3.11

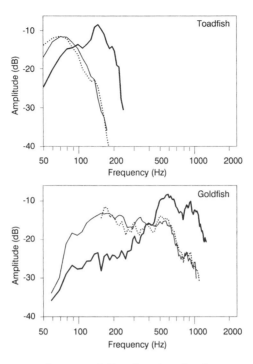

FIGURE 3.12. Upper panel: averaged filter functions for the two major categories of saccular afferents of the toadfish *Opsanus tau* (thick and thin solid lines) and the average for the high-frequency afferents (dotted line) after a −24 dB/octave tilt around a pivot point of ~100 Hz (from Fay 1997). Lower panel: a similar comparison of the two categories of saccular afferents in the goldfish *Carassius auratus*. For the goldfish, the high-frequency filter function (thick solid line) matches the low-frequency function (thin solid line) best after a −15 dB/octave tilt around a pivot point of ~400 Hz (dotted line). Although the goldfish hears in a wide frequency range compared with toadfish, the hearing range is similarly divided between two afferent types in both species. (Modified from Fay and Edds-Walton 1997b.)

and 3.12). As Lewis (1992) has pointed out (see also Lewis and Narins, Chapter 4), second-order resonances result in filter functions having decelerating roll-off slopes (concave filter skirts), yet primary afferents tend to have filter functions with accelerating slopes (i.e., have convex filter skirts), especially above the center frequency. These and other features of the filter functions and revcors indicate filtering of high dynamic order (Lewis 1992) and suggest that electrical resonances of hair cells are absorbed into the complex dynamics of an entire system that could include bidirectional transduction (mechanical to electrical and the reverse).

For both the goldfish and *Opsanus*, spiking hair cells appear to provide input to their respective low-frequency afferents. If true, how can the frequency selectivity of the low-frequency afferents be explained? A possible explanation is illustrated in Fig. 3.12 showing averaged revcor filter functions for the two classes of saccular afferents in goldfish (Fay 1997) and *Opsanus* (Fay and Edds-Walton 1997b). The dotted line in each panel is the high-frequency filter function that has been spectrally "tilted" about the point at which the two filter functions intersect (400 Hz for goldfish, 100 Hz for *Opsanus*). The best-fitting spectral tilt is −15 dB per octave for the goldfish and −24 dB per octave for *Opsanus*. Although the magnitudes of the tilts are difficult to interpret, this analysis suggests that the low-frequency filter function in each species may be the result of a simple transformation of the high-frequency filter. One mechanism that could help explain these correspondences is that low-CF afferents may innervate hair cells responding to otolith displacement, whereas high-CF afferents contact hair cells responding to otolith acceleration. These differing response properties could arise from differences among hair cells in hair bundle stiffness and the friction of coupling to the otolith (cf. Rogers and Cox 1988). In this context, we note that such a spectral tilt also operates in the mechanosensory lateral line system: the acceleration-coupled canal neuromasts respond to higher frequencies than the velocity-coupled superficial neuromasts (Denton and Gray 1983). In any case, micromechanical processes combined with hair cell resonance (and possibly bidirectional transduction) apparently contribute to peripheral frequency selectivity in fishes as well as in amphibians (Lewis 1992; see also Lewis and Narins, Chapter 4), reptiles (Köppl and Manley 1992), and birds (Manley and Gleich 1992).

The goldfish saccule is crudely tonotopically organized; high-CF afferents originate primarily from the rostral region (Furukawa and Ishii 1967). However, there is no evidence that the saccule of *Opsanus* is tonotopically organized. The frequency response functions of afferents from the rostral, middle, and caudal regions of the saccular epithelium show no tendency for tonotopy (Fay and Edds-Walton 1997a). A similar lack of tonotopy was reported for the saccule of *Gadus* (Horner et al. 1981). Furthermore, Steinacker and Romero (1992) found resonant, nonresonant, and spiking hair cells in all regions of the *Opsanus* saccular epithelium. It appears that although individual elements of the saccule and its nerve show diverse frequency selectivity, the epithelium may not be topographically organized with respect to frequency. The sort of frequency analysis observed in saccular afferents of goldfish and *Opsanus* (Fig. 3.12) may be the most primitive basis for parsing the acoustic spectrum yet observed in vertebrates (Fay 1997).

Fishes were probably the first vertebrates to solve problems of frequency analysis, and these solutions may have formed a model for those of their tetrapod descendants. However, both goldfish and *Opsanus* differ from other vertebrate species in having a small number of differently tuned channels (two or three) compared with the continuously variable tuning observed in the auditory nerves of anuran amphibians, reptiles, birds, and mammals. We note that many cells of the torus semicircularis of goldfish show sharper tuning and a more continuous distribution of CF than primary afferents do (Lu and Fay 1993). This sharpening and dispersion in CF is most likely created by inhibition (Lu and Fay 1996), and the result is a set of filters with continuously distributed CF synthesized in the brain (see Feng and Schellart, Chapter 6). These sorts of computations may underlie many of the behavioral capacities for frequency analysis revealed in psychophysical studies (reviewed by Fay 1988; see also Fay and Megela Simmons, Chapter 7).

5.5 Suppression in Saccular Afferents

Saccular afferents of the goldfish reveal a set of nonlinear phenomena that are common to auditory afferents in most vertebrate classes and thus may be primitive functional characteristics of the peripheral neural code underlying the vertebrate sense of hearing. The most striking of these phenomena are two-tone rate suppression and single-tone suppression.

In a linear system, adding two stimuli results in a response that is equal to or greater than the response to either stimulus presented alone. Two-tone rate suppression (TTRS) is a nonlinear effect defined as the reduction in evoked spike rate to one stimulus as a result of the addition of a second stimulus. In a typical TTRS experiment, the response to a stimulus just above threshold at an afferent's CF can be reduced by the simultaneous presentation of a second tone in frequency regions either below or above CF. Because this is a physiologically vulnerable response observable at the level of basilar membrane motion in mammals (reviewed in Patuzzi 1996), it has been hypothesized to arise from stimulus-evoked movements of outer hair cells that feed back onto the mechanical response of the cochlear partition. TTRS has been observed in primary auditory afferents of anuran amphibians (Capranica and Moffat 1980; see also Lewis and Narins, Chapter 4), reptiles (Manley 1990), birds (Hill et al. 1989b), and mammals (Sachs and Kiang 1968) and most recently in saccular afferents of the goldfish (Lu and Fay 1996). As in the amphibian papilla of anurans, TTRS in the goldfish saccule occurs only in a subpopulation of low-frequency afferents. However, unlike the case for anurans, TTRS occurs in goldfish only for suppressive frequencies well above the CF. Although the functional significance of TTRS is not clear at present, its appearance in goldfish suggests that it is a primitive characteristic of vertebrate auditory systems and may be caused by different mechanisms in different taxa.

In goldfish, what appears to be spontaneous activity can also be suppressed in some low-CF saccular afferents by single tones presented at frequencies well above the CF (Fay 1990, 1991). This controversial sort of suppression has been termed "single-tone suppression" to distinguish it from TTRS (see also Lewis and Narins, Chapter 4). Single-tone suppression has also been reported for mammals (Henry and Lewis 1992), some reptiles

(Manley 1990), and birds (e.g., Hill et al. 1989a). Figure 3.13 illustrates suppression of apparently spontaneous activity in the response areas of two saccular afferents (units 242 and 91) with CFs between 200 and 300 Hz; as the stimulus frequency rises above the CF, the response transitions from excitation to suppression. This is most evident during the second 25-msec epoch of the 50-msec tone burst (bottom panels). In the region of transition between declining excitation and suppression, it is difficult to distinguish frequency-dependent adaptation (Coombs and Fay 1985, 1987) from suppression. Single-tone suppression and frequency-dependent adaptation are observed in many low-CF saccular afferents but in not all of them (see unit 38 in Fig. 3.13). Lewis (1986) has made similar observations for afferents of the anuran amphibian papilla (see also Lewis and Narins, Chapter 4).

The origin of single-tone suppression is not clear. Hill et al. (1989c) proposed that it could result from hypopolarization at the spike initiation zone due to positive, extracellular fields produced by receptor currents through nearby hair cells. This is plausible for goldfish because low-CF saccular afferents must pass near rostral saccular hair cells that respond best at higher frequencies. One of the consequences of single-tone suppression is that the frequency RAs for some saccular afferents are sharpened above CF as time progresses throughout a brief tone burst (compare upper and lower panels in Fig. 3.13).

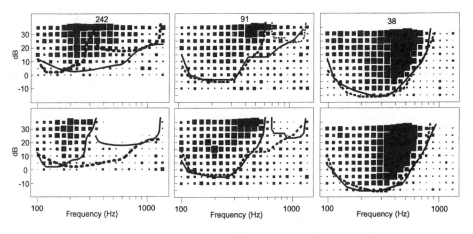

FIGURE 3.13. Frequency-by-level RAs for three saccular afferents of the goldfish *Carassius auratus* tuned near 250 Hz. The filled areas indicate the number of spikes recorded during the first (top panels) and second (bottom panels) 25-msec epochs of a 50-msec tone burst presented at the frequencies and levels indicated. The area of each square is proportional to the number of spikes evoked. The solid lines outline the tuning curves based on both increases and decreases of spike rate. The dotted lines show the tuning curves from the opposite (upper or lower) panels. (From Fay 1990.)

5.6 Phase Locking

In goldfish (Fay 1978b), *Gadus* (Horner et al. 1981), and *Opsanus* (Fay and Edds-Walton 1997a,b), all sound-responsive saccular afferents phase lock, or synchronize, to all acoustic waveforms within the frequency range of hearing including noise (Fay et al. 1983; Fay 1997). Because phase locking is ubiquitous among low-frequency afferents (for effective frequencies below ~4 kHz in most cases) in all vertebrate auditory systems investigated, it is apparently a primitive characteristic of the neural code for hearing.

Phase locking is used in encoding interaural time differences in ongoing sounds and in improving signal-to-noise ratios for binaural signal detection in humans (Jeffress 1948) and presumably in all vertebrates including fishes (e.g., Schuijf 1975; Rogers et al. 1988; Fay and Coombs 1983; see also Section 5.7). In addition, phase locking has been hypothesized to play a role in pitch perception and other aspects of frequency analysis in tetrapods (Wever 1949) and fishes (Fay et al. 1983). In goldfish, the temporal error with which saccular afferents synchronize to tones predicts behavioral frequency discrimination acuity (Fay 1978b), suggesting that central computation based on a phase-locked peripheral code is a primitive strategy for signal analysis in vertebrate auditory systems.

Saccular afferents also synchronize to the envelopes of amplitude-modulated tones and noise (Fay 1980). For modulated tones, each afferent has a modulation rate to which it is most sensitive, ranging between 20 and over 200 Hz. Sensitivity to amplitude-modulated tones can be quite high (Furukawa et al. 1982), with significant responses to increments and decrements of sound pressure as small as 0.1 dB (Fay 1985). Responses of goldfish saccular afferents to temporally asymmetrical envelopes exhibit asymmetries in the phase angle of phase locking and spike rate that are qualitatively predicted by Furukawa's (1986) model of the hair cell synapse (see Section 5.1) and that account for the perceptual distinctiveness of envelope shapes (Fay et al. 1996; see also Fay and Megela Simmons, Chapter 7).

5.7 Directional Responses to Whole Body Acceleration

As predicted by de Vries (1950), the primitive and shared mode of sound detection in fishes results from the otolith organs responding to acoustic particle motion as inertial accelerometers (Fay and Olsho 1979). Whole body acceleration activates primary afferents from all otolith organs (saccule, lagena, and utricle) in goldfish, with the most sensitive afferents having thresholds as low as 0.1 nm at 140 Hz (Fay 1984). Saccular afferents in *Opsanus* have similar thresholds, and most afferents are saturated for displacements greater than 1 µm (Fay and Edds-Walton 1997a,b). This displacement sensitivity is remarkable. At 0-dB sound pressure level (near the human threshold of hearing at 1 kHz), basilar membrane displacement in the guinea pig is about 0.2 nm (Allen 1996). In studies on the vibration sensitivity of the bullfrog sacculus, Koyama et al. (1982) reported acceleration thresholds as low as 10^{-6} g at 50 Hz, corresponding to displacements of about 0.1 nm. This sensitivity to whole body motion predicts that behavioral detection thresholds for hearing generalists such as *Astronotus* (Lu et al. 1996) and *Opsanus* (Fish and Offutt 1972) are probably determined by the direct detection of acoustic particle motion (see Fay and Megela Simmons, Chapter 7, for a more complete discussion of behavioral detection thresholds).

In goldfish, all otolithic afferents responding to acceleration of the head tend to have low-pass frequency-threshold curves when the signal level is expressed in acceleration units, with some saccular afferents responding to higher frequencies than lagenar and utricular afferents (Fay 1981). In response to whole body acceleration, saccular afferents of *Opsanus* are frequency selective, with one class of afferents responding best at 74 Hz another class responding best at 140 Hz, and a third population with response peaks at both frequencies (Fay and Edds-Walton 1997b). In general, the frequency-response functions for lagenar afferents of the goldfish resemble those for saccular afferents of *Opsamus* and *Gadus* (Horner et al. 1981) in having the best frequencies between 100 and 200 Hz. Based on these data, it is suggested that all otolith organs in fishes will have best displacement thresholds in the region of 0.1 nm and a frequency response extending up to at least 200 Hz.

In goldfish (Fay 1984), *Opsanus* (Fay and Edds-Walton 1997a), and *Gadus* (Hawkins and Horner 1981), the response of most otolithic afferents vary as a function of the axis of translatory motion according to a cosine function. Because a cosinusoidal directional response function has been measured for individual hair cells (Hudspeth and Corey 1977), it appears that most otolithic afferents probably receive effective input only from homogeneously oriented hair cells. Figure 3.14 illustrates the directional response of a typical afferent of the saccule of *Opsanus* (Fay and Edds-Walton 1997a). In polar coordinates, synchronized spike rate functions of the stimulation axis are cosinusoidal functions (panel A) and threshold functions are straight lines (panel B), as would be expected in a linear system. The line that passes through the origin and represents the long axis of the dipole figure in panel A is the axis producing the greatest synchronized spike rate. The line that passes through the origin and is perpendicular to the threshold functions in panel B is coincident with the line in panel A. This line defines the axis of stimulation at which threshold is lowest, or the characteristic axis (CA). It is generally assumed that directional hearing by fishes depends on central computations using the directionality of individual neural channels as inputs. The directionality observed in primary afferents is preserved at least to the level of the midbrain in trout (*Oncorhynchus mykiss*) (e.g., Wubbles et al. 1993) and in goldfish (Ma and Fay 1996). See Feng and Schellart, Chapter 6, for further detail on directional representations in the central auditory system.

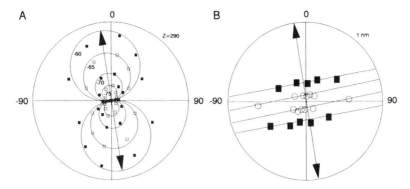

FIGURE 3.14. A: polar plot of the responsiveness (Z) of a saccular afferent of the toadfish *Opsanus tau* as a function of the axis of 100-Hz translatory motion on the horizontal plane. $Z = r^2n$, where r is the coefficient of synchronization and n is the number of spikes recorded. Data points are plotted twice to facilitate interpretation: once at the nominal stimulus axis angle and once again at the nominal angle plus 180°. Filled and open symbols serve to visually separate the data obtained at four stimulus levels (given as root mean square displacement in dB re: 1 μm). The double-ended arrow indicates the axis on the horizontal plane that is most excitatory. B: polar plot of displacement thresholds from the same afferent in A. The symbols are displacement thresholds for the criteria of $Z = 20$ (open squares) and $Z = 100$ (filled squares; the highest threshold is off scale and not plotted). Again, thresholds are plotted twice: once at the nominal stimulus axis angle and once again at this angle plus 180°. The straight lines through the symbols are the best-fitting linear functions to the threshold points. The double-ended arrow perpendicular to the threshold lines indicates the axis at which the threshold is lowest. The radius of the circle is 1 nm.

In goldfish (Fay 1984) and *Opsanus* (Fay and Edds-Walton 1997a), otolithic afferents are widely distributed with respect to orientation of their CA in spherical coordinates. Figure 3.15 shows these data for both species. In this view, each symbol represents a single afferent. The symbol's location on the northern hemisphere of the globe represents the location at which its CA would penetrate the globe's surface. In the goldfish right saccule (panel A), afferents are tightly grouped (40–60° elevation, 0–30° azimuth). The CAs of lagenar afferents are more widely scattered in elevation but are still loosely grouped in azimuth between 0 and 90° (panel B), and those of utricular afferents tend to fall near the equator on the horizontal plane (panel C). In the left saccule of *Opsanus* (panel D), CAs tend to cluster on an axis near −45°. This is qualitatively consistent with the oblique angle of the saccular epithelium in the horizontal plane (Fay et al. 1996a).

The diverse patterns of hair cell orientation over the surface of the epithelium (see Section 4.4) primarily determine the azimuth of utricular CAs and the elevation of saccular and lagenar CAs. Utricular afferents of

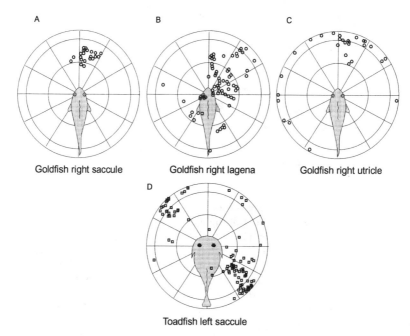

FIGURE 3.15. A–C: distributions of the characteristic axis (CA) for afferents of the right saccular (n = 22), utricular (n = 29), and lagenar (n = 85) nerves of the goldfish (Fay 1984). The view is down onto a northern hemisphere with the fish's ears at the globe's center. The symbols locate the points at which CAs penetrate the globe's surface (from Fay 1984). D: distribution of CAs for the left saccular nerve (n = 96) of *Opsanus*. (From Fay and Edds-Walton 1997a.)

Opsanus tend to have CAs clustering at 0° elevation, corresponding to the essentially horizontal orientation of the utricular epithelium. For the saccule and lagena of goldfish and the saccule of *Opsanus*, CA azimuths tend to correspond with the overall azimuthal orientations of the otoliths and sensory epithelia of the respective organs. This correspondence results from the nearly vertical orientation of the saccular and lagenar sensory epithelia. Thus saccular and lagenar hair cell orientation patterns appear to have relevance only for vertical sound localization. In this case, the elevation of the axis of acoustic particle motion could be resolved through central computations across the array of afferent input. It is interesting to consider that sound source elevation in humans and other terrestrial tetrapods is thought to be represented, at least in part, in terms of spectral shape (Wightman et al. 1991). Spectral shape is a stimulus feature represented monaurally in mammals as an across-afferent (tonotopic) profile of activity. The saccule of *Opsanus* is not tonotopically organized but rather is organized directly with respect to elevation. Thus there appears to be a parallel between fishes and tetrapods in the strategy for encoding sound source elevation.

Encoding sound source azimuth in fishes could well depend on binaural processing as it does in tetrapods. For many terrestrial animals, azimuth is encoded through interaural stimulus differences of time and intensity brought about by the physical separation of the external ears by the head. In fishes, there probably are no usable interaural stimulus differences in time and intensity. However, given the azimuthal orientations of the right and left saccular and lagenar epithelia and the corresponding clustering of CAs in azimuth, direction-dependent interaural differences in response magnitude will result. These response differences are not due to interaural stimulus differences but to response differences that arise from the inherent directional characteristics of hair cells. Differences between the two ears in terms of overall response magnitude naturally create interaural response-time differences among low-spontaneous saccular afferents. As noted above, the angle at which these afferents phase lock is level dependent over a 90° range, amounting to time differences of 0–2.5 msec at 100 Hz (Fay and Edds-Walton 1997a). It is possible that these interaural response-time differences are used in binaural processing for resolving sound source azimuth in fishes. Thus there appear to be parallels between fishes and tetrapods in the use of binaural strategies for encoding sound source azimuth as well as in a monaural strategy for encoding elevation.

In the saccular nerves of *Opsanus* (Fay and Edds-Walton 1997a) and *Gadus* (Horner et al. 1981), there is sufficient diversity in CA elevation to account for directional hearing in the vertical planes. However, this is not true for the saccule of the goldfish in which afferents are tightly clustered with respect to best elevation. Thus goldfish (and probably other hearing specialists) may require inputs from both the saccule and lagena, or the lagena alone, for encoding the elevation of sound sources.

For goldfish and other hearing specialists that are particularly sensitive to sound pressure, input to the otolith organs via the swim bladder is likely to be identical for corresponding elements of both ears. This input is also probably independent of sound source direction because the swim bladder provides input to both ears equally, independent of sound source location (van Bergeijk 1967). Thus the sound pressure waveform probably produces responses of the saccule that are highly correlated interaurally. Any additional direct particle motion input to the otolith organs (i.e., acceleration of the head) that is not on the midsagittal plane will tend to produce responses with interaural differences (Fay et al. 1982). Subtracting the response from the two ears (cf. Colburn and Durlach 1978; a sort of common-mode rejection) will leave information about interaural response differences (amplitude and phase) caused by the head-acceleration component of the stimulus. At the same time, adding the responses from both ears would tend to emphasize the sound pressure waveform (common to both ears) and minimize interaural response differences (Popper et al. 1988).

Finally, we note that the directional response of the saccule to acoustic particle motion gives fishes the equivalent of monaural "pressure-gradient" receivers that have been demonstrated for the ears of some amphibians and birds (reviewed by Fay and Feng 1987; see also Lewis and Narins, Chapter 4) and in the lateral line system (Denton and Gray 1983; see also Coombs and Montgomery, Chapter 8). The accelerometer mode of otolith stimulation in fishes could be viewed as pressure-gradient detection because particle motions normally result from pressure gradients (Rogers and Cox 1988).

One issue of directional hearing by fishes remains an enigma: a system of particle motion receivers is subject to an ambiguity about the direction of sound propagation (e.g., Schuijf 1975; Schellart and de Munck 1987; Rogers et al. 1988). The axis of motion may be resolved, but it is not clear how an accelerometer array alone can represent the vector toward the source. Schuijf (1975) has shown that this ambiguity can be resolved in the nearfield through the encoding of the phase relationships between particle motion and sound pressure. Present understanding of the otolithic ears and the peripheral neural codes data does not shed sufficient light on this persistent question; studies at the behavioral and central levels are required.

6. Summary and Conclusions

Tremendous strides have been made in our understanding of fish hearing since the pioneering work of Karl von Frisch (1936, 1938) and his students (e.g., von Frisch and Stetter 1932). This has been particularly the case since the behavioral studies of fish hearing by Tavolga and Wodinsky (1963) demonstrated differences in hearing capacities of different fish species and the first volumes on fish hearing (Tavolga 1964, 1967). Since that time, we have gained a reasonable understanding of the auditory periphery and, as discussed in other chapters in this volume, the central auditory system and the sense of hearing revealed through behavioral experiments.

Still, there are many significant questions that remain in order for us to achieve an understanding of the auditory periphery in fishes that is comparable to that already gained for many amniotes (see Popper and Fay 1993). The following are some of the most important questions that need to be answered.

1. Are the known patterns of similarity and diversity in the structures and functions of otolithic ears characteristic of fishes in general, or are there important dimensions of the patterns that have been missed? The view we now have is based on a few species, but there are many thousands of extant species whose peripheral auditory systems have not been described or analyzed. Functional studies on nonteleosts including Chondrichthyes and Agnatha are needed, along with a broader survey of the bony fishes, to

more clearly determine the functions that are primitive and those that are derived.

2. What is the functional significance of species differences in ear structures? These differences include species-specific hair cell orientation patterns, innervation patterns, and the relative sizes, shapes, and orientations of the otoliths and their respective organs. It is generally believed that species-specific structures imply different functions, but it is not presently clear that fishes differ in hearing functions other than in bandwith and in relative sensitivity to sound pressure and acoustic particle motion.

3. What are the peripheral requirements and the neural codes for sound source localization, and do they differ between hearing generalists and hearing specialists? Is the encoding of the sound pressure waveform required for the unambiguous representation of sound source location? Is the localization code distributed across otolith organs, or is one organ sufficient? What, if any, is the role of the utricle in hearing and source localization in nonclupeid fishes?

4. What is the role of the lagena in hearing? The lagena is large among otophysans but diminutive among many hearing generalists. Does the otophysan lagena play a role in directional hearing comparable to that generally accepted for the saccule of the generalists?

5. What is the role of the swim bladder in hearing among generalists? Is the distinction between specialists and generalists with respect to sound pressure encoding a matter of degree, or is it the case that sound pressure is simply ineffective in normal hearing by generalists?

6. What are the efficiency and filtering characteristics of the Weberian ossicles? What is the nature of the energy-transmission pathway between the output of the Weberian ossicles and the sacculi (and possibly other otolith organs)?

7. What is the origin of peripheral frequency selectivity, and how do species differ in a labeled-lines representation of the acoustic spectrum? What are the consequences of hair cell resonance and spiking for the afferent code in vivo? What hair cell mechanisms underlie species differences in hearing bandwidth?

8. What are the peripheral neural codes underlying the analysis of sound source characteristics other than source location? What are the relative roles of temporal and spatial codes in spectral analysis? What dimensions of neural activity are used to synthesize pitchlike and timbrelike perceptions? Do these differ among species, or are there widely shared general principles for encoding sound source characteristics?

9. What is the functional significance of hair cell addition throughout the life span of fish? Do fishes with more hair cells hear differently than fishes with fewer hair cells?

10. What are the relative roles of the saccule, lagena, and utricle in vestibular and auditory functions? Are these functions mixed within an organ or parsed among organs?

Acknowledgments. We are grateful to our colleagues and students over the years who have participated with us in many of these studies and who have provided insight and fruitful discussions. In particular, we are grateful to our mentors, William N. Tavolga (A.N.P.) and Ernest Glen Wever (R.R.F.) for their support, guidance, and friendship. We also thank Dr. Peggy Edds-Walton for providing a detailed review of the manuscript. The writing of this chapter was supported in part by Program Project Grant 1-P01-DC-00293-11 from the National Institute on Deafness and Other Communication Disorders, National Institutes of Health, to the Parmly Hearing Institute.

We dedicate this chapter to the memory of our friend and colleague Antares Parvuleseu who passed away at the age of 74 in July, 1998. Antares is certainly best known in the acoustics community for many major contributions to the physics of underwater sound. However, he also made pivotal contributions to marine bio-acoustics by helping our community understand and appreciate the very complex nature of sound in tanks. Antares demonstrated to us in two very important papers (1964, 1967) that tank acoustics are far more complex than any biologist had previously appreciated, and he made us think through far more clearly the need to control, and account for, tank acoustics. Indeed, his papers continue to guide our field even in 1999.

Though Antares only had one additional paper in fish bioacoustics (Popper et al. 1973), he continued to follow the field with genuine joy and excitement. He was always delighted to participate in meetings, talk with ourselves or our students, and contribute his very unique blend of interests and great intellect to whatever problems we brought to him. Our community will miss him greatly.

References

Alexander RMcN (1962) The structure of the Weberian apparatus in the *Cyprini*. Proc Zool Soc Lond 139:451–473.

Allen J (1997) Out hair cells shift the excitation pattern via basement membrane tension. In: Lewis ER, Long GR, Lyon RF, Narins PM, Steele CR and Hecht-Poinar E (eds) Diversity in Auditory Mechanics. Singapore: World Scientific Publishers, pp. 167–175.

Barber VC, Emerson CJ (1980) Scanning electron microscopic observations on the inner ear of the skate *Raja ocellata*. Cell Tissue Res 205:199–215.

Blaxter JHS (1981) The swimbladder and hearing. In: Tavolga WN, Popper AN, Fay RR (eds) Hearing and Sound Communication in Fishes. New York: Springer-Verlag, pp. 61–71.

Blaxter JHS, Denton EJ, Gray JAB (1981) Acoustico-lateralis systems in clupeid fishes. In: Tavolga WN, Popper AN, Fay RR (eds) Hearing and Sound Communication in Fishes. New York: Springer-Verlag, pp. 39–59.

Buwalda RJA (1981) Segregation of directional and non-directional acoustic information in the cod. elasmobranchs. In: Tavolga WN, Popper AN, Fay RR (eds) Hearing and Sound Communication in Fishes. New York: Springer-Verlag, pp. 139–171.

Buwalda RJA, Schuijf A, Hawkins AD (1983) Discrimination by the cod of sounds from opposing directions. J Comp Physiol 150:175–184.

Capranica R, Moffat AJM (1980) Nonlinear properties of the peripheral auditory system of amphibians. In: Popper AN, Fay RR (eds) Comparative Studies of Hearing in Vertebrates. New York: Springer-Verlag, pp. 139–166.

Carlström D (1963) A crystallographic study of vertebrate otoliths. Biol Bull 125:441–463.

Chang JSY, Popper AN, Saidel WM (1992) Heterogeneity of sensory hair cells in a fish ear. J Comp Neurol 324:621–640.

Colburn S, Durlach H (1978) Models of binaural interaction. In: Carterette E, Friedman M (eds) Handbook of Perception, Vol. IV. New York: Academic Press, pp. 461.

Coombs SL, Fay RR (1985) Adaptation effects on amplitude modulation detection: behavioral and neurophysiological assessment in the goldfish auditory system. Hear Res 19:57–71.

Coombs SL, Fay RR (1987) Response dynamics of goldfish saccular nerve fibers: effects of stimulus frequency and intensity on fibers with different tuning, sensitivity and spontaneous activity. J Acoust Soc Am 81:1025–1035.

Coombs SL, Popper AN (1979) Hearing differences among Hawaiian squirrelfishes (family Holocentridae) related to differences in the peripheral auditory system. J Comp Physiol 132:203–207.

Coombs S, Hastings M, Finneran J (1996) Measuring and modeling lateral line excitation patterns to changing dipole source locations. J Comp Physiol 178:359–371.

Corwin JT (1977) Morphology of the macula neglecta in sharks of the genus *Carcharhinus*. J Morphol 152:341–362.

Corwin JT (1978) The relation of inner ear structure to the feeding behavior in sharks and rays. In: Johari OM (ed) Scanning Electron Microscopy/1978, vol. 2. Chicago: SEM, pp. 1105–1112.

Corwin JT (1981a) Audition in elasmobranchs. In: Tavolga WN, Popper AN, Fay RR (eds) Hearing and Sound Communication in Fishes. New York: Springer-Verlag, pp. 81–105.

Corwin JT (1981b) Postembryonic production and aging in inner ear hair cells in sharks. J Comp Neurol 201:541–553.

Corwin JT (1983) Postembryonic growth of the macula neglecta auditory detector in the ray, *Raja clavata*: continual increases in hair cell number, neural convergence, and physiological sensitivity. J Comp Neurol 217:345–356.

Corwin JT (1985) Perpetual production of hair cells and maturational changes in hair cell ultrastructure accompany postembryonic growth in an amphibian ear. Proc Natl Acad Sci USA 82:3911–3915.

Corwin JT (1989) Functional anatomy of the auditory system in sharks and rays. J Exp Zool Suppl 2:62–74.

Corwin JT, Warchol ME (1991) Auditory hair cells: structure, function, development and regeneration. Annu Rev Neurosci 14:301–333.

Crawford AC, Fettiplace R (1981) An electrical tuning mechanism in turtle cochlear hair cells. J Physiol (Lond) 312:377–412.

Dallos P (1996) Overview: cochlear neurobiology. In: Dallos P, Popper AN, Fay RR (eds) The Cochlea. New York: Springer-Verlag, pp. 1–43.

De Boer E, de Jongh HR (1978) On cochlear coding: potentialities and limitations of the reverse correlation technique. J Acoust Soc Am 63:115–135.

Denton EJ, Gray JAB (1980) Receptor activity in the utriculus of the sprat. J Mar Biol Assoc UK 60:717–740.

Denton EJ, Gray JAB (1983) Stimulation of the acoustico-lateralis system of clupeid fish by external sources and their own movements. Philos Trans R Soc Lond [B] 341:113–127.

Denton EJ, Gray JAB (1989) Some observations on the forces acting on neuromasts in fish lateral line canals. In: Coombs S, Görner P, Münz P (eds) The Mechanosensory Lateral Line—Neurobiology and Evolution. New York: Springer-Verlag, pp. 229–246.

De Vries HL (1950) The mechanics of the labyrinth otoliths. Acta Otolaryngol 38:262–273.

Dunkelberger DG, Dean JM, Watabe N (1980) The ultrastructure of the otolithic membrane and otolith of juvenile mummichog. J Morphol 163:367–377.

Edds PL, Presson J, Popper AN (1989) A comparison of saccular innervation in the goldfish (Carassius auratus), an otophysan, and the oscar (Astronotus ocellatus), an acanthopterygian. Abstr Assoc Res Otolaryngol 12:77–78.

Edds-Walton PL, Highstein SM, Fay RR (1996) Characteristics of directional auditory afferents in the toadfish (Opsanus tau). Soc Neurosci Abstr 178:447.

Enger PS (1963) Single unit activity in the peripheral auditory system of a teleost fish. Acta Physiol Scand 59(suppl 210):9–48.

Fay RR (1978a) Coding of information in single auditory nerve fibers of the goldfish. J Acoust Soc Am 63:136–146.

Fay RR (1978b) Phase-locking in goldfish saccular nerve fibers accounts for frequency discrimination capacities. Nature 275:320–322.

Fay RR (1980) Psychophysics and neurophysiology of temporal factors in hearing by the goldfish: amplitude modulation detection. J Neurophysiol 44:312–332.

Fay RR (1981) Coding of acoustic information in the eighth nerve. In: Tavolga WN, Popper AN, Fay RR (eds) Hearing and Sound Communication in Fishes. New York: Springer-Verlag, pp. 189–222.

Fay RR (1984) The goldfish ear codes the axis of particle motion in three dimensions. Science 225:951–953.

Fay RR (1985) Sound intensity processing by the goldfish. J Acoust Soc Am 78:1296–1309.

Fay RR (1988) Hearing in Vertebrates: A Psychophysics Databook. Winnetka, IL: Hill-Fay Associates.

Fay RR (1990) Suppression and excitation in auditory nerve fibers of the goldfish, Carassius auratus. Hear Res 48:93–110.

Fay RR (1991) Masking and suppression in auditory nerve fibers of the goldfish, Carassius auratus. Hear Res 55:177–187.

Fay RR (1997) Frequency selectivity of saccular afferents of the goldfish revealed by revcor analysis. In: Lewis ER, Long GR, Lyon RF, Narins PM, Steele CR, Hecht-Poinar E (eds) Diversity in Auditory Mechanics. Singapore: World Scientific Publishers. pp. 69–75.

Fay RR, Coombs SL (1983) Neural mechanisms in sound detection and temporal summation. Hear Res 10:69–92.

Fay RR, Edds-Walton PL (1997a) Directional response properties of saccular afferents of the toadfish, *Opsanus tau*. Hear Res, 111:1–21.

Fay RR, Edds-Walton PL (1997b) Diversity in frequency response properties of saccular afferents of the toadfish (*Opsanus tau*). Hear Res, 113:235–246.

Fay RR, Feng A (1987) Directional hearing among nonmammalian vertebrates. In: Yost WA, Gourevitch G (eds) Directional Hearing. New York: Springer-Verlag, pp. 179–213.

Fay RR, Olsho LW (1979) Discharge patterns of lagenar and saccular neurons of the goldfish eighth nerve: displacement sensitivity and directional characteristics. Comp Biochem Physiol 62:377–386.

Fay RR, Ream TJ (1986) Acoustic response and tuning in saccular nerve fibers of the goldfish (*Carassius auratus*). J Acoust Soc Am 79:1883–1895.

Fay RR, Ahroon WA, Orawski AA (1978) Auditory masking patterns in the goldfish (*Carassius auratus*): psychophysical tuning curves. J Exp Biol 74:83–100.

Fay RR, Hillery CM, Bolan K (1982) Representation of sound pressure and particle motion information in the midbrain of the goldfish. Comp Biochem Physiol 71:181–191.

Fay RR, Yost WA, Coombs SL (1983) Repetition noise processing by a vertebrate auditory system. Hear Res 12:31–55.

Fay RR, Edds-Walton PL, Highstein SM (1994) Directional sensitivity of saccular afferents of the toadfish to linear acceleration at audio frequencies. Biol Bull 187:258–259.

Fay RR, Edds-Walton PL, Highstein SM (1996a) Tuning in saccular afferents of the toadfish revealed by the reverse correlation method. Biol Bull 191:255–257.

Fay RR, Chronopoulos M, Patterson RD (1996b) The sound of a sinusoid: perception and neural representations in the goldfish (*Carassius auratus*). Aud Neurosci 2:377–392.

Fish JF, Offutt GC (1972) Hearing thresholds from toadfish, *Opsanus tau*, measured in the laboratory and field. J Acoust Soc Am 51:1318–1321.

Flock Å (1965) Electron microscopic and electrophysiological studies on the lateral line canal organ. Acta Otolaryngol Suppl 199:1–90.

Flock Å (1971) Sensory transduction in hair cells. In: Lowenstein WR (ed) Handbook of Sensory Physiology, vol. 1. Principles of Receptor Physiology. Berlin: Springer-Verlag, pp. 396–411.

Forge A, Li L, Nevill G (1995) Mammalian vestibular hair cell regeneration. Science 267:706–707.

Furukawa T (1978) Sites of termination on the saccular macula of auditory nerve fibers in the goldfish as determined by intracellular injection of procion yellow. J Comp Neurol 180:807–814.

Furukawa T (1981) Effects of efferent stimulation of the saccule of goldfish. J Physiol (Lond) 315:203–215.

Furukawa T (1986) Sound reception and synaptic transmission in goldfish hair cells. Jpn J Physiol 36:1059–1077.

Furukawa T, Ishii Y (1967) Neurophysiological studies on hearing in goldfish. J Neurophysiol 30:1377–1403.

Furukawa T, Ishii Y, Matsuura S (1972) Synaptic delay and time course of post synaptic potentials at the junction of hair cells and eighth nerve fibers of the goldfish. Jpn J Physiol 22:617–635.

Furukawa T, Hayashida Y, Matsuura S (1978) Quantal analysis of the size of excitatory postsynaptic potentials at synapses between hair cells and afferent nerve fibers in goldfish. J Physiol (Lond) 276:211–226.

Furukawa T, Kuno M, Matsuura S (1982) Quantal analysis of a decremental response at hair cell-afferent fibre synapses in the goldfish sacculus. J Physiol (Lond) 322:181–195.

Gauldie RW (1996) Fusion of otoconia: a stage in the development of the otolith in the evolution of fishes. Acta Zool 77:1–23.

Grozinger B (1967) Elektro-physiologische Untersuchungen an der Hörbahn der Schleie (*Tinca tinca* [L.]). Z Vergl Physiol 57:44–76.

Guinan JJ Jr (1996) Physiology of olivocochlear efferents. In: Dallos P, Popper AN, Fay RR (eds) New York: Springer-Verlag, pp. 435–502.

Hawkins AD, Chapman CJ (1975) Masked auditory thresholds in the cod *Gadus morhua* L. J Comp Physiol 103A:209–226.

Hawkins AD, Horner K (1981) Directional characteristics of primary auditory neurons from the codfish ear. In: Tavolga WN, Popper AN, Fay RR (eds) Hearing and Sound Communication in Fishes. New York: Springer-Verlag, pp. 311–328.

Henry KH, Lewis ER (1992) One-tone suppression in the cochlear nerve of the gerbil. Hear Res 63:1–6.

Hill K, Mo J, Stange G (1989a) Induced suppression in spike response to tone-on-noise stimuli in the auditory nerve of the pigeon. Hear Res 39:49–62.

Hill K, Mo J, Stange G (1989b) Excitation and suppression of primary auditory fibers in the pigeon. Hear Res 39:37–48.

Hill K, Stange G, Gummer A, Mo J (1989c) A model proposing synaptic and extra-synaptic influences on the responses of cochlear nerve fibers. Hear Res 39:75–90.

Holton T, Weiss TF (1983) Frequency selectivity of hair cells and nerve fibers in the alligator lizard cochlea. J Physiol (Lond) 345:241–260.

Horner K, Hawkins A, Fraser P (1981) Frequency characteristics of primary auditory neurons from the ear of the cod, *Gadus morhua* L. In: Tavolga WN, Popper AN, Fay RR (eds) Hearing and Sound Communication in Fishes. New York: Springer-Verlag, pp. 223–242.

Horner K, Sand O, Enger P (1980) Binaural interaction in the cod. J Exp Biol 85:323–331.

Hoshino T (1975) An electron microscopic study of the otolithic maculae of the lamprey (*Entosephenus japonicus*). Acta Otolaryngol 80:43–53.

Hudspeth AJ (1983) Transduction and tuning by vertebrate hair cells. Trends Neurosci 6:366–369.

Hudspeth AJ, Corey DP (1977) Sensitivity, polarity and conductance change in the response of vertebrate hair cells to controlled mechanical stimuli. Proc Natl Acad Sci USA 74:2407–2411.

Ishii Y, Matsuura S, Furukawa T (1971) Quantal nature of transmission at the synapse between hair cells and eight nerve fibers. Jpn J Physiol 21:79–89.

Jarvick E (1980) Basic Structure of Evolution of the Vertebrates, Vol. 1. New York: Academic Press.

Jeffress LA (1948) A place theory of sound localization. J Comp Physiol Psychol 41:35–39.

Jørgensen JM, Mathiesen C (1988) The avian inner ear: continuous production of hair cells in vestibular sensory organs, but not in the auditory papilla. Naturwissenschaften 75:319–320.

Kalmijn AJ (1988a) Hydrodynamic and acoustic field detection. In: Atema J, Fay RR, Popper AN, Tavolga WN (eds) Sensory Biology of Aquatic Animals. New York: Springer-Verlag, pp. 83–130.

Kalmijn AJ (1988b) Detection of weak electric fields. In: Atema J, Fay RR, Popper AN, Tavolga WN (eds) Sensory Biology of Aquatic Animals. New York: Springer-Verlag, pp. 151–186.

Kalmijn AJ (1989) Functional evolution of lateral line and inner ear systems. In: Coombs S, Görner P, Münz P (eds) The Mechanosensory Lateral Line— Neurobiology and Evolution. New York: Springer-Verlag, pp. 187–216.

Köppl C, Manley G (1992) Functional consequences of morphological trends in the evolution of lizard hearing organs. In: Webster DB, Fay RR, Popper AN (eds) The Evolutionary Biology of Hearing. New York: Springer-Verlag, pp. 489–510.

Koyama H, Lewis ER, Leverenz EL, Baird RA (1982) Acute seismic sensitivity of the bullfrog ear. Brain Res 250:168–172.

Kros CJ (1996) Physiology of mammalian cochlear hair cells. In: Dallos P, Popper AN, Fay RR (eds) The Cochlea. New York: Springer-Verlag, pp. 318–385.

Lanford PJ, Popper AN (1994) Heterogeneity of hair cell populations in the otic endorgans of the goldfish, *Carassius auratus*. Abstr Assoc Res Otolaryngol 17:96.

Lanford PJ, Popper AN (1996) A unique afferent structure in the crista ampullaris of the goldfish. J Comp Neurol 366:572–579.

Lanford PJ, Presson JC, Popper AN (1996) Cell proliferation and hair cell addition in the ear of the goldfish, *Carassius auratus*. Hear Res 100:1–9.

Lewis ER (1986) Adaptation, suppression, and tuning in amphibian acoustical fibers. In: Moore BCJ, Patterson R (eds) Auditory Frequency Selectivity. New York: Plenum, pp. 129–136.

Lewis E (1992) Convergence and design in vertebrate acoustic sensors. In: Webster DB, Fay RR, Popper AN (eds) The Evolutionary Biology of Hearing. New York: Springer-Verlag, pp. 163–184.

Lombarte A, Popper AN (1994) Quantitative analyses of postembryonic hair cell addition in the otolithic endorgans of the inner ear of the European hake, *Merluccius merluccius* (Gadiformes, Teleostei). J Comp Neurol 345:419–428.

Long JA (1995) The Rise of Fishes. Baltimore, MD: The Johns Hopkins University Press.

Lowenstein O (1970) The electrophysiological study of the responses of the isolate labyrinth of the lamprey (*Lampetra fluviatilis*) to angular acceleration, tilting and mechanical vibration. Proc R Soc Lond [B] 174:419–434.

Lowenstein O, Thornhill RA (1970) The labyrinth of Myxine, anatomy, ultrastructure and electrophysiology. Proc R Soc Lond [B] 176:21–42.

Lowenstein O, Osborne MP, Wersäll J (1964) Structure and innervation of the sensory epithelia of the labyrinth in the thornback ray (*Raja clavata*). Proc R Soc Lond [B] 160:1–12.

Lowenstein O, Osborne MP, Thornhill RA (1968) The anatomy and ultrastructure of the labyrinth of the lamprey (*Lampetra fluviatilis* L.). Proc R Soc Lond [B] 170:113–134.

Lu Z, Fay RR (1993) Acoustic response properties of single units in the torus semicircularis of the goldfish. J Comp Physiol A 173:33–48.

Lu Z, Fay RR (1996) Two-tone interaction in auditory nerve fibers and midbrain neurons of the goldfish, *Carassius auratus*. Aud Neurosci 2:257–273.

Lu Z, Popper AN, Fay RR (1996) Behavioral detection of acoustic particle motion by a teleost fish, *Astronotus ocellatus*. J Comp Physiol A 179:227–233.

Lu Z, Popper AN (1997) Encoding of acoustic particle motion by saccular ganglion cells of a fish: Intracellular recording and tracing. Soc Neurosci Abst 23:180.

Lu Z, Song J, Popper AN (1998) Encoding of acoustic directional information by saccular afferents of the sleeper goby, *Dormitator latifrons*. J Comp Physiol A, in press.

Ma D, Fay RR (1996) Directional response properties of cells of the goldfish midbrain. J Acoust Soc Am 100:2786.

Manley G (1990) Peripheral Hearing Mechanisms of Reptiles and Birds. New York: Springer-Verlag.

Manley GA, Gleich O (1992) Evolution and specialization of function in the avian auditory periphery. In: Webster DB, Fay RR, Popper AN (eds) Evolutionary Biology of Hearing. New York: Springer-Verlag, pp. 561–580.

Mann DA, Lu Z, Popper AN (1977) Ultrasound detection by a teleost fish. Nature 389:341.

Mathiesen C, Popper AN (1987) The ultrastructure and innervation of the ear of the gar, *Lepisosteus osseus*. J Morphol 194:129–142.

McCormick CA (1978) Central projections of the lateralis and eighth nerves in the bowfin, *Amia calva*. Doctoral Thesis, University of Michigan.

McCormick CA (1982) The organization of the octavolateralis area in actinopterygian fishes: a new interpretation. J Morphol 171:159–181.

McCormick CA, Popper AN (1984) Auditory sensitivity and psychophysical tuning curves in the elephant nose fish, *Gnathonemus petersii*. J Comp Physiol 155:753–761.

Moeng RR, Popper AN (1984) Auditory response of saccular neurons of the catfish, *Ictalurus punctatus*. J Comp Physiol 155:615–624.

Nelson, JS (1994) Fishes of the World, 3rd ed. New York: Wiley.

Northcutt RG (1980) Central auditory pathways in anamniotic vertebrates. In: Popper AN, Fay RR (eds) Comparative Studies of Hearing in Vertebrates. New York: Springer-Verlag, pp. 79–118.

Parvulescu A (1964) Problems of propagation and processing. In: Tavolga WN (ed) Marine Bio-Acoustics. Oxford: Pergamon Press, pp. 87–100.

Parvulescu A (1967) The acoustics of small tanks. In: Tavolga WN (ed) Marine Bio-Acoustics II. Oxford: Pergamon Press, pp. 7–14.

Patuzzi R (1996) Cochlear micromechanics and macromechanics. In: Dallos P, Popper AN, Fay RR (eds) The Cochlea. New York: Springer-Verlag, pp. 186–257.

Platt C (1977) Hair cell distribution and orientation in goldfish otolith organs. J Comp Neurol 172:283–297.

Platt C (1994) Hair cells in the lagenar otolith organ of the coelacanth are unlike those in amphibians. J Morphol 220:381.

Platt C, Popper AN (1981) Structure and function in the ear. In: Tavolga WN, Popper AN, Fay RR (eds) Hearing and Sound Communication in Fishes. New York: Springer-Verlag, pp. 3–38.

Platt C, Popper AN (1984) Variation in lengths of ciliary bundles on hair cells along the macula of the sacculus in two species of teleost fishes. SEM/1984 1915–1924.

Platt C, Popper AN (1996) Sensory hair cell arrays in lungfish inner ears suggest retention of the primitive patterns for bony fishes. Soc Neurosci Abstr 22:1818.

Popper AN (1971) The effects of size on the auditory capacities of the goldfish. J Aud Res 11:239–247.

Popper AN (1977) A scanning electron microscopic study of the sacculus and lagena in the ears of fifteen species of teleost fishes. J Morphol 153:397–418.

Popper AN (1978) Scanning electron microscopic study of the otolithic organs in the bichir (*Polypterus bichir*) and shovel-nose sturgeon (*Scaphirhynchus platorynchus*). J Comp Neurol 18:117–128.

Popper AN (1981) Comparative scanning electron microscopic investigations of the sensory epithelia in the teleost sacculus and lagena. J Comp Neurol 200:357–374.

Popper AN (1983) Organization of the inner ear and processing of acoustic information. In: Northcutt RG, Davis RE (eds) Fish Neurobiology and Behavior. Ann Arbor, MI: University of Michigan Press, pp. 125–178.

Popper AN, Coombs S (1982) The morphology and evolution of the ear in Actinopterygian fishes. Am Zool 22:311–328.

Popper AN, Fay RR (1977) Structure and function of the elasmobranch anditory system. Am Zool 17:443–452.

Popper AN, Fay RR (1993) Sound detection and processing by fish: critical review and major research questions. Brain Behav Evol 41:14–38.

Popper AN, Fay RR (1997) Evolution of the ear and hearing: Issues and questions. Brain Behav Evol 50:213–221.

Popper AN, Hoxter B (1984) Growth of a fish ear: 1. Quantitative analysis of sensory hair cell and ganglion cell proliferation. Hear Res 15:133–142.

Popper AN, Hoxter B (1987) Sensory and nonsensory ciliated cells in the ear of the sea lamprey, *Petromyzon marinus*. Brain Behav Evol 30:43–61.

Popper AN, Northcutt RG (1983) Structure and innervation of the inner ear of the bowfin, *Amia calva*. J Comp Neurol 213:279–286.

Popper AN, Platt C (1979) The herring ear has a unique receptor pattern. Nature 280:832–833.

Popper AN, Platt C (1983) Sensory surface of the saccule and lagena in the ears of ostariophysan fishes. J Morphol 176:121–129.

Popper AN, Platt C (1993) Inner ear and lateral line of bony fishes. In: Evans DH (ed) The Physiology of Fishes. Boca Raton, FL: CRC Press, pp. 99–136.

Popper AN, Saidel WM (1990) Variations in receptor cell innervation in the saccule of a teleost fish ear. Hear Res 46:211–227.

Popper AN, Tavolga WN (1981) Structure and function of the ear of the marine catfish, *Arius felis*. J Comp Physiol 144:27–34.

Popper AN, Salmon M, Parvulescu A (1973) Sound localization by two species of Hawaiian squirrelfish, *Myripristis berndti* and *M. argyromus*. Anim Behav 21:86–97.

Popper AN, Rogers PH, Saidel WM, Cox M (1988) The role of the fish ear in sound processing. In: Atema J, Fay RR, Popper AN, Tavolga WN (eds) Sensory Biology of Aquatic Animals. New York: Springer-Verlag, pp. 687–710.

Popper AN, Platt C, Edds P (1992) Evolution of the vertebrate inner ear: an overview of ideas. In: Webster DB, Fay RR, Popper AN (eds) Comparative Evolutionary Biology of Hearing. New York: Springer-Verlag, pp. 49–57.

Popper AN, Saidel WM, Chang JSY (1993) Two types of sensory hair cell in the saccule of a teleost fish. Hear Res 66:211–216.

Presson JC, Popper AN (1990) A ganglionic source of new eighth nerve neurons in a post-embryonic fish. Hear Res 46:23–38.

Presson JC, Edds PE, Popper AN (1992) Central-peripheral and rostral-caudal organization of the innervation of the saccule in a cichlid fish. Brain Behav Evol 39:196–207.

Presson JC, Lanford PJ, Popper AN (1996) Hair cell precursors are ultrastructurally indistinguishable from mature support cells in the ear of a postembryonic fish. Hear Res 100:10–20.

Retzius G (1881) Das Gehörorgan der Wirbelthiere, vol. 1. Stockholm: Samson and Wallin.

Roberts BL, Meredith, GE (1992) The efferent innervation of the ear: variations on an enigma. In: Webster DB, Fay RR, Popper AN (eds) Comparative Evolutionary Biology of Hearing. New York: Springer-Verlag, pp. 185–210.

Roberts BL, Russell IJ (1972) The activity of lateral-line efferent neurones in stationary and swimming dogfish. J Exp Biol 57:435–448.

Rogers PH, Cox M (1988) Underwater sound as a biological stimulus. In: Atema J, Fay RR, Popper AN, Tavolga WN (eds) Sensory Biology of Aquatic Animals. New York: Springer-Verlag, pp. 131–149.

Rogers PH, Popper AN, Cox M, Saidel WM (1988) Processing of acoustic signals in the auditory system of bony fish. J Acoust Soc Am 83:338–349.

Rubel EW, Dew LA, Roberson DW (1995) Mammalian vestibular hair cell regeneration. Science 267:701–703.

Ruggero M (1992) Physiology and coding of sound in the auditory nerve. In: Popper AN, Fay RR (eds) The Mammalian Auditory Pathway: Neurophysiology. New York: Springer-Verlag, pp. 34–93.

Sachs M, Kiang NY-S (1968) Two-tone inhibition in auditory nerve fibers. J Acoust Soc Am 43:1120–1128.

Saidel WM, Crowder JA (1997) Expression of cytochrome oxidase in hair cells of the teleost utricle. Hear Res 109:63–77.

Saidel WM, Popper AN (1983a) Spatial organization in the saccule and lagena of a teleost: hair cell pattern and innervation. J Morphol 177:301–317.

Saidel WM, Popper AN (1983b) The saccule may be the transducer for directional hearing of nonostariophysine teleosts. Exp Brain Res 50:149–152.

Saidel WM, Popper AN (1987) Sound reception in two anabantid fishes. Comp Biochem Physiol 88A:37–44.

Saidel WM, Presson JC, Chang JS (1990) S-100 immunoreactivity identifies a subset of hair cells in the utricle and saccule of a fish. Hear Res 47:139–146.

Saidel WM, Lanford PJ, Yan HY, Popper AN (1995) Hair cell heterogeneity in the goldfish saccule. Brain Behav Evol 46:362–370.

Sans A, Highstein SM (1984) New ultrastructural features in the vestibular labyrinth of the toadfish, Opsanus tau. Brain Res 308:191–195.

Schellart NAM, de Munck JC (1987) A model for directional and distance hearing in swimbladder-bearing fish based on the displacement orbits of the hair cells. J Acoust Soc Am 82:822–829.

Schellart NAM, Popper AN (1992) Functional aspects of the evolution of the auditory system of Actinopterygian fish. In: Webster DB, Fay RR, Popper AN (eds) Comparative Evolutionary Biology of Hearing. New York: Springer-Verlag, pp. 295–322.

Schuijf A (1975) Directional hearing of cod (*Gadus morhua*) under approximate free field conditions. J Comp Physiol 98:307–332.

Schuijf A, Buwalda RJA (1980) Underwater localization—a major problem in fish acoustics. In: Popper AN, Fay RR (eds) Comparative Studies of Hearing in Vertebrates. New York: Springer-Verlag, pp. 43–77.

Schuijf A, Hawkins AD (1983) Acoustic distance discrimination by the cod. Nature 302:143–144.

Sento S, Furukawa T (1987) Intra-axonal labeling of saccular afferents in the goldfish, *Carassius auratus*: correlations between morphological and physiological characteristics. J Comp Neurol 258:352–367.

Slepecky NB (1996) Structure of the mammalian cochlea. In: Dallos P, Popper AN, Fay RR (eds) The Cochlea. New York: Springer-Verlag, pp. 44–129.

Song J, Yan HYY, Popper AN (1996) Damage and recovery of hair cells in fish canal (but not superficial) neuromasts after gentamicin exposure. Hear Res 91:63–71.

Spoendlin H (1964) Organization of the sensory hairs in the gravity receptors in the utricle and saccule of the squirrel monkey. Z Zellforsch 62:701–716.

Steinacker A, Romero A (1992) Voltage-gated potassium current resonance in the toadfish saccular hair cell. Brain Res 574:229–236.

Stensiö E (1927) The Downtonian and Devonian vertebrates of Spitzberg, I. Cephalaspidae. Skr Svalb Ish 12:1–391.

Stipetić E (1939) Über des Gehörorgan der Mormyriden. Z Vergl Physiol 26:740–752.

Sugihara I, Furukawa T (1989) Morphological and functional aspects of two different types of hair cells in the goldfish sacculus. J Neurophysiol 62:1330–1343.

Suzue T, Wu G-B, Furukawa T (1987) High susceptibility to hypoxia of afferent synaptic transmission in the goldfish sacculus. J Neurophysiol 58:1066–1079.

Tavolga WN (ed) (1964) Marine Bio-Acoustics. Oxford, UK: Pergamon Press.

Tavolga WN (ed) (1967) Marine Bio-Acoustics, II. Oxford, UK: Pergamon Press.

Tavolga WN, Wodinsky J (1963) Auditory capacities in fishes. Pure tone thresholds in nine species of marine teleosts. Bull Am Mus Nat Hist 126:177–240.

Tester AL, Kendall JI, Milisen WB (1972) Morphology of the ear of the shark genus *Carcharhinus*, with particular reference to the macula neglecta. Pac Sci 26:264–274.

Tricas TC, Highstein SM (1990) Visually mediated inhibition of lateral line primary afferent activity by the octavolateralis efferent system during predation in the free-swimming toadfish, *Opsanus tau*. Exp Brain Res 83:233–236.

Tricas TC, Highstein SM (1991) Action of the octavolateralis efferent system upon the lateral line of free-swimming toadfish, *Opsanus tau*. J Comp Physiol 169:25–37.

Van Bergeijk WA (1967) The evolution of vertebrate hearing. In: Neff WD (ed) Contributions to Sensory Physiology. New York: Academic Press, pp. 1–49.

Von Békésy G (1960) Experiments in Hearing. New York: McGraw-Hill.

Von Frisch K (1936) Über den Gehorsinn der Fische. Biol Rev 11:210–246.

Von Frisch K (1938) Über die Bedeutung des Sacculus und der Lagena fár den Gehorsinn der Fische. Z Vergl Physiol 25:703–747.

Von Frisch K, Stetter H (1932) Untersuchungen über den Sitz des Gohörsinnes bei der Elritze. Z Vergl Physiol 17:686–801.

Weber EH (1820) De Aure et Auditu Hominis et Animalium. Pars I. De Aure Animalium Aquatilium. Leipzig: Gerhard Fleischer.

Wegner N (1982) A qualitative and quantitative study of a sensory epithelium of the inner ear of a fish (*Colisa labiosa*; Anabantidae). Acta Zool (Stockh) 63:33–146.

Wegner NT (1979) The orientation of hair cells in the otolithic organs and papilla neglecta in the inner ear of the Anabantid fish *Colisa labiosa* (Day). Acta Zool (Stockh) 60:205–216.

Weiss TF, Mulroy MJ, Turner RG, Pike CL (1976) Tuning of single fibers in the cochlear nerve of the alligator lizard: relation to receptor morphology. Brain Res 115:71–90.

Wersäll J, Flock Å, Lundquist P-G (1965) Structural basis for directional sensitivity in cochlear vestibular sensory structures. Cold Spring Harbor Symp 30:115–132.

Wever EG (1949) Theory of Hearing. New York: Wiley.

Wever EG (1974) The evolution of vertebrate hearing. In: Keidel WD, Neff WD (eds) Handbook of Sensory Physiology, vol. V/1, Auditory System. Berlin: Springer-Verlag, pp. 423–454.

Wightman FL, Kistler DJ, Arruda M (1991) Monaural localization, revisited. J Acoust Soc Am 89:1995.

Wohlfahrt TA (1933) Anatomische Untersuchungen über das Labyrinth der Elritze (*Phoxinus laevis* L.). Z Vergl Physiol 17:659–685.

Wubbels RJ, Kroese ABA, Schellart NAM (1993) Response properties of the lateral line and auditory unites in the medulla oblongata of the rainbow trout (*Oncorhynchus mykiss*). J Exp Biol 179:77–92.

4
The Acoustic Periphery of Amphibians: Anatomy and Physiology

Edwin R. Lewis and Peter M. Narins

1. Introduction

According to current classification, the living amphibians are distributed among three orders—Caudata (newts and salamanders, or urodeles), Gymnophiona (caecilians), and Anura (frogs and toads)—which often are grouped in a single subclass—Lissamphibia. A current summary of the biology of the Lissamphibia is found in Duellman and Trueb (1994). Among the morphological features common to the three orders of Lissamphibia, but lacking in fish, are four evidently related to acoustic sensing (see Bolt and Lombard 1992; Fritzsch 1992 for recent reviews): (1) a hole (the oval window) in the bony wall of the otic capsule; (2) the insertion of one or two movable skeletal elements, the columella and the operculum, into that hole from its lateral side; (3) a periotic labyrinth, part of which projects into the hole from its medial side; and (4) two extraordinarily thin membranes (contact membranes), comprising locally fused epithelial linings of the periotic and otic labyrinths, each contact membrane forming part of the wall of a separate papillar recess in the otic labyrinth. The two papillae themselves may be homologues of two sensors found in fish—the macula neglecta and the basilar papilla. In amphibians, the putative homologue of the macula neglecta is called the amphibian papilla. Among fish, the basilar papilla has been found only in the coelacanth fish, *Latimeria* (Fritzsch 1987). Another morphological feature common to the Lissamphibia, but absent in fish, is a large periotic cistern separated from the saccular recess of the otic labyrinth by a large, extraordinarily thin membranous wall (like the contact membranes). The periotic cistern is the part of the periotic labyrinth that protrudes into the oval window and contacts the columella and operculum (Lombard 1980).

The frogs and toads (anurans) exhibit further acoustic specializations (Fig. 4.1). In many terrestrial anuran species, the columella spans an air-filled space (middle ear) between the inner ear and a tympanum (ear drum), and provides relatively rigid coupling between the tympanum and the wall of the periotic labyrinth at the oval window—analogous to the coupling

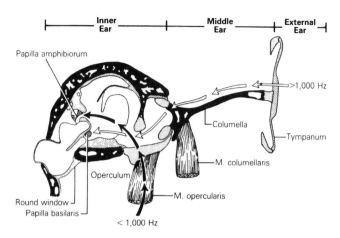

FIGURE 4.1. Diagram of the ear of a frog or toad, depicting the elements associated with the oval window and the relative positions of two of the acoustic sensors (the amphibian and basilar papillae). Arrows show the putative paths of acoustic energy in two frequency ranges according to the observations of Lombard and Straughan (1974) (see Section 3). (From Duellman and Trueb. *Biology of Amphibians* © 1994. The Johns Hopkins University Press.)

provided by the ossicular chain in mammals (Wever 1985; Jaslow et al. 1988). Physiological experiments have demonstrated conclusively that the anuran basilar papilla and the anuran amphibian papilla both are acoustic sensors, especially responsive to airborne sound (Frishkopf and Goldstein 1963; Capranica 1965; Frishkopf and Geisler 1966; Feng et al. 1975; Lewis, Baird et al. 1982). In the more derived anurans, the saccule is uniquely specialized morphologically and has been shown physiologically to be an acoustic sensor—exquisitely sensitive to substrate vibration (seismic signals) (Ashcroft and Hallpike 1934; Cazin and Lannou 1975; Koyama et al. 1982). These various physiological results have led to the following commonly accepted inferences: (1) The periotic labyrinth in anurans channels acoustic signals from the oval window to the acoustic sensors (saccule, amphibian papilla, basilar papilla) in the otic labyrinth (Harrison 1902; de Burlet 1935; van Bergeijk and Witschi 1957). (2) Acoustic coupling of seismic signals from the substrate and low-frequency sound from the air to the oval window is enhanced by the presence of the operculum and its associated muscles (Lombard and Straughan 1974; Hetherington et al. 1986). (3) In terrestrial anurans, acoustic coupling between the air and the oval window is enhanced by the presence of the tympanum/columella system (Frishkopf et al. 1968; Lombard and Straughan 1974). Thus, the four morphological features listed in the first paragraph—columella/operculum, oval window, periotic labyrinth, and contact membranes—are considered to be successive elements in two paths for acoustic signal from the animal's periphery to the sensors of the otic labyrinth. By analogy, these structures commonly are presumed to provide similar functions in those Lissamphibia (ceacilians and urodeles) for which acoustic physiological data are sparse (Kingsbury and Reed 1909; Monath 1965; Smith 1968; Wever and Gans 1976; Ross and Smith 1980). Among the living Lissamphibia, only anurans possess a tympanum and air-filled middle ear. Physiological and behavioral studies, which have been carried out in many anuran species, both primitive and derived, show clearly that the anuran nervous system and the animal itself respond conspicuously to auditory and seismic signals. The frog or toad thus has its ear to the ground and to the air at the same time. The urodeles (newts and salamanders) and caecilians, on the other hand, may have their ears largely to the ground.

2. End Organs and Homologies

The amphibian inner ear is unusual in the number of end organs (separate sensory surfaces) it contains (Retzius 1881; Wever 1985; Lewis et al. 1985). Recall that the mammalian inner ear contains six: three semicircular canals, a saccule, a utricle, and a cochlea. The inner ears of some caecilians contain nine: three semicircular canals, a lagena, a utricle, a saccule, an amphibian papilla, a papilla neglecta, and a basilar papilla (Sarasin and Sarasin 1892; White and Baird 1982). The inner ears of some urodeles and all anurans contain eight: three semicircular canals, a lagena, a utricle, a saccule, an amphibian papilla, and a basilar papilla (de Burlet 1928, 1934a,b; van Bergeijk 1957; Geisler et al. 1964; Mullinger and Smith 1969; Wever 1973; Lombard 1977; White 1986). Some species of caecilians and urodeles lack a basilar papilla; this lack is considered a derived state by Lombard and White (Wever 1975; Lombard 1977; White and Baird 1982). The usual presumption, based on morphological and developmental criteria, is that inner-ear end organs of the same name, but in different taxa, are homologous. Thus the semicircular canals, the utricle, and the saccule of amphibians are considered to be homologous to the corresponding end organs (of the same names) in fish, reptiles, birds, and mammals; and the lagenar macula of amphibians can be considered homologous to the maculae of the same name in fish, reptiles, birds, and monotremes (Fritzsch 1992). The presumption is brought into question, from time to time, however. Fritzsch (1992), for example, has argued compellingly that a recess for the lagena, separate from that of the saccule, has arisen independently at least three times (i.e., the macula may be homologous from taxon to taxon, but its recess not).

2.1 Basilar Papilla

The basilar papilla of amphibians often is taken to be homologous to the end organs of the same name in reptiles and birds, which in turn are considered to be homologous to the mammalian cochlea (Baird 1974b; Fritzsch 1992). In reptiles, birds, and monotremes, however, the basilar papilla lies within the lagenar recess, close to its union with the saccular recess (Baird 1974), and among the Lissamphibia, it occurs in that position in the urodeles and caecilians (White 1978; White and Baird 1982; White 1986). In the anurans, it lies in a separate recess that opens directly into the saccular recess (van Bergeijk and Witschi 1957; Geisler et al. 1964; Wever 1973). This led Wever (1974, 1985) to propose that the end organ labeled "basilar papilla" in anurans arose independently of that in urodeles. Structural differences between the basilar papilla in amphibians and that in amniotes led Lombard and Bolt (1979) to suggest that it was independently derived in those two groups. In *Latimeria*, Fritzsch (1987) identified a sensory papilla in the lagenar recess, close to its union with the saccular recess. Its location, structure and innervation led him to conclude that it was homologous to the basilar papillae of tetrapods. Subsequently, he argued that the basilar papilla arose just once, in the sarcopterygians, and was retained in somewhat different forms in *Latimeria*, the amphibians, and the amniotes (Fritzsch 1992).

2.2 Amphibian Papilla

The amphibian papilla was considered by Retzius (1881) to be homologous to the papilla neglecta of fish. The papilla neglecta, however, normally is associated either with the upper part of the inner ear (pars superior) and the utricle, or with the duct connecting the pars superior to the lower part of the ear (pars inferior). In the anurans and the caecilians, the amphibian papilla resides in an outpocketing of the pars inferior and thus is associated with the saccule (de Burlet 1934b). Furthermore, the papilla neglecta (in an appropriate position in the utricle) and the amphibian papilla both are found in the inner ears of caecilians, suggesting that the amphibian papilla may have arisen independently (Sarasin and Sarasin 1890). The issue is not resolved, however (Baird 1974a,b; Corwin 1977; White and Baird 1982). In many fish, the papilla neglecta comprises two sensory patches (Platt 1977, 1983; Lewis et al. 1985). In urodeles, caecilians, and the most primitive living frogs (genera *Ascaphus* and *Leiopelma*), the amphibian papilla comprises just one patch (Mullinger and Smith 1969; White 1978; White and Baird 1982; White 1986; Lewis 1981a). In the remaining anurans, it comprises two patches that develop separately and merge during maturation (Li and Lewis 1974; Lewis 1981b, 1984). White and Baird (1982) argued that the amphibian papilla in all cases was derived from a primitive, two-patch papilla neglecta, with one patch being lost in urodeles and primitive anurans but retained in the caecilians and the remaining anurans. In the caecilians, they argue, the two patches migrated to opposite ends of the large-diameter duct separating the pars inferior and pars superior. Their argument is strengthened by the location of the amphibian papilla in many urodeles—in an outpocketing from the duct itself (Lombard 1977).

Some investigators have proposed that the amphibian orders arose independently of one another (see Duellman and Trueb 1994, pp. 437–443). If the amphibian papilla did arise independently in the amphibians, independent origin of any amphibian order would imply that the amphibian papilla of its members is not homologous to the amphibian papilla of the other amphibian orders. Based on considerable evidence, the generally accepted view currently is that of a common origin of the three amphibian orders (but see Duellman and Trueb 1994, p. 443), in which case the amphibian papilla could be homologous among all amphibian taxa. Several shared morphological features (described in a later section of this chapter) among the amphibian papillae of caecilians, urodeles, and the most primitive living anurans (*Ascaphus* and *Leiopelma*) strengthen the argument for homology at least within the Lissamphibia (Lewis 1981a,b; White and Baird 1982).

2.3 Functional Overview

Although there have been a few such studies of the acoustic periphery in urodeles (Ross and Smith 1977, 1978, 1980) and caecilians (Wever and Gans 1976; Wever 1985), most physiological and biophysical studies of amphibian auditory and seismic senses have been carried out on anurans. Four of the inner-ear end organs (saccule, lagena, amphibian papilla, and basilar papilla) of the frog seem to serve acoustic functions (auditory and/or seismic), and all of these end organs except the lagena appear to be dedicated entirely to acoustic function (Ashcroft and Hallpike 1934; Frishkopf et al. 1968; Feng et al. 1975; Caston et al. 1977; Lewis, Baird et al. 1982; Baird and Lewis 1986). The acoustic frequency range seems to be divided among the saccule, the amphibian papilla, and the basilar papilla. Where its frequency range has been studied, in the American toad (*Bufo americanus*), the American bullfrog (*Rana catesbeiana*), the European grass frog (*R. temporaria*), the leopard frog (*R. pipiens*), and the white-lipped frog (*Leptodactylus albilabris*), the anuran saccule exhibits best excitatory frequencies (BEFs) typically below 100 Hz (Moffat and Capranica 1976; Koyama et al. 1982; Yu et al. 1991; Jørgensen and Christensen-Dalsgaard 1991; Christensen-Dalsgaard and Narins 1993; Christensen-Dalsgaard and Jørgensen 1996a). Where their frequency ranges have been studied, in *B. americanus*, *R. catesbeiana*, *R. temporaria*, *R. pipiens*, *L. albilabris*, the green frog (*R. clamitans*), the Puerto Rican coqui (*Eleutherodactylus coqui*), the green treefrog (*Hyla cinerea*), the spring peeper (*H. crucifer*), the barking treefrog (*H. gratiosa*), cricket frogs (*Acris gryllus*, *Acris crepitans*), the little green toad (*B. debilis*), Couch's spadefoot toad (*Scaphiopus couchi*), the fire-bellied toad (*Bombina orientalis*), and the tailed frog (*Ascaphus truei*), the anuran amphibian papilla has exhibited BEFs ranging from approximately 80 Hz to between 600 Hz and 1600 Hz, depending on the species, and the anuran basilar papilla has exhibited BEFs higher than those of the amphibian papilla (Frishkopf and Goldstein 1963; Sachs 1964; Capranica 1965; Liff 1969; Capranica et al. 1973; Capranica and Moffat 1974a,b; Moffat and Capranica 1974; Capranica and Moffat 1975; Feng et al. 1975; Capranica 1976; Narins and Capranica 1976; Wilczynski et al. 1983, 1984; Hillery and Narins 1987; Zakon and Wilczynski 1988; Ronken 1991).

In all anurans in which its physiological properties have been studied, the basilar papilla has been found to be tuned to some component of the animal's call (Capranica 1965; Frishkopf et al. 1968; Loftus-Hills and Johnstone 1970; Loftus-Hills 1973; Narins and Capranica 1976; Wilczynski et al. 1983, 1984; Zakon and Wilczynski 1988). The sacculus and amphibian papilla both seem to be more general-purpose acoustic sensors, providing tuning over broad ranges of frequencies. Amphibian papillar units have been found to be tuned relatively sharply, providing especially good spectral resolution, while saccular units have been found to be much more broadly tuned, apparently sacrificing some spectral resolution to achieve greater temporal resolution. The lagena has a vibration-sensitivity (Narins 1975; Caston et al. 1977), which was found to be distributed along the very center of its central (striolar) band (Lewis, Baird et al. 1982; Baird and Lewis 1986). That region is surrounded by large areas of orientation and postural-motion sensing regions (yielding traditional adapting and nonadapting vestibular responses) (Narins 1975; Caston et al. 1977; Baird and Lewis 1986; Cortopassi and Lewis 1995, 1996). Where it has been studied, in *R. catesbeiana*, the vibratory frequency range of the lagena overlaps those of the sacculus and the low-frequency regions of the amphibian papilla, and lagenar units generally are broadly tuned and less sensitive than saccular units (Cortopassi and Lewis 1996).

2.4 Amphibian Papilla and Cochlea: Analogies and Distinctions

Whether or not the frog basilar papilla is a homologue of the mammalian cochlea, the frog amphibian papilla seems to be a very much closer analogue to the cochlea. The frog amphibian papilla and mammalian cochlea have several, remarkable functional similarities: (1) The shapes of the frog amphibian papillar tuning curves are very similar to those of cochlear units with comparable BEFs, both kinds of units exhibiting high dynamic order, which gives them steep tuning band edges and allows them to achieve high spectral resolution and still maintain high temporal resolution (Frishkopf and Goldstein 1963; Kiang et al. 1965; Narins and Hillery 1983; see Lewis 1992). (2) The tuning of frog amphibian papillar units and cochlear units undergo very similar adjustments to changes in background sound intensity—trading spectral resolution for temporal resolution as background sound levels increase, and vice versa (Evans 1977; Møller 1977, 1978; Carney and Yin 1988; Dunia and Narins 1989; Yu 1991; Lewis and Henry 1994). (3) The sensitivity of amphibian-papillar units and cochlear units also undergoes similar adjustments to changes in background sound intensity—both exhibiting a gain-control (nonlinear) form of adaptation (Abbas 1981; Costalupes et al. 1984; Megela 1984; Narins 1987; Narins and Zelick 1988; Zelick and Narins 1985; Narins and Wagner 1989; Yu 1991; Lewis and Henry 1995). (4) The frog amphibian papilla and the cochlea both exhibit tonotopy, which could facilitate neural computations involving acoustic spectra (von Békésy 1960; Lewis and Leverenz 1979; Lewis et al. 1982). (5) The individual units of the frog amphibian papilla and the cochlea both are subject to suppression by stimuli at frequencies above and below BEF (Goldstein et al. 1962; Katsuki et al. 1962; Frishkopf and Goldstein 1963; Sachs and Kiang 1968; Capranica and Moffat 1980; Ehret, Moffat and Capranica 1983; Lewis 1986; Henry and Lewis 1992; Benedix et al. 1994; Christensen-Dalsgaard and Jørgensen 1996b); this includes both suppression of spontaneous activity (e.g., one-tone suppression) and suppression of driven activity (e.g., two-tone suppression). (6) Units of the frog amphibian papilla and the cochlea both exhibit responses to simultaneously applied tone pairs that imply nonlinearities even at low stimulus levels; amphibian papillar units exhibit especially strong responses at the f_2-f_1 frequency, implying quadratic distortion, and cochlear units exhibit especially strong responses at the $2f_1$-f_2 frequency, implying cubic distortion (Goldstein and Kiang 1968; Capranica and Moffat 1980). (7) The cochlea and the frog amphibian papilla both evidently are sources of otoacoustic emissions (Kemp and Martin 1976; Kemp 1979; Palmer and Wilson 1982), suggesting that electromechanical transduction as well as mechanoelectric transduction occurs in both.

The frog amphibian papilla and mammalian cochlea also have several remarkable structural similarities: (1) Each has a long, narrow, tonotopically organized sensory epithelium that is compressed into a short space by coiling (in two or three dimensions) (Lewis et al. 1985; Lewis et al. 1992). (2) Both are tectorial (as opposed to otoconial or otolithic) end organs, and in each the tectorial membrane covers all of the sensory surface (Lim 1972; Lewis 1976; Kronester-Frei 1978; Shofner and Feng 1983). (3) In both, the hair cell-body length decreases as one moves from the low-frequency end to the high-frequency end (Bohne and Carr 1985; Evans 1988; Simmons et al. 1994).

The frog amphibian papilla and the mammalian cochlea also have conspicuous differences: (1) The tuning range of the cochlea extends to much higher frequencies: the range of BEFs of amphibian papillar units extends over approximately 2.5 to 4 octaves, ranging upward from approximately 100 Hz; the range of BEFs of cochlear units extends over approximately 8 to 10 octaves, also ranging upward from approximately 100 Hz. (2) The higher-frequency units of the cochlea exhibit sharp tuning tips superimposed on broad tuning tails; amphibian papillar units exhibit no such tips or tails [e.g., see Kiang et al. (1965) and Zakon and Wilczynski (1988)]. (3) The tendency of spikes in cochlear afferent axons to be phase-locked to stimulus wave-forms extends to higher frequencies than does that of spikes in amphibian papillar afferent axons: approximately 5 kHz for cochlear axons, and approximately 900 Hz for amphibian papillar axons (Kiang et al. 1965; Anderson et al. 1970; Narins and Hillery 1983). (4) The organizations of supporting cells in the two organs are profoundly different. The supporting cells surrounding the hair cells of the amphibian papilla appear to be simple columnar epithelial cells (Geisler et al. 1964); those surrounding the hair cells in the cochlea are highly modified, forming the elaborate organ of Corti (e.g., see Iurato 1962). (5) The amphibian papilla is mounted on the thick labyrinthine wall overlying the semicircular canals and lacks any structure resembling the basilar membrane of the cochlea (Geisler et al. 1964; Lewis 1976). (6) The ability of hair cells to proliferate and regenerate is different: hair-cell proliferation continues in adult frogs (Alfs and Schneider 1973; Lewis and Li 1973); it is not known to do so in the cochlea.

3. The Far Periphery

The conduction of acoustic signals from the environment to the amphibian inner ear is a subject of considerable uncertainty at this time. In anurans and urodeles, the simplest picture is that of a dual system comprising a columellar subsystem that primarily conducts acoustic signals that were airborne or water-borne and an opercular subsystem that conducts acoustic signals that were borne through the ground (seismic signals). In urodeles, for example, the columellar subsystem evidently serves as the predominant acoustic pathway in aquatic larvae, and the opercular subsystem (Fig. 4.2) develops as the animal emerges to a terrestrial life, at which time the columella becomes part of the otic capsule—fused to the other skeletal elements and no longer movable (Monath 1965; Wever 1974, 1985; Lombard 1977; Hetherington 1988). Adult anurans typically possess both subsystems. In some burrowing species, however, the columellar subsystem is lost, leaving only the opercular subsystem; and in some aquatic species, the opercular subsystem is lost or very much reduced, leaving the columel-lar subsystem (de Villiers 1932, 1934; Henson 1974). Larval forms of at least some members of the family Ranidae possess a third subsystem, the bron-chial columella, which connects the pulmonary bronchus to the inner ear (Witschi 1949, 1955; Henson 1974). Most caecilians possess a columellar subsystem, but all evidently lack an opercular subsystem (Taylor 1969; Hetherington 1988).

In most anurans, the columellar subsystem comprises a tympanum that covers a largely air-filled middle ear, a bony structure (the columella) that spans that air-filled space, the linkages at each end of that bony structure,

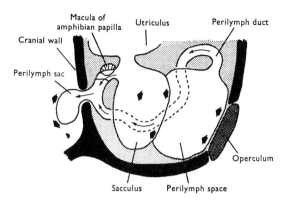

FIGURE 4.2. Putative paths (arrows) of fluid displacement in the inner ear of a salamander. Perilymph space = periotic cistern. (From Smith 1968, with permission.)

and a small muscle (*M. columellaris*) connected between the bony structure and the pectoral girdle (Wever 1979). The distal end of the columella is linked by cartilage to the tympanum; the proximal end contacts the inner ear at the oval window. Distally, where it spans the middle-ear space, the columella is slender, rod-like; proximally, where it is attached to the oval window, it is expanded into a thin plate (the footplate). *M. columellaris* inserts on an extension of the footplate.

According to Duellman and Trueb (1994), the tympanum in most anuran species is either the same diameter in males and females or slightly larger in females. In certain species of the family ranidae, however, the tympanum is conspicuously larger in males. Frishkopf et al. (1968) found no correlation between tympanum size and sensitivity to airborne sound in one such species, *R. catesbeiana*. In one group of West African frogs (genus *Petropedetes*) the tympanum of adult males during mating season is decorated with a fleshy papilla (Lawson 1993). It is not clear how this affects the acoustic properties of the columellar subsystem. An especially interesting example of the complexity and uncertainty of our current picture of the acoustic periphery in anurans is provided by the work of Purgue (1997), who has demonstrated that the peripheral path for emissions of the higher-frequency components of vocalizations by males of numerous species of *Rana* (which also possess tympanae conspicuously larger than the females of the same species) is through the tympanum rather than the gular sac. These observations are especially interesting in light of the evidence found by Narins, Ehret, and colleagues strongly implying that a significant path for low-frequency sound from the environment to the inner ear passes through the frog's body wall rather than through the tympanum (Narins et al. 1988; Ehret et al. 1990; Jørgensen et al. 1991; Ehret et al. 1994). This has been demonstrated in a wide range of species—*Eleutherodactylus coqui, Smilisca baudini*, *Hyla cinerea*, *Osteopilus septentrionalis*, *Dentrobates tinctorius*, *D. histrionicus*, *Epipedobates tricolor*, and *E. azureiventris*.

The columellar subsystem in urodeles lacks the tympanum and air-filled middle-ear volume, yet the columella still contacts the oval window and still includes a movable rod-like element (oriented approximately perpendicular to the oval window). Lombard and Bolt (1979) proposed that the urodeles and anurans shared a common ancestor that possessed an anuran-like columellar subsystem, with a tympanum and air-filled middle ear, and that the absence of those elements in living urodeles is a derived condition. In caecelians, the columella comprises a massive footplate and a stubby rod-like extension that articulates with the quadrate bone of the skull (Marcus 1935; Brand 1956; Taylor 1969; Duellman and Trueb 1994, p. 306).

In anurans and urodeles, the opercular subsystem comprises a bony (some newts and salamanders) or cartilaginous (other newts and salamanders and frogs and toads) plate (the operculum) attached to the oval window, plus one or more small muscles connected between the plate and skeletal elements of the pectoral girdle. Evidently, in some urodeles (Monath 1965) and possibly in the anuran family pipidae (de Villiers 1932), the footplate of the columella is fused with the operculum.

The view presented at the beginning of this subsection, namely of a dual system comprising a columellar subsystem that primarily conducts acoustic signals that were airborne or water-borne and an opercular subsystem that conducts acoustic signals that were borne through the ground, clearly is overly simplified. In 1974, Lombard and Straughan presented evidence suggesting that both subsystems serve as paths for sounds that were airborne, with the opercular system serving for low-frequency sound components and the columellar system serving for high-frequency components. This work should be revisited in light of more recent studies of extratympanal acoustic pathways. Wever (1979) suggested that the dual muscle system in anurans (one separately innervated muscle associated with the operculum, one with the columella) allows the animal to adjust the efficacy of each subsystem independently by adjusting the tensions on the muscles.

4. The Periotic System

The conduction of acoustic signals from the oval window (where the columellar and opercular subsystems terminate) to the sensory surfaces of the inner ear is presumed to be accomplished, at least in part, by the periotic system. The periotic system includes tissue that lines the wall of the otic capsule, and it includes fluid-filled chambers and ducts surrounded by thin squamous epithelium (Fig. 4.3). These chambers and ducts and their thin walls, together, are known as the periotic labyrinth. The periotic labyrinth surrounds part of the otic labyrinth, which comprises the inner-ear cham-

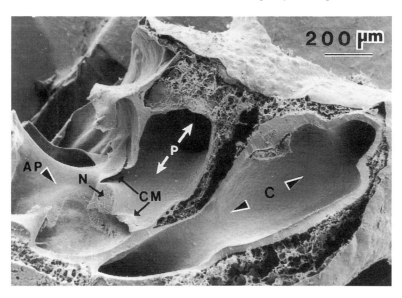

FIGURE 4.3. Scanning electron micrograph showing a dorsad view of the transected membranous labyrinth (combined otic and periotic labyrinths) of the Colorado River toad, *Bufo alvarius*, in the vicinity of the amphibian papilla (AP). The periotic duct (P) is separated from the recess of the amphibian papilla by a contact membrane (lower CM) and from the saccular space by another contact membrane (upper CM). The two contact membranes are separated by the amphibian papillar twig of the eighth nerve. The amphibian papilla resides on a thick wall of limbic tissue that separates it from the confluence (C) of the posterior and horizontal semicircular canals. Fibrous periotic tissue is seen lying between the dense walls of the periotic duct and canal confluence and the thin outer wall of the membranous labyrinth.

bers in which the various sensory surfaces reside, and the ducts connecting them, including the semicircular canals (Fig. 4.4). The solute composition of the fluid (perilymph) in the periotic labyrinth is very similar to that of the extracellular fluids in the animal as a whole, with sodium and chloride being the predominant electrolytes (Rauch and Rauch 1974). The walls of the periotic labyrinth flex easily in response to forces applied perpendicular to the wall, but are rigid and tough in response to tangential forces.

4.1 Periotic Cistern and the Saccule

One chamber, the periotic cistern, lies between the lateral wall of the recess in which the saccule resides and the lateral wall of the otic capsule, where the oval window resides. The periotic cistern is in direct contact with the columellar and opercular subsystems, through the oval window. Where they are in direct contact, the squamous epithelial linings of the saccular

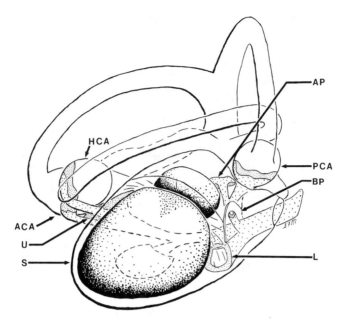

FIGURE 4.4. Diagram of the inner ear of the American bullfrog (*Rana catesbeiana*), showing the relative size of the saccular otoconial mass (depicted with darkly shaded margins) and the relative positions of the various sensors (L, lagena; BP, basilar papilla; AP, amphibian papilla; U, utricle; ACA, HCA, PCA, ampules of the anterior, horizontal, and posterior semicircular canals; S, saccule. Beneath the saccular mass, the outline of the eighth nerve and macular pad of the saccule are depicted with dashed lines. (Illustration by Steven F. Myers, Biology Department, University of Michigan–Flint.)

recess and the periotic cistern are fused, forming a thin lateral wall over the saccular recess. In *Latimeria* there is a similar space between the lateral wall of the otic capsule and the saccular recess, but it is filled with fat rather than perilymph (Millot and Anthony 1965; Fritzsch and Wake 1988). In the more derived anurans, the periotic cistern and the thin wall associated with it nearly surround the saccular recess (McNally and Tait 1925; Wever 1985; Lewis and Lombard 1988). The wall is thickened slightly in one place, forming a small epithelial pad (Fig. 4.5) on which the saccular macula and its hair cells reside. Between this macular pad and the wall of the otic capsule is the fluid-filled space of the periotic cistern. Spanning that space are several very thin, cylindrical struts of periotic connective tissue as well as the much thicker saccular twig of the eighth cranial nerve. Blood flow enters the macular pad through arterioles in the nerve twig and exits through venules in the struts. The hair-cell side of the macular pad faces the inside of the saccular recess, part of the otic labyrinth. In being a relatively thin sensory surface (with hair cells) lying directly between the periotic and

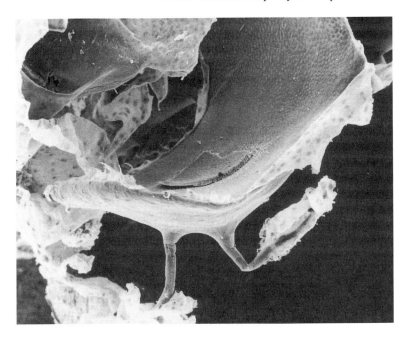

FIGURE 4.5. Macular pad of the saccule of a small Puerto Rican frog, *Eleuthero-dactylus portoricensis*. From this pad, only two struts (bottom center) spanned the periotic fluid space (extension of the periotic cistern) beneath the pad. In larger frogs, such as *R. catesbeiana*, there can be as many as seven struts. Remnants of the extremely thin tissue formed by fusion of the epithelial linings of the periotic cistern and the saccular recess is seen around the edges of the pad and elsewhere. The dark spots in this tissue are the nuclei of epithelial cells. On the top (endolymphatic surface) of the pad, one can seen the gelatinous membrane, stuck to the surface of the macula. The saccular twig of the nerve, covered by epithelial cells, is at the top of the micrograph.

otic labyrinths, the anuran macular pad is analogous to the basilar papillae in reptiles and birds and to the organ of Corti in the mammalian ear (Wever 1974, 1976). Undoubtedly more than any other feature, this configuration of the macular pad in the anuran saccule led to the hair cell of the saccule of *R. catesbeiana* becoming a standard model for studies of cellular and molecular mechanisms of transduction (Hudspeth and Corey 1977; Hudspeth and Lewis 1988). After cutting the nerve twig and the struts, one can remove the macular pad of *R. catesbeiana* from a living frog for use in in vitro experiments.

4.2 Periotic Canal and Sac and the Auditory Papillae

The periotic cistern is connected through a duct (periotic canal) to a second chamber—the periotic sac, which lies at least partially outside of the otic capsule, in the brain case. En route, the periotic canal passes next to the fluid-filled recess in which the amphibian papilla resides (see Fig. 4.3). As it does so, the walls of the otic and periotic labyrinths are fused into a patch of extraordinarily thin membrane (contact membrane) that forms one wall of the amphibian-papillar recess. That wall protrudes conspicuously into the canal, partially occluding it (Purgue and Narins 1997b). The papilla itself, in which the hair cells reside, is not part of that wall. Instead, it is embedded in thick limbic tissue with no underlying periotic fluid space. In this way, the amphibian papilla is distinctly not analogous to the basilar papillae of reptiles and birds, or to the organ of Corti in mammals (Wever 1974, 1976). In the more derived anurans, close to the contact membrane of the amphibian-papillar recess, is a second patch of extraordinarily thin membrane that lies between the periotic canal and the saccular recess (Lewis 1984). In some more derived urodeles, the periotic canal also passes close to the recess of the basilar papilla, with the walls of the periotic and otic labyrinths again fusing to form a contact membrane. In primitive urodeles and apparently in all anurans, the periotic canal bypasses the recess of the basilar papilla. In those animals, a short, second duct emerges from the periotic sac and terminates at a contact membrane that forms one wall of the basilar-papillar recess. In all amphibians, the basilar papilla is not part of the wall between periotic and otic labyrinths. Like the amphibian papilla, it is embedded in thick limbic tissue with no underlying periotic fluid space, and in this way is distinctly not analogous to the basilar papillae of reptiles and birds, or to the organ of Corti in mammals. The same thing is true of the saccular maculae in all amphibians except the derived anurans (Lombard 1970; Lewis and Nemanic 1972; Lombard 1977; Wever 1985). In amniotes, the periotic duct is rostral to the otic labyrinth; in amphibians it is caudal. This difference led Baird (1974a) and Lombard (1977) to suggest independent origins of the amphibian and amniote periotic systems (see also Fritzsch 1992).

4.3 Putative Acoustic Paths

It is interesting to contemplate, in detail, the transfer of acoustic energy through the columella or operculum to the periotic system. In solid mechanical systems, energy flow occurs when there is force accompanied by motion. If a force is applied through the skeletal element (columella or operculum) to the oval window, and that force is accompanied by motion of the oval window (in the same direction as the force), then energy is being delivered through the element to the oval window. At any instant, the rate of energy flow (SI unit $1.0\,\text{J/s} = 1.0\,\text{W}$) equals the product of the force (SI unit $1.0\,\text{N} = 1.0\,\text{J/m}$) and the velocity (SI unit $1.0\,\text{m/s}$) of that motion.

In fluid systems, energy flow occurs when there is pressure accompanied by fluid (volume) flow. Most of the force applied by the skeletal element to the oval window membrane will be passed through to the fluid, where it will be translated to pressure. In that case, the combination of skeletal element and oval window membrane serve as parts of a (piston-like) transducer that takes energy across from the realm of solid mechanics to the realm of fluids, and vice versa. If the direction of the force were perpendicular to the oval window membrane, and the contact between the skeletal element and the membrane were flat, then the pressure (SI unit $1.0\,N/m^2$) developed by the transducer would equal the force passed through to the fluid divided by the area (SI unit $1.0\,m^2$) of the contact. Under those same circumstances, the flow (SI unit $1.0\,m^3/s$) of the fluid would equal the velocity of the skeletal element in the direction of the force multiplied by the area of the contact. At any instant, the rate of energy flow (again SI unit $1.0\,W$) to the fluid would equal the product of the pressure and the fluid flow. Some of that energy will be absorbed by the periotic labyrinth, and some will be transferred to the otic labyrinth, where the sensory surfaces reside.

There are two fundamentally different ways that the perilymph will flow in response to the pressure developed by the transducer: (1) it will compress and expand (decreasing and increasing the total volume of perilymph); and (2) the walls of the periotic system will flex, allowing the perilymph to move from place to place, but leaving its total volume constant. Owing to the large value of the elastic modulus (bulk modulus) for compression and expansion of water (approximately $2.1 \times 10^9\,N/m^2$) and the limited volume of perilymph, the amplitude of flow from the first mode will be exceedingly small (e.g., for a peak sound pressure level of $1.0\,Pa$, which is extremely loud, the peak volume displacement of perilymph would be approximately 0.48×10^{-9} times the total perilymph volume). If the otic capsule were sealed, and its walls rigid, then the walls of the periotic system could flex only through compression or expansion of the tissues and fluid spaces adjacent to it within the otic capsule; and the amplitude of flow from the second mode also would be exceedingly small. In all amphibians, however, the periotic system extends into the cranial cavity, through one or more holes in the medial wall of the otic capsule (e.g., see Frishkopf and Goldstein 1963; Smith 1968; Fritzsch 1992). This provides a pathway by which the second mode of perilymph flow can be greatly enhanced, allowing perilymph to move in and out of the cranial cavity in response to pressure changes at the oval window (van Bergeijk 1957; Smith 1968). Thus one can envision perilymph flow as being directed across the entire otic capsule, between the oval window on its lateral side and the cranial cavity on its medial side (see Fig. 4.2).

The wall of the periotic sac within the cranial cavity is thin and flexible. As it flexes to accommodate the acoustically driven flow of perilymph from the otic capsule into or out of the cranial cavity, it will displace or be displaced by other tissues within the cavity. There are three fundamentally different ways that such displacement will occur: (1) bulk compression or expansion of the surrounding cranial tissues, (2) expansion or compression of the cranial wall, and (3) displacement of tissue (e.g., blood) out of or into the cranial cavity through various openings (e.g., blood vessels). Again, owing to the large value of the bulk modulus of water, displacements resulting from the first mode will be exceedingly small (although they now are limited by the fluid volume of the cranial cavity rather than that of the otic capsule). The second and third modes require acoustically driven pressure differences between the perilymph within the periotic sac and the environments immediately outside the cranial cavity. Even for airborne or water-borne sounds with wavelengths that are very long relative to the dimensions of the ear, so that acoustic pressure gradients outside the head are very small, such acoustically driven pressure differences could be created, for example, by making the amplitude of the sound pressure developed by the pistion-like transducer at the oval window greater than the sound pressure impinging on the animal. This would require that the middle-ear subsystem conducting the sound provide pressure amplification, that is, serve as an acoustic transformer (see Lewis 1996 for basic transformer theory).

Frogs and toads, including the most primitive living frogs (*Ascaphus*), possess a compliant window in the cranial wall immediately adjacent to the periotic sac (van Bergeijk and Witschi 1957; Wever 1985). It is called the round window (Fig. 4.1) and is considered to be an analogue but not a homologue of the structure of the same name in the mammalian ear (Henson 1974). It is covered by a taut, but compliant membrane, with muscle tissue on the outside, and it clearly provides a pathway by which the acoustically driven flow of perilymph through the otic capsule is further enhanced.

5. The Acellular Systems

Within the otic labyrinth, the extracellular fluid is endolymph, in which the predominant cation is the potassium ion rather than the sodium ion (Simon et al. 1973; Rauch and Rauch 1974; Lewis et al. 1985). In addition to this fluid, one finds a fascinating variety of acellular structures, including elaborate gelatinous structures and collections of calcium carbonate crystals. Together, the endolymph and these structures are presumed to provide the path for acoustic energy flow from the periotic system to the hair bundles on the surfaces of the sensory maculae and papillae.

5.1 Saccular Otoconial Suspension and Otoconial Membrane

When one opens the otic capsule of an amphibian, the most conspicuous structure usually is the viscous suspension of calcium carbonate crystals (otoconia) that occupies almost the entire saccular recess (e.g., see Lewis and Nemanic 1972). The crystal form and shape evidently are determined by a protein matrix at the core of each otoconium (Pote and Ross 1991; Pote, Hauer et al. 1993; Pote, Weber, and Kretsinger 1993). In adult amphibians, the crystal form in the saccule is aragonite, as it is in bony fishes, *Latimeria*, and adult reptiles (Carlström and Engström 1955; Carlström 1963; Lewis and Nemanic 1972; Marmo et al. 1981; Marmo et al. 1983). In birds and mammals, and in amphibian utricles, it is calcite. Associated with recent research in space and gravitational biology, there has been extensive investigation on otoconial development in the Japanese red-bellied newt, *Cynops pyrrhogaster* (see Steyger et al. 1995; Wiederhold et al. 1995). In that animal, the crystal form in the larval saccule is calcite, becoming aragonite in the adult saccule. Otoconia in the adult amphibian saccule appear to come in two shapes (Lewis and Nemanic 1972), which Steyger and colleagues (1995) call prismatic and fusiform (Fig. 4.6). The latter are considerably larger than the former. The suspension of mixed aragonite crystals is bounded laterally by the thin wall formed by fusion of epithelial linings of the periotic cistern and the saccular recess of the otic labyrinth, and medially by the thick medial wall of the saccular recess. In unfixed preparations, if the thin wall is ruptured, the otoconial suspension will flow readily out of the saccular recess. In the more recently derived anurans, in which the thin wall is extended, the suspension is bounded ventrally and medially (in part) as well as laterally by the thin wall (Wever 1985; Lewis

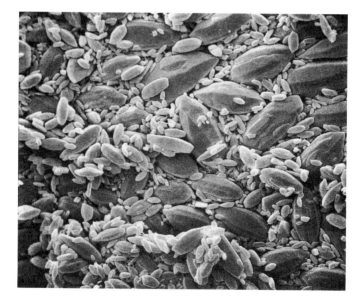

FIGURE 4.6. Saccular otoconia from the American bullfrog (*Rana catesbeiana*). Width of micrograph = 70 µm.

and Lombard 1988). In those animals, the macular pad forms an island on the ventromedial part of the thin wall bounding the otoconial suspension. In all of the amphibians, the otoconial suspension is clearly visible (white) through the thin wall.

Immediately adjacent to the macular surface lies a gelatinous pad (the gelatinous membrane) presumably comprising a variety of highly hydrated glycoproteins or mucopolysaccharides (Wislocki and Ladman 1955; Tachibana et al. 1973; Steel 1985; Fermin et al. 1987; Fermin and Lovett 1989). This structure (Fig. 4.7) covers the entire macula and is contiguous to and evidently continuous with the organic matrix of the otoconial suspension (Hillman 1969, 1976; Lewis and Nemanic 1972). It contains no otoconia, but appears to provide a linkage between the otoconial suspension and the hair bundles of the receptor cells. Over

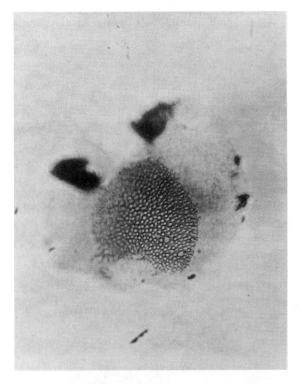

FIGURE 4.7. Gelatinous membrane from the saccule of the *R. catesbeiana*. This acellular structure was photographed in a buffer solution after fixation and decalcification. The gelatinous membrane is the porous, flat structure. Background material includes the matrix in which the otoconia were embedded.

each hair cell is a cylindrical, smooth-walled, fluid-filled pore that is oriented perpendicular to the macular surface and penetrates almost all the way through the gelatinous membrane (Hillman 1969; Lewis and Nemanic 1972). In fixed preparations, the gelatinous membrane appears to be connected to the tips of the microvilli of the supporting cells by fibers, which form a layer between the gelatinous membrane and the epithelial surface (Hillman 1969; Lewis and Nemanic 1972). The thickness of the fibrous layer is approximately equal to the length of the longest stereocilia in each hair bundle, so that only the top of the bundle contacts the gelatinous membrane. The connection between the hair bundle and the gelatinous membrane appears to be formed entirely between the tip of the kinocilium and the rim of the adjacent, fluid-filled pore (Hillman 1969). The stereocilia lie directly beneath the pore itself and appear to have no direct connections to the solid part of the gelatinous membrane (Fig. 4.8).

5.2 The Tectoria

The basilar and amphibian papillae lack otoconia, but possess tectorial membranes. Each papilla lies in a separate recess, one wall of which is formed by a contact membrane. With the exception of the tectorium and the papilla, which protrudes slightly into the recess, the recess is filled with endolymph.

5.2.1 Basilar Papilla

To date, basilar papillae have been found to be present in two species of caecilians, *Ichthyophis kohtaoensis* and *I. glutinosus* (Sarasin and Sarasin 1890; Jørgensen 1981; White and Baird 1982), both of which belong to a family (Ichthyophiidae) considered to be primitive (Duellman and Trueb 1994). They are absent in *Caecilia occidentalis* (Lombard 1977) and *Dermophis mexicanus* (White and Baird 1982), both members of the family Caeciliidae, and in *Typhlonectes natans* (Typhlonectidae) (White and Baird 1982). Caeciliidae and Typhlonectidae are considered to be derived families. Wever (1985) shows no basilar papilla in *I. glutinosus*, *I. orthoplicatus*, *D. mexicanus*, and *Geotrypes seraphini* (Caeciliidae) (see also Wever 1975; Wever and Gans 1976). The presence of the basilar papilla in *I. glutinosus*, however, has been confirmed (Sarasin and Sarasin 1890; Jørgensen 1981).

Basilar papillae are known to occur in six of the nine urodele families (Ambystomatidae, Dicamptodontidae, Salamandridae, Hynobiidae, Cryptobranchidae, and Amphiunidae) and so far have been found to be absent in members of the other three (Proteidae, Sirenidae, and Plethodontidae) (Lombard 1977). The Sirenidae are considered to be primitive, the Proteidae to be intermediate, and the Plethodontidae to

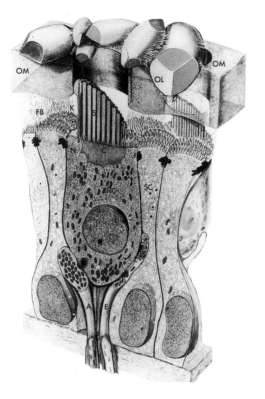

FIGURE 4.8. Diagram of a cross-section through the macula of the bullfrog saccule, the bulbed kinocilium (K) attached to the gelatinous membrane (OM) at the edge of a pore. Also shown are otoconial crystals (OL), the stereocilia bundle (S) and the dense cuticular plate (C) to which they are attached, the fibrous layer (FB) between the cell surfaces and the gelatinous membrane, and afferent (A) and efferent (E) innervation to the hair-cell body (RC). (From Hillman 1976, with permission.)

be derived (Duellman and Trueb 1994). To date, a basilar papilla has been found in anurans ranging from the most primitive (*Ascaphus* and *Leiopelma*) to the most recently derived; no frog or toad has been found to lack the sensor.

In all of these amphibians, the recess of the basilar papilla is tubular (Fig. 4.9). In caecilians and urodeles, the end of the recess is backed by thick periotic tissue; but its ventromedial wall is formed by the contact membrane and backed by the periotic canal (derived urodeles) or by a short extension of the periotic sac (caecilians and primitive urodeles) (Lombard 1977; White 1978; White 1986; White and Baird 1982). The papilla lies on the anterolateral wall at the entrance to the recess, sometimes extending well outside of it—into the lagenar recess. The tectorium comprises a bulky structure that overlies the distal-most part of the papilla (relative to the

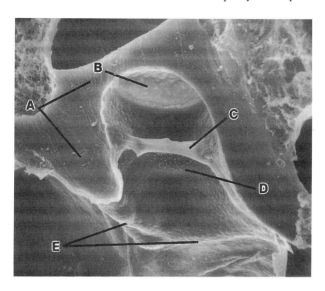

FIGURE 4.9. Scanning electron micrograph showing the basilar papillar recess from the toad, *Bufo americanus*. The wall (A), comprising dense limbic tissue, has been transected to expose the interior of the recess. The contact membrane (B) has a pebbly appearance owing to nuclei bulging from within the extremely flat epithelial cells. The tectorium (C) overlies the sensory surface (D) on which the hair cells reside. The entrance (E) to the recess from the saccular space is at the bottom.

recess entrance) and thin strands that extend to microvilli and cilia of the more proximal cells. The bulky part of the tectorium contains fluid-filled pores, similar to those in the gelatinous membrane of the sacculus. Immediately over each hair cell covered by this bulky structure is the entrance to one of these pores. In the tiger salamander, *Ambystoma tigrinum*, thin strands have been seen connecting the bulky part of the tectorium to the opposite wall of the chamber (White 1978).

In anurans, the recess of the basilar papilla extends directly from the large, saccular recess to the contact membrane, where it ends (van Bergeijk 1957; van Bergeijk and Witschi 1957; Frishkopf and Flock 1974; Lewis 1978; Lewis et al. 1985). The contact membrane is backed by a long, tubular extension of the periotic sac (e.g., see Fritzsch 1992). The papilla lies well within the recess. In contrast with the bulky structure in urodeles and caecilians, the tectorium forms a thin, semicircular membrane that extends from the distal-most part of the papilla to an evidently strong, tense strand of tectorial material that is suspended across the recess, anchored at opposite sides. Around the curved part of its perimeter, the semicircular membrane (Fig. 4.10) is connected to densely microvillous epithelial cells on the wall of the recess. Part of this connection lies over the papilla itself, and the densely microvillous epithelial cells in that case are supporting cells

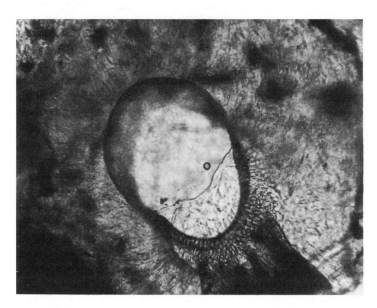

FIGURE 4.10. Phase-contrast micrograph showing the papillar recess of a leopard frog, *Rana pipiens*, in cross section. The tissue was fixed and photographed in a buffer solution. The outline of the tectorium is clearly visible in the lower right portion of the recess. The pores in the tectorium are visible close to the papillar surface, extending perpendicular from it. The osmium-stained basilar papillar twig of the eighth nerve extends from the papilla toward the lower right of the micrograph.

surrounding the two or three distal-most rows of hair cells (Fig. 4.11). Over each of these hair cells is a cylindrical, smooth-walled, fluid-filled pore in the tectorium. The remaining hair cells seem to lack direct association with the tectorium (Fig. 4.12). Being essentially a semidiaphragm, the anuran basilar-papilla tectorium extends directly into the path of any axial fluid flow in the basilar papillar recess (see Section 5.3).

5.2.2 Amphibian Papilla

No amphibian has been found to lack an amphibian papilla. In all three amphibian orders, the amphibian papilla resides on the dorsal surface (ceiling) of its associated recess (Retzius 1881; Geisler et al. 1964; Mullinger and Smith 1969; Wever 1975; Lewis 1976, 1978; Wever and Gans 1976). In urodeles, caecilians, and the most primitive living anurans (*Ascaphus* and *Leiopelma*), the recess is straight, extending medially from the saccular recess and ending at the contact membrane, which forms its medial wall. In the derived anurans it extends medially, then bends to extend caudally (Fig. 4.13). It ends at the contact membrane, which forms its caudal wall. The tectorium of the caecilian amphibian papilla has not been examined

FIGURE 4.11. Micrograph of part of the distal region of the basilar papilla of the African clawed frog, *Xenopus laevis*, showing a view directly down on four hair bundles with bulbed kinocilia. The microvilli of supporting cells in the lower left are clearly visible. To the distal edge (upper right) of the papilla, the remaining microvilli are obscured by remnants of the fibrous layer of the tectorium to which they were attached. Width of micrograph = 18.4 μm.

carefully. In urodeles and the most primitive frogs (*Ascaphus* and *Leiopelma*) it is a bulky structure that hangs from the papilla and fills much of the papillar recess (White 1978; Lewis 1981a, 1984). In the other frogs and toads, the papilla comprises two patches of sensory epithelium that arise separately and grow together during development (Li and Lewis 1974); and the tectorium (Fig. 4.14) has three distinct parts: (1) a bulky structure that hangs from the anterior sensory patch and fills much of the volume of the papillar recess adjacent to that patch, (2) a thinner structure that hangs from the posterior sensory patch and fills only a small fraction of the recess volume adjacent to that patch, and (3) a diaphragm-like structure that lies in a plane perpendicular to that of the adjacent sensory epithelium and extends across the entire papillar recess (Wever 1973, 1985; Lewis 1976, 1981a,b, 1984; Yano et al. 1990). The diaphragm-like structure is connected around its entire margin, to the other two parts of the tectorial membrane dorsally, and ventrally to a row of densely microvillous epithelial cells that rings the tubular recess. This thin, gelatinous diaphragm lies between the contact membrane and the opening of the amphibian papillar recess to the saccular recess, standing directly across the path of axial flow of endolymph

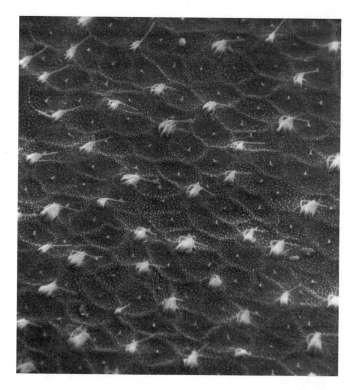

FIGURE 4.12. Micrograph of part of the nontectorial (proximal) region of the basilar papilla of the toad *Bufo mauritanicus*. Here the kinocilia are not bulbed and extend well beyond the bundles of stereocilia.

between those two places. For that reason, Wever (1973) called it the "sensing membrane," although it also is known as the tectorial curtain.

In urodeles and anurans, the entire amphibian papilla is covered by tectorium. As in the saccular gelatinous membrane, over each hair cell is a cylindrical, smooth-walled, fluid-filled pore (Fig. 4.15). It extends in a direction perpendicular to the papillar surface and then bends. A large proportion of the volume of the amphibian-papillar tectorium is occupied by these pores, all running very precisely in parallel. In fixed preparations, the tectorial membrane appears to be connected to the tips of the microvilli of the supporting cells by fibers (Lewis 1976; Lewis and Leverenz 1983). As in the saccule, these fibers form a layer between the tectorium and the epithelial surface. Again, the thickness of the fibrous layer is approximately equal to the length of the longest stereocilia in each hair bundle, so that only the top of the bundle contacts the gelatinous membrane; and the connection between the hair bundle and the gelatinous membrane appears to be formed entirely between the tip of the kinocilium and the rim of the adjacent, fluid-filled pore (Lewis and Leverenz 1983). As in the saccule, the

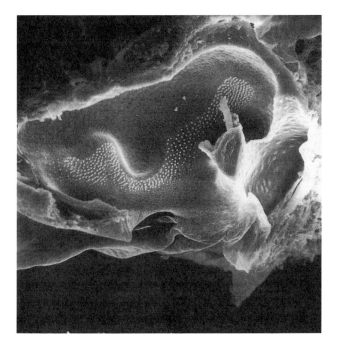

FIGURE 4.13. Micrograph of a transected amphibian papillar recess of the green tree frog, *Hyla cinerea*, showing a dorsad view of the papilla. The individual hair bundles appear as white dots over the sensory surface. The caudal (high-frequency) region is at the upper right, rostral (low-frequency) region at the lower left.

stereocilia lie directly beneath the pore and appear to have no direct connections to the solid part of the tectorium.

5.3 Putative Acoustic Paths Revisited

A miniature scuba diver exploring the periotic labyrinth could start in the periotic cistern and swim, in perilymph, unimpeded through the entire labyrinth—along the thin lateral wall of the saccule, through the long, narrow periotic canal, to the contact membranes of the amphibian and basilar papillae, and on into the periotic sac. The diver would find no path into the chambers and tubes of the otic labyrinth. The fluid systems of the two labyrinths are entirely separate. Exploring the otic labyrinth, the diver could start in the saccular recess and swim unimpeded through endolymph paths into the recesses of the basilar and amphibian papillae, and on to the contact membranes of the amphibian and basilar papillae.

As it was with the periotic system, it is interesting to contemplate the flow of acoustic energy within the otic labyrinth. Again, there are two fundamentally different ways that the endolymph will flow in response to the pressure

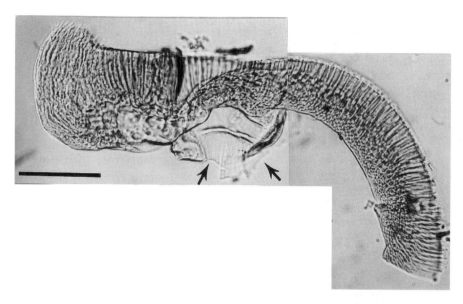

FIGURE 4.14. Montage of two phase-contrast micrographs showing a view from the medial side of the amphibian-papillar tectorium of the Puerto Rican coqui frog, *Eleutherodactylus coqui*. The tissue was fixed and photographed in buffer solution. The upper edge of the tectorium was attached to the papilla and follows the surface contour of that structure. The extent of the downward curve of the caudal region of the papilla surface in this animal can be seen in the micrograph on the right. Arrows in the micrograph on the left point to the tectorial curtain (Wever's sensing membrane). The elaborate network of fluid-filled pores in the tectorium is clearly visible. Horizontal line = 100 µm.

transmitted through the perilymph: (1) it will compress and expand (decreasing and increasing the total volume of endolymph), and (2) its pressure changes will cause the wall of the otic labyrinth system to expand and contract (shifting the endolymph, but leaving its total volume constant). Again, owing to the large value of the bulk modulus of water, the amplitude of the flow from the first mode will be extremely small. For each acoustic recess in the otic labyrinth, the second mode of flow requires two flexible walls (e.g., see van Bergeijk 1957; Smith 1968).

The most obvious flexible-wall candidates are (1) the very thin wall separating the saccular recess of the otic labyrinth from the periotic cistern, and (2) the very thin walls (contact membranes) separating the periotic canal or periotic sac from the recesses of the amphibian and basilar papillae and (in the more derived anurans) from the saccular recess (Smith 1968). The presence of these very thin walls suggests the following causal chain: (1) the piston-like transducer at the oval window is pushed inward, displacing perilymph; (2) part of the displaced perilymph deforms the lateral wall

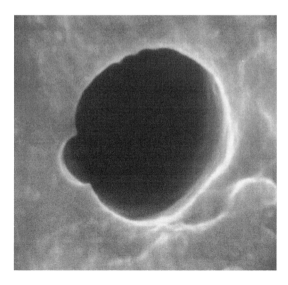

FIGURE 4.15. Scanning electron micrograph showing the entrance to a single pore in the amphibian-papillar tectorium from the American bullfrog, *Rana catesbeiana*. The semicircular notch at the edge of the pore entrance is the place where the tectorium was attached to the kinociliary bulb of the hair bundle associated with the pore.

of the saccular recess, displacing endolymph in that recess; (3) the displaced endolymph flows through the cylindrical recesses of the basilar and amphibian papillae, deforming the contact membranes (which are at the far ends of those recesses), displacing perilymph in the periotic sac; (4) the wall of the periotic sac expands to accept the displaced perilymph. The other part of the displaced perilymph in step 2 flows through the periotic canal and directly displaces perilymph in the periotic sac. The fraction of the fluid flow through each of these paths will be inversely proportional to the fluid impedance of the path (e.g., see Lewis 1996). If the periotic canal were cylindrical, of length L and cross-sectional area A, it would behave as a resistive impedance to fluid flows at low sound frequencies, with the resistance being proportional to L/A^2. At high frequencies, it would behave as an inertial impedance (a mass-like element), with the impedance being directly proportional to frequency and to L/A. The transition frequency (f_0) is given approximately by $f = 4\eta/\rho A$, where η is the viscosity of the fluid and ρ is the density of the fluid. Because the periotic canal is longer and much narrower than the other fluid paths in the amphibian ear, regardless of the frequency, its fluid impedance must be high relative to those of the other paths, strongly suggesting that the flow owing to displacement of the perilymph at the oval window will be directed largely through the recesses of the two papillae (see Smith 1968; Purgue and Narins 1997a). This will lead to displacements of the structures (tectorial membranes and free-standing hair bundles) protruding into the flow paths from the walls of those recesses.

6. The Sensory Epithelia

6.1 Acoustic Hair Cells

As in all vertebrates, the hair bundles of amphibians comprise a large group of stereocilia (sometimes called stereovilli) arrayed on the apical surface of the hair cell in rows of successively greater length, with a single kinocilium (a true cilium, with paired microtubules and a basal body) immediately adjacent to the row of longest stereocilia. The (supporting) cells adjacent to each papilla or macula exhibit a single, very short cilium and microvilli (Lewis and Li 1973). Geisler and colleagues (1964) noted that along the lateral edge of the anuran amphibian papilla the "hair cells gradually fade out into undifferentiated cells," suggesting that hair cells might arise by direct transformation of the supporting cells adjacent to the papilla. Studies of morphogenetic sequences in tadpoles and adults of *R. catesbeiana* supported this proposition (Lewis and Li 1973; Li and Lewis 1974, 1979) and suggested that direct transformation of a supporting cell to an acoustic hair cell in that animal begins with elongation of the cilium, accompanied by shifting of the microvilli into an array of rows and transformation of the microvilli into stereocilia (stereovilli). Baird and colleagues (1996) provide strong evidence that the formation of hair cells in *R. catesbeiana* occurs in two ways: (1) from direct transformation (transdifferentiation) of supporting cells to hair cells, and (2) from mitotic division of a progenitor cells. It is clear that hair cells continue to be formed in adult anuran acoustic sensors (Alfs and Schneider 1973; Lewis and Li 1973; Corwin 1985; Baird et al. 1996). Because mechanisms of hair cell formation in anurans (and other submammalian vertebrates) may have profound implications with respect to the development of ways to promote hair cell regeneration in humans made deaf by hair cell loss, considerable effort currently is aimed at understanding those mechanisms (Corwin et al. 1993; Cotanche and Lee 1994; Baird et al. 1996).

In anurans other than *Ascaphus* and *Leiopelma*, the typical acoustic hair cell has several (usually five or more) rows of stereocilia and a kinocilium with its distal end expanded to form a distinct bulb (Hillman 1969; Lewis and Li 1975; Lewis 1977, 1978; Baird and Lewis 1986). The bulb is connected to the tips of the stereocilia in the row adjacent to the kinocilium, so that the kinocilium does not extend, distally, beyond that row. It also is connected to one side of the rim of the fluid-filled pore in the tectorium or gelatinous membrane overlying the hair cell, and evidently provides a focused mechanical linkage between the array of stereocilia and the overlying gelatinous structure (Hillman 1969; Hillman and Lewis 1971; Lewis 1976). In these same animals, the nonacoustic hair cells typically have fewer rows of stereocilia; they all lack the kinocilia bulb, and the unbulbed kinocilium typically extends distally well beyond the row of longest stereocilia (Lewis and Li 1975; Hillman 1976).

Typical anuran acoustic hair cells occur along the midline (seismic region) of the anuran lagenar macula (Baird and Lewis 1986). They also occur throughout most of the anuran saccule and amphibian papilla. At the growing edges of those two end organs, however, one finds a gradation of morphology, between the typical acoustic type and a small hair-cell type that Lewis and Li (1973) considered to be an undifferentiated, juvenile form. This juvenile hair cell has fewer rows of stereocilia and an unbulbed kinocilium that is conspicuously longer than the longest stereocilia. The hair cells of the distal-most two or three rows in the anuran basilar papilla (those hair cells with direct connection to the tectorial membrane) also are the typical acoustic type; the more proximal hair cells have unbulbed kinocilia and resemble nonacoustic types (Lewis 1977).

In caecilians, urodeles, and *Ascaphus* and *Leiopelma*, the typical acoustic hair cell has many rows of stereocilia and a kinocilium that is neither conspicuously bulbed nor conspicuously longer than the longest stereocilia (Lewis and Nemanic 1972; Lewis 1981a; White and Baird 1982; White 1986). White (1986) found slight swellings at the tips of kinocilia in the distal-most hair cells of the basilar papillae of some urodeles. One sees no evidence of such swellings in the saccular hair cells of the mudpuppy (*Necturus maculosus*), however, but there is evidence of a specialized connection between the distal tip of the kinocilium and the longest stereocilia (Lewis and Nemanic 1972). The distribution of typical acoustic hair cells in these amphibians seems to follow the pattern of the anurans, including a gradation to juvenile forms along the putative growing edges of the saccule and amphibian papilla (Lewis and Nemanic 1972).

6.2 Hair-Bundle Orientation Patterns—Sensitivity Maps

Comparing micromorphology with observed physiological responses in semicircular canals, Lowenstein and Wersäll (1959) concluded that hair cells are functionally polarized, responding in an excitatory manner to hair-bundle strain directed along a vector oriented parallel to the surface of the sensory epithelium and directed across the center of the apical surface of the cell, toward the kinocilium. This sort of strain would be produced by an appropriately directed shearing action between the epithelial surface and the overlying acellular structures. Lowenstein and Wersäll's conclusion has been substantiated by direct observation of stimulus-response properties of individual hair cells (Hudspeth and Corey 1977). Thus, by placing an appropriately oriented arrow over each hair cell in a micrograph of an inner-ear sensory surface, one can construct a sensitivity map—a map of the local shearing motion that leads to maximum excitation. Shotwell and colleagues (1981) demonstrated that a given strain yields hair-cell depolarization that closely matches a cosine function of the angle of the strain vector relative to the arrow: strain aligned in the direction of the arrow yields maximum positive voltage change; strain in the opposite direction yields maximum negative voltage change; and strain perpendicular to the arrow yields no voltage change.

An excellent compilation of sensitivity maps can be found in Leverenz's comprehensive review of inner-ear morphology (Lewis et al. 1985, Chapter 3). In most amphibians that have been observed (the caecilian *Ichthyophis kohtaoensis*; among urodeles—five species of *Ambystoma* plus *Dicamptodon ensatus* and *Rhycotriton olympicus*; and among anurans— one leiopelmatid species, three discoglossids, two pipids, one paleoarctic pelobatid, 11 hylids, two dendrobatids, four ranids, two hyperoliids, one rhacophorid, and one paleoarctic microhylid) arrows for the basilar papilla are aligned predominantly parallel with the axis of the papillar recess (toward or away from the saccular or lagenar recess), implying sensitivity predominantly to axial motion of the endolymph or tectorium (Lewis 1978; White and Baird 1982; White 1986). Some anurans (12 bufonid species, five neoarctic pelobatids, and two neoarctic microhylids) exhibit conspicuous deviation from this pattern (Lewis 1978). In those species, the arrows for the hair cells attached to the tectorium are aligned in parallel with the axis of the recess, but the arrows for the other hair cells form inward or outward vortex-like patterns (Fig. 4.16), implying sensitivity to rotating (eddy) patterns of flow proximal to the tectorial membrane.

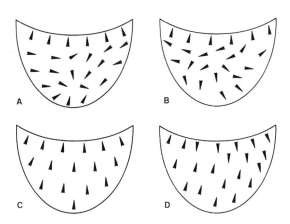

FIGURE 4.16. Sensitivity maps (showing hair cell polarization patterns) of the basilar papillae of four anurans: (A) *Scaphiopus hammondi* (Pelobatidae), (B) *Hypopachus variolosus* (Microhylidae), (C) *Kassina* sp. (Hyperoliidae), and (D) *Bombina orientalis* (Discoglossidae).

In the caecilians and urodeles in which it has been constructed, and in *Ascaphus* and *Leiopelma*, the sensitivity map for the amphibian papilla has arrows directed parallel to the axis of the papillar recess (toward or away from the saccule), implying sensitivity to axial motion of the tectorial membrane. In the other anurans for which maps have been constructed, the papillar recess bends and extends caudally, and sensitivity maps are more complicated but consistent (Lewis 1976, 1978, 1981a,b). In anurans considered to be most ancient (three species of discoglossids, four pipids, six pelobatids) the maps imply sensitivity to tectorial motion parallel to the long, curved axis of the posterior patch. In anurans considered to have arisen more recently (one sooglossid, nine leptodactylids, 15 hylids, two dendrobatids, seven ranids, three hyperoliids, one rhacophorid, three microhylids), the maps imply sensitivity to tectorial motion parallel to the long axis over the anterior part of the posterior patch, and tectorial motion perpendicular to that axis over the posterior part of the patch (Fig. 4.17). The transition between these two modes of sensitivity occurs directly under the gelatinous diaphragm that Wever called the "sensing membrane."

6.3 Transduction to Neuroelectric Signals

So far, we have followed the putative paths that the energy from external acoustic signals takes from the periphery to the acellular structures overlying acoustic hair cells, and we have used sensitivity maps to infer the directions of motion in those structures that will be effective in exciting the hair cells themselves. The combination of force and velocity in the strain applied to the hair bundle evidently is the last instance in the causal chain of acoustic sensing that involves energy directly from the acoustic signals. Davis (1965) promoted the idea that transduction of an acoustic signal to a neuroelectric signal in the cochlea was accomplished by strain-modulated electrical resistance in combination with the electrical battery provided by the inner-ear fluids (in the mammalian cochlea there is an electrical potential of approximately 0.1 V between endolymph and perilymph). Strelioff and his colleagues corroborated this hypothesis with direct measurement of sound-induced electrical resistance changes in the guinea pig cochlea (Strelioff et al. 1972; Honrubia et al. 1976). In the saccule of *R. catesbeiana*, Hudspeth and his colleagues identified a cellular basis for this mode of transduction and one of the cellular elements responsible for it—a strain-gated ion channel, located at or near the tip of each stereocilium (Hudspeth and Corey 1977; Hudspeth and Jacobs 1979; Shotwell et al. 1981; Hudspeth 1982; Jaramillo and Hudspeth 1991).

In transduction of this sort, the acoustic energy reaching the transducer is used to modulate the resistance, but the energy in the resulting electrical signal comes from the battery, and the energy of the battery comes from metabolic sources, through ion pumps. From this point on, through the hair

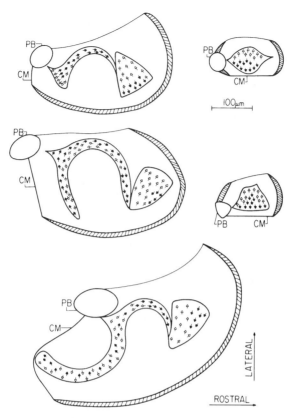

FIGURE 4.17. Sensitivity maps of the amphibian papillae of four anurans and one urodele [*Ambystoma maculatum* (Ambystomatidae), upper right]. Three of the anurans, *Bombina orientalis* (Discoglossidae, upper left), *Scaphiopus couchi* (Pelobatidae, left center), and *Ascaphus truei* (Leiopelmatidae, right center), are members of families considered to be ancient. The other anuran, *Kassina senegalensis* (Hyperoliidae, bottom of figure) is considered to be a recently derived frog. The contact membrane (CM) and cross-section of the papillar branch (PB) of the eighth nerve are depicted in each map.

cells and their synapses, through the eighth cranial nerve, through the brain stem and beyond, the signals representing environmental sounds and vibrations do not include energy from those sounds and vibrations (see McCormick, Chapter 5, for a discussion of the amphibian CNS).

There still is one thing that must happen to the energy from the sounds and vibrations before we move past it. It must be coupled to the gates of the channel molecules in such a way that it can modulate their opening and closing. The forces and motions of the tectorial or gelatinous membrane evidently are coupled through the distal end of the kinocilium to the distal end of the row of tallest stereocilia. Pickles and colleagues (1984) identified structures, which they called tip links, that seem to be ubiquitous in vertebrate hair bundles and evidently serve to couple forces and motions from row to row through the entire array of stereocilia (Assad et al. 1991). Owing to the gradation of stereocilia length from one row to the next, as the hair bundle is tilted (by shear forces applied through the distal end of the kinocilium), the members of one row will slide relative to members of adjacent rows. If one imagines that each tip link provides a spring-like coupling that is under tension at rest, then that tension should increase for shearing motion aligned with the sensitivity arrow (i.e., directed across the luminal surface of the hair cell toward the kinocilium) and decrease for shearing motion in the opposite direction. Presuming that the probability that a given strain-gated channel is open increases as the tension in the associated tip link increases, one has in this structure not only the means for coupling the acoustic energy to its final target, but also a mechanism for the functional polarization first observed by Lowenstein and Wersäll (Corey et al. 1989; Pickles 1993).

6.4 Reverse Transduction: Otoacoustic Emissions

Using engineering theory, Gold (1948) argued that if tuning in the mammalian cochlea were accomplished largely by mechanical elements, and if it were as sharp as he believed it was, on the basis of psychophysical experiments, then it would require feedback with energy augmentation. He suggested that this is accomplished by a loop involving a reverse transduction process, allowing the mechanical energy from the external acoustic source to be augmented by energy derived from metabolism. The reverse transducer would convert the metabolically derived energy to mechanical energy.

Presumably, under some circumstances, perhaps pathological, some of this metabolically derived mechanical energy could find its way back through the acoustical paths of the inner and middle ears to the periphery, and could be emitted into the air or water as sound. Clinical studies of tinnitus led to the discovery of spontaneous acoustic emissions from human ears (Citron 1969; see Lewis et al. 1985), implying that a reverse transduction mechanism does exist in the cochlea. Kemp (1979) showed that such emissions could be evoked by externally applied acoustic stimuli. Subsequently, spontaneous and evoked acoustic emissions were observed in *Rana temporaria* and in the hybrid species, *Rana esculenta* (Palmer and Wilson 1982; Whitehead et al. 1986), implying that reverse transduction also occurs in amphibians. The frequencies of the emissions observed so far from the frogs correspond to the BEFs of the basilar papilla and the posterior patch of the amphibian papilla (see Section 2.3), suggesting that the emissions may arise from those places (Whitehead et al. 1986; van Dijk et al. 1989; van Dijk et al. 1996; Long et al. 1996). Working with frog saccular hair cells in vitro, Assad and colleagues (1989) observed reverse transduction directly—

changing the membrane potential in isolated saccular hair cells produced pivoting (tilting) movement of the hair bundle (but no apparent movement of other parts of the cell). Benser and colleagues (1996) found that this hair-bundle motion could be very rapid. In response to mechanical stimuli, some hair bundles twitched (in the sense that a muscle-fiber twitches), doing work against the mechanical stimulation device; some hair bundles exhibited spontaneous twitches and some produced stimulus-evoked mechanical oscillations. Evidence gathered by Assad and Corey (1992) implies that the site of this reverse transduction (which they call the "adaptation motor") is the tip link.

6.5 Innervation of the Hair Cells

Among amphibians, detailed eighth nerve fiber counts are available only for adult *R. catesbeiana* (Dunn 1978). The saccular branch averages approximately 960 myelinated axons and 170 unmyelinated axons; the basilar papilla branch averages approximately 400 myelinated axons and 360 unmyelinated axons; and the amphibian papilla branch averages approximately 1,550 myelinated axons and 60 unmyelinated axons. In adult *R. catesbeiana*, the saccular macula has approximately 2,500 hair cells, the basilar papilla has approximately 100 hair cells, and the amphibian papilla has approximately 1,500 hair cells (Lewis and Li 1973; Lewis 1976, 1978). In adult *R. catesbeiana*, saccular afferent axons have been found to innervate from two to approximately 200 hair cells, amphibian papillar afferent axons from one to approximately 15 hair cells, and basilar papillar afferent axons from one to four hair cells with one being typical (Lewis, Baird et al. 1982; Lewis, Leverenz, and Koyama 1982). In the *R. catesbeiana* amphibian papilla, afferent axons to the anterior patch tended to innervate more hair cells than did those to the posterior patch. Simmons and colleagues (1992) found a similar pattern in adult *R. pipiens*. In that animal, however, they found that a high proportion of the afferent axons innervating the basilar papilla were branched, and that the average number of hair cells innervated by the branched afferents was five. The afferent axons traced into the basilar papillar of *R. catesbeiana* all innervated hair cells in the rows directly underlying the tectorial membrane. Those in *R. pipiens* were not so limited, possibly accounting for the observed difference in the innervation pattern. In *R. catesbeiana*, efferent innervation was found to be absent in the basilar papilla and present in the saccule and amphibian papilla (Flock and Flock 1966; Robbins et al. 1967; Frishkopf and Flock 1974). In the amphibian papilla, efferent terminals are more abundant in the anterior patch than they are in the posterior patch (Flock and Flock, 1966). In *R. catesbeiana*, Flock and Flock (1966) found tight junctions between axons in the amphibian papilla, suggesting the possibility of electrical connections; and Dunn (1980) found reciprocal synapses between hair cells and afferent axons in the sensory surfaces (cristae) of the semicircular canals. These observations raise the possibility of interesting neural circuitry in the acoustic end organs of amphibians, a possibility that is largely unexplored.

7. Physiology of Afferent Axons

Physiological studies of acoustic afferent axons in anurans are aided by the fact that the fiber bundle from each end organ maintains its integrity and relative position in the eighth nerve as the latter spans the gap between the inside of the cranial wall and the brain (Boord et al. 1970). One can conceive of the spike activity over all of the axons of the various acoustic sensors of the amphibian inner ear as representing a dynamic image of the acoustic ambiance of the animal, with each axon bearing an element of that image. Many of the properties of these individual elements were summarized in Section 2.3 and 2.4, above. Based on these properties, one can think of the image as being spectrographic—reflecting both time and frequency. Terms commonly applied to the elements of spectrographic images are *tuning* and *sensitivity*. Tuning, interpreted broadly as selective responsiveness to the shapes of acoustic waveforms, has been the focus of many physiological studies of auditory and seismic axons in the amphibian ear.

7.1 Tuning and Putative Tuning Mechanisms

Eighth-nerve tuning properties have been studied in two fundamentally different ways: (1) variants of the frequency threshold tuning curve (FTC), and (2) variants of the Volterra-Wiener systems of descriptive models. FTCs and their variants are based entirely on changes in mean spike rate in response to tonal (sinusoidal) stimuli, and thus inherently reflect the action of an underlying nonlinear process akin to rectification. The Volterra-Wiener descriptions may include components based on mean spike-rate changes, but they also include components based on linear modulation of the spike rate by the stimulus waveform. When it is less than 0.5, the widely used synchronization index (vector strength) is such a measure of linear modulation (Goldberg and Brown 1969; Hillery and Narins 1987). Other such measures are the reverse-correlation (revcor) function, which is equivalent to the first-order Wiener kernel (de Boer and de Jongh 1978; Yu et al. 1991); and the phase-tuning curve—the phase of the linear spike-rate modulation as a function of frequency (Hillery and Narins 1984). The current state of the theory of dynamic processes makes linear descriptions much easier to interpret than nonlinear descriptions in terms of underlying physical processes. The linear descriptions of tuning as seen through individual afferent axons of the frog saccule and the frog amphibian papilla strongly imply underlying dynamics of high dynamic order (Lewis et al. 1990), meaning dynamics involving a large number of independent energy-storage elements (e.g., many mass elements, each moving independently, and many elastic elements, each compressed or expanded independently).

Studying FTCs in frog auditory axons, Moffat and Capranica (1976) found that the BEFs of amphibian-papillar axons are strongly dependent on temperature and that those for basilar papillar axons are not. Because the properties of mechanical elements (masses and springs) typically are not strongly dependent on temperature, they concluded that tuning in the amphibian papilla might involve electrochemical (molecular) mechanisms, which typically do exhibit strong temperature dependence. Subsequently, Pitchford and Ashmore (1987) found electrochemical resonances in the low-frequency region of the frog amphibian papilla, and Lewis and Hudspeth (1983) found such resonances in the frog saccule. Hudspeth and Lewis (1988) showed that, in the frog saccule, the resonance arises from an interaction of voltage-dependent calcium channels and calcium-dependent potassium channels, all in the individual hair cells. In the frog saccule, however, the frequency of electrochemical resonance typically is above the range of tuning seen in the axons. Furthermore, tuning in the saccular axons is only weakly dependent on temperature (Egert and Lewis 1995), while the frequency of the electrochemical resonance is strongly dependent on temperature (Smotherman and Narins 1998). This evidence suggests that the tuning seen in the saccular axons does not involve the electro-chemical resonance. Similar resonances are found in the hair cells of frog semicircular-canal cristae, where they also exhibit frequencies well above the tuning range of the canal axons and seem not to be involved at all in that tuning (Houseley et al. 1989).

Smotherman and Narins (1997) found that the resonance frequencies of hair cells taken from various places along the low-frequency region of the frog amphibian papilla, on the other hand, match very well the known tonotopy of the afferent axons from the same region (Lewis, Leverenz, and Koyama 1982). This finding, along with the strong temperature dependence of the resonance (Smotherman and Narins 1998) and that of the axonal tuning (van Dijk et al. 1990; Stiebler and Narins 1990), strongly implies that the electrochemical resonances observed in the hair cells participate in the tuning seen in the axons. The electrochemical resonance, however, has a dynamic order of two. The dynamic order observed in amphibian-papillar axonal tuning typically is eight to ten, but sometimes as high as 16 (Lewis et al. 1990). Electrochemical resonance, with its low dynamic order, is observed in isolated hair cells and in hair cells in situ, but with the overlying tectorium removed. Noting the observations of frog otoacoustic emissions by Palmer and Wilson (1982), Lewis (1988) argued that if the tuning in axons is a consequence of the electrochemical resonance, then the high dynamic order seen in axons implies electromechanical coupling between hair cells, through the tectorium, which in turn implies the presence of reverse transduction. Although it is now clear that reverse transduction is present, it has not yet been shown definitively that the high dynamic order of axonal tuning actually is achieved by electromechanical coupling of electrochemical resonances. Nevertheless, the evidence to date suggests that the reverse transducer (the tip-link/adaptation-motor) may play a key role in tuning observed in the low-frequency amphibian papillar axons.

Lacking strong temperature sensitivity, tuning in the frog saccule and basilar papilla as well as that in the higher-frequency regions of the amphibian papilla may be accomplished largely by mechanical elements. At the high-frequency (caudal) end of the amphibian papilla, correlations between morphology and tonotopy suggest involvement of the tectorium as mass and the hair bundles as springs (Lewis and Leverenz 1983; Shofner and Feng 1983). The mechanisms underlying mechanical tuning, however, including the roles of the elaborate and beautifully regular pores in the tectorium, remain an open topic for future studies. Van Bergeijk and Witschi (1957) proposed that the tectorium of the basilar papilla is a tuned, two-dimensional wave structure; Hillery and Narins (1984), among others, made a similar proposal regarding the tectorium of the amphibian papilla. Although the details of both of these proposals require modification in light of subsequent observations, they provide interesting starting points for further study. Whatever the mechanisms may be, it is clear that they give rise to tuning of high dynamic order.

An aspect of tuning that has been under increasing scrutiny is the responsiveness to temporal aspects of the acoustic waveform (Rose and Capranica 1985; Schwartz and Simmons 1990; Simmons and Ferragamo 1993; Bodnar and Capranica 1994; Simmons et al. 1996). An insightful approach to the study of such responsiveness is the use of Wiener kernels as descriptive models of tuning (de Boer and de Jongh 1978; Wolodkin et al. 1997). The power of such an approach can be seen in Figure 4.18. Presently, it is being extended successfully to incorporate nonlinear aspects of tuning (Wit et al. 1994; Yamada et al. 1997).

7.2 Adaptation

As implied by its name, studies of the tip-link/adaptation-motor have been motivated by interest in its possible role in a phenomenon labeled *adaptation* (Eatock et al. 1987; Howard and Hudspeth 1987; Shepherd and Corey 1994). Whereas adaptation in vision science produces an adjustment of photosensitivity to changes in ambient light intensity, adaptation associated with tip links seems to be related to tuning rather than sensitivity. It makes the saccular hair cells, in which it has been studied, unresponsive to acoustic stimuli of very low frequency. As mentioned in the previous section, it also may account for the high dynamic order in tuning in low-frequency amphibian papillar axons.

Acoustic axons are known to exhibit sensitivity adjustments to ambient acoustic levels, however (see Lewis and Henry 1995 for a discussion about mechanisms). This type of adaptation (akin to adaptation in vision science) is conspicuous in afferent axons from the frog amphibian papilla and in afferent axons from the frog basilar papilla (Ehret and Capranica 1980; Megela 1984; Zelick and Narins 1985; Penna and Narins 1989; Yu 1991). It is not conspicuous in frog saccular axons, however, although it does occur to a slight extent in some of them (Lewis 1986; Egert 1993). In response to an increase in the ambient sound level, the sensitivities of amphibian-papillar axons and basilar papillar axons gradually decrease (Megela and Capranica 1981; Megela 1984; Yu 1991). As mentioned in Section 2.4, frog amphibian-papillar axons also exhibit phenomena known as one-tone suppression and two-tone suppression (Lewis 1986; Christensen-Dalsgaard and Jørgensen 1996b)—reduction in sensitivity to intrinsic noise in the presence of a background (suppressing) tone, and reduction in sensitivity to a tone at a frequency close to BEF in response to a second (suppressing) tone at another frequency. Showing that the response of the *R. catesbeiana* amphibian-

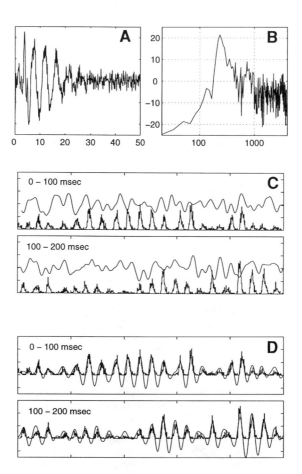

papillar axon to a single tone can be divided into a linear (AC) component and a nonlinear (DC) component, reflecting AC and DC components of the underlying generator potential at the spike trigger, Lewis (1986) presented evidence implying that adaptation and suppression both are consequences of the DC component of generator potential being negative, biasing the spike trigger away from threshold and making the axon less responsive to the AC component (see Lewis and Henry 1995 for the basic theory). Recently, Christensen-Dalsgaard and Jørgensen (1996b) drew the same conclusions for *R. temporaria*. The biophysical basis of the negative DC response remains an open question for future research.

7.3 Directional Sensitivity

Phonotaxis tests with female treefrogs have demonstrated that anuran amphibians are able to accurately localize sound in both azimuth and elevation (Feng et al. 1976; Rheinlaender et al. 1979; Gerhardt and Rheinlaender 1980, 1982; Passmore et al. 1984; Rheinlaender and Klump 1988; Klump and Gerhardt 1989; Jørgensen and Gerhardt 1991). Measurements of many species have shown that the frog ear is an inherently directional, asymmetri-

◄————————————————————————————————

FIGURE 4.18. Characterization and responses of a single afferent nerve fiber from the amphibian papilla of *Rana catesbeiana*. A: First-order Wiener kernel, derived by the revcor method (horizontal axis shows time in milliseconds). This is an estimate of the impulse response of the peripheral tuning structure associated with the fiber. It is noisy because it is the time average of the noise stimuli occurring immediately prior to each spike. The noise (random acoustic stimulus) had a gaussian amplitude distribution and a uniform power spectral density from 100 Hz to 5.0 kHz. B: Tuning curve, derived by discrete Fourier transform of the first-order Wiener kernel in A (horizontal axis shows frequency in Hz; vertical axis shows relative amplitude in decibels). The center frequency is 230 Hz, and the peak of the tuning curve extends approximately 20 dB above the background noise level (inherent in the noisiness of the waveform in A). C: The upper (smooth) line in each panel shows the time course of an acoustic stimulus waveform applied repeatedly to the ear. The stimulus was a segment of noise with gaussian amplitude distribution and uniform power spectral density from 70 Hz to 600 Hz. The lower (ragged) line shows a peristimulus time histogram of spikes (i.e., the spike rate) in response to the waveform. Note that the peaks and troughs of the spike-rate response are not a very good match to those of the stimulus. D: The smooth line shows the result of passing the stimulus waveform in C through a filter with the impulse response of A (computed simply by convolution of the impulse response in A and the smooth waveform in C). The smooth line was scaled to allow detailed comparison with the spike rate response of C, which is displayed again in D. Note how well the peaks and troughs of the spike rate response are matched by the output of the filter. The goodness of the match attests to the fidelity with which the waveform in A represents the tuning properties associated with this axon. (Courtesy of Walter Yamada, Graduate Group in Neurobiology, University of California, Berkeley.)

cal, pressure-gradient receiver (Rheinlaender et al. 1979, 1981; Feng 1980; Chung et al. 1981; Feng and Shofner 1981; Pinder and Palmer 1983; Palmer and Pinder 1984; Vlaming et al. 1984; Aertsen et al. 1986; Eggermont 1988; Michelsen et al. 1986; Wang et al. 1996). In such a receiver the eardrum is driven by instantaneous differences between sound pressures arriving at the external and internal surfaces of the tympanic membrane (TM). Thus it has been proposed that sound localization in frogs depends on acoustic pathways to the internal surfaces of the TMs and the resultant phase differences across the TMs (Eggermont 1988; Narins et al. 1988).

Several studies have clearly demonstrated directional sensitivity of single auditory nerve fibers in the frog (Feng 1980; Feng and Shofner 1981; Schmitz et al. 1992; White et al. 1992). Laser Doppler vibrometric (LDV) measurements of the eardrum also suggest that the TM response is highly directional (Jorgensen and Gerhardt 1991; Jorgensen et al. 1991). Sound transmission to the inner ear through pathways other than through the TM has been demonstrated in anuran amphibians. Among the identified pathways are the mouth, the opercular-opercularis muscle complex, the lateral body walls via the lung and glottis, and substrate-borne vibrations through the entire body (Lombard and Straughan 1974; Feng and Shofner 1981; Narins and Lewis 1984; Wilczynski et al. 1987; Narins et al. 1988; Ehret et al. 1990; Jorgensen et al. 1991; Christensen-Dalsgaard and Narins 1993; Ehret et al. 1994). Simultaneous single fiber recordings and LDV measurements of the eardrum revealed that statistically significant correlations between eardrum velocity and nerve fiber firing rate were present in about 45% of the fibers recorded, implicating a significant role for extratympanic sound transmission to the inner ear (Wang et al. 1996).

The directional properties of phase-locking in the auditory have also been investigated. Schmitz et al. (1992) showed that sound direction affected vector strength very little, whereas it had a strong influence on the preferred firing phase of eighth nerve fibers. Thus, phase-locking of amphibian papillar neurons can potentially provide intensity-independent information for sound localization. In addition, the consistent differences between the directional masking patterns for vector strength and for the preferred phase functions suggest that different mechanisms underlie noise masking of these two measures of phase-locking in the amphibian auditory nerve (Wang and Narins 1996).

8. Summary

In the present climate of deep reductionism in biology, there is a tendency among many scientists to believe that the only important issues in sensory biology center around molecular devices and cellular mechanisms. Surely, such devices and mechanisms are intriguing, and they are important. The authors of this chapter hope, however, that we have convinced the reader that knowledge about molecular devices and cellular mechanisms forms only a small part of what we understand about the amphibian ear, and that without the integrative context provided by our knowledge of morphology, physiology, and neuroethology, the significance of that part would be reduced to zero. We also hope that we have demonstrated to the reader that there are fascinating open questions regarding the amphibian ear at all levels—morphological, physiological, neuroethological, and molecular—and that the comparative approach clearly is not only the path of choice but the path promising the greatest adventures.

Acknowledgments. More than 30 years ago, at MIT in Cambridge, Massachusetts, and the Bell Telephone Laboratories in Murray Hill, New Jersey, a small group of researchers combined engineering and biology to bring astounding new clarity to the study of the amphibian ear. Their work opened the way to the current evolutionary and neuroethological approaches to hearing research; without them and the scientific approaches they pioneered, this chapter would not have been worth writing. It is most appropriate that this volume is dedicated to one of the members of that MIT/Bell-Labs team—Bob Capranica, our long-time friend and mentor. We wish to acknowledge the other members, Bob's own mentor, and his colleagues at Cambridge and Murray Hill: Moise H. Goldstein, Jr, Lawrence S. Frishkopf, Willam A. van Bergeijk, C. Daniel Geisler, Gerard Harris, and Charles E. Molnar.

During the preparation of this manuscript, the work of the authors was supported by grants DC00112 (to E.R.L.) and DC00222 (to P.M.N.) from the National Institutes of Health, National Institute on Deafness and other Communicative Disorders.

References

Abbas PJ (1981) Auditory-nerve fiber responses to tones in a noise masker. Hear Res 5:69–80.

Aertsen AMHJ, Vlaming MSMG, Eggermont JJ, Johannesma PIM (1986) Directional hearing in the grassfrog (*Rana temporaria* L.). II. Acoustics and modelling of the auditory periphery. Hear Res 21:17–40.

Alfs B, Schneider H (1973) Vergleichend-anatomische Untersuchungen am Labyrinth zentraleuropäischer Froschlurch-Arten (Anura). Z Morph Ökol Tiere 76:129–143.

Anderson DJ, Rose JE, Hind JE, Brugge JF (1970) Temporal position of discharges in single auditory nerve fibers within the cycle of a sine-wave stimulus: frequency and intensity effects. J Acoust Soc Am 49:1131–1139.

Ashcroft DW, Hallpike CS (1934) Action potentials in the saccular nerve of the frog. J Physiol (Lond) 81:23P–24P.

Assad JA, Corey DP (1992) An active motor model for adaptation by vertebrate hair cells. J Neurosci 12:3291–3309.

Assad JA, Hacohen N, Corey DP (1989) Voltage dependence of adaptation and active bundle movement in bullfrog saccular hair cells. Proc Natl Acad Sci USA 86:2918–2922.

Assad JA, Shepherd GMG, Corey DP (1991) Tip-link integrity and mechanical transduction in vertebrate hair cells. Neuron 7:985–994.

Baird IL (1974a) Anatomical features of the inner ear in submammalian vertebrates. In: Keidel WD, Neff WD (eds) Handbook of Sensory Physiology, vol. 1: Auditory System. Berlin: Springer-Verlag, pp. 159–212.

Baird IL (1974b) Some aspects of the comparative anatomy and evolution of the inner ear in submammalian vertebrates. Brain Behav Evol 10:11–36.

Baird RA, Lewis ER (1986) Correspondences between afferent innervation patterns and response dynamics in the bullfrog utricle and saccule. Brain Res 369:48–64.

Baird RA, Steyger PS, Schuff NR (1996) Hair cell regeneration in the bullfrog vestibular otolith organs. Ann NY Acad Sci 781:59–70.

Békésy G von (1960) Experiments in Hearing. Wever EG (ed). New York: McGraw-Hill.

Benedix JH Jr, Pedemonte M, Yelluti R, Narins PM (1994) Temperature dependence of two-tone rate suppression in the northern leopard frog, *Rana pipiens pipiens*. J Acoust Soc Am 96:2738–2745.

Benser ME, Marquis RE, Hudspeth AJ (1996) Rapid, active hair bundle movements in hair cells from the bullfrog's sacculus. J Neurosci 16:5629–5643.

Bergeijk WA van (1957) Observations on models of the basilar papilla of the frog's ear. J Acoust Soc Am 29:1159–1162.

Bergeijk WA van (1967) The evolution of vertebrate hearing. In: Neff WD (ed) Contributions to Sensory Physiology. New York: Academic Press, pp. 1–49.

Bergeijk WA van, Witschi E (1957) The basilar papilla of the anuran ear. Acta Anat 30:81–91.

Bodnar DA, Capranica RR (1994) Encoding of phase spectra by the peripheral auditory system of the bullfrog. J Comp Physiol 174:157–171.

Boer E de, de Jongh HR (1978) On cochlear encoding: potentialities and limitations of the reverse-correlation technique. J Acoust Soc Am 63:115–135.

Bohne BA, Carr CD (1985) Morphometric analysis of hair cells in the chinchilla cochlea. J Acoust Soc Am 77:153–158.

Bolt JR, Lombard EL (1992) Nature and quality of the fossil evidence for otic evolution in early tetrapods. In: Webster DB, Fay RR, Popper AN (eds) The Evolutionary Biology of Hearing. New York: Springer-Verlag, pp. 377–403.

Boord RL, Grochow LB, Frishkopf LS (1970) Organization of the posterior ramus and ganglion of the VIIIth cranial nerve of the bullfrog *Rana catesbeiana*. MIT Res Lab Electr Quart Prog Rep 99:180–182.

Brand DJ (1956) On the cranial morphology of *Scolecomorphus* uluguruensis (Barbour and Loveridge). Ann Univ Stellenbosch 32(A):1–25.

Burlet HM de (1928) Über die Papilla neglecta. Anat Anz 66:199–209.

Burlet HM de (1934a) Zur vergleichenden Anatomie und Physiologie des perilymphatischen Raumes. Acta Otolaryngol 13:153–187.

Burlet HM de (1934b) Vergleichende Anatomie des statoakustichen Organs. In: Bolk L, Göppoert E, Kallius E, Lubosch W (eds) Handbuch der Vergleichenden Anatomie der Wirbeltiere, 2/2. Berlin: Urban and Schwarzenberg, pp. 1293–1432.

Burlet HM de (1935) Über endolymphatische und perilymphatische Sinnesendstellen. Acta Otolaryngol 22:287–305.

Capranica RR (1965) The evoked vocal responses of the bullfrog. Massachusetts Inst Tech Res Monog 33:1–106.

Capranica RR (1976) Morphology and physiology of the auditory system. In: Llinas R, Precht W (eds) Frog Neurobiology. Berlin: Springer-Verlag, pp. 551–575.

Capranica RR, Frishkopf LS, Nevo E (1973) Encoding of geographic dialects in the auditory system of the cricket frog. Science 182:1272–1275.

Capranica RR, Moffat AJM (1974a) Excitation, inhibition, and disinhibition in the inner ear of the toad (*Bufo*). J Acoust Soc Am 55:480.

Capranica RR, Moffat AJM (1974b) Frequency sensitivity of auditory fibers in the eighth nerve of the spadefoot toad, *Scaphiopus couchi*. J Acoust Soc Am 55:S85.

Capranica RR, Moffat AJM (1975) Selectivity of peripheral auditory system of spadefoot toads. J Comp Physiol 100:231–249.

Capranica RR, Moffat AJM (1980) Nonlinear properties of the peripheral auditory system of anurans. In: Popper AN, Fay RR (eds) Comparative Studies of Hearing in Vertebrates. New York: Springer-Verlag, pp. 139–165.

Carlström D (1963) A crystallographic study of vertebrate otoliths. Biol Bull 125:441–463.

Carlström D, Engström H (1955) The ultrastructure of statoconia. Acta Otolaryngol 45:14–18.

Carney LH, Yin TCT (1988) Temporal coding of resonances by low-frequency auditory nerve fibers: single-fiber responses and a population model. J Neurophysiol 60:1653–1677.

Caston J, Precht W, Blanks RHI (1977) Response characteristics of frog's lagena afferents to natural stimulation. J Comp Physiol 118:273–289.

Cazin L, Lannou J (1975) Response du saccule à la stimulation vibratoire directe de la macule, chez la grenouille. C R Soc Biol (Paris) 169:1067–1071.

Christensen-Dalsgaard J, Jørgensen MB (1996a) Sound and vibration sensitivity of VIIIth nerve fibers in the grassfrog *Rana temporaria*. J Comp Physiol 179:437–445.

Christensen-Dalsgaard J, Jørgensen MB (1996b) One-tone suppression in the frog auditory nerve. J Acoust Soc Am 100:451–457.

Christensen-Dalsgaard J, Narins PM (1993) Sound and vibration sensitivity of VIIIth nerve fibers in the frogs *Leptodactylus albilabris* and *Rana pipiens pipiens*. J Comp Physiol 172:653–662.

Chung S-H, Pettigrew A, Anson M (1981) Hearing in the frog: dynamics of the middle ear. Proc R Soc Lond [B] 212:459–485.

Citron L (1969) Observations on a case of objective tinnitus. Excerpta Medica Int Conf Ser 189:91.

Corey DP, Hacohen N, Huang PL, Assad JA (1989) Hair cell stereocilia bend at their bases and touch at their tips. Soc Neurosci Abstr 15:208.

Cortopassi KA, Lewis ER (1995) Tuning properties of acoustic and equilibrium axons: a comparison of frequency response curves in the bullfrog lagena. In: Burrows M, Matheson T, Newland PL, Shuppe H (eds) Nervous Systems and Behaviour. New York: Georg Thieme Verlag, p. 344.

Cortopassi KA, Lewis ER (1996) High-frequency tuning properties of bullfrog lagenar vestibular afferent fibers. J Vestib Res 6:105–119.

Corwin JT (1977) Morphology of the macula neglecta in sharks of the genus *Carcharhinus*. J Morphol 152:341–362.

Corwin JT (1985) Perpetual production of hair cells and maturational changes in hair-cell ultrastructure accompany postembryonic growth in an amphibian ear. Proc Natl Acad Sci USA 82:3911–3915.

Corwin JT, Warchol ME, Kelley MT (1993) Hair cell development. Curr Opin Neurobiol 3:32–37.

Costalupes JA, Young ED, Gibson DJ (1984) Effects of continuous noise backgrounds on rate response of auditory-nerve fibers in cat. J Neurophysiol 51:1326–1344.

Cotanche DA, Lee KH (1994) Regeneration of hair cells in the vestibulocochlear system of birds and mammals. Curr Biol 4:509–513.

Davis H (1965) A model for transducer action in the cochlea. Cold Spring Harb Symp Quant Biol 30:181–190.

Dijk P van, Lewis ER, Wit HP (1990) Temperature effects on auditory nerve fiber response in the American bullfrog. Hear Res 44:231–240.

Dijk P van, Wit HP, Segenhout JM (1989) Spontaneous otoacoustic emissions in the European edible frog (*Rana esculenta*): spectral details and temperature dependence. Hear Res 42:273–282.

Dijk P van, Narins PM, Wang J (1996) Spontaneous otoacoustic emissions in seven frog species. Hear Res 101:102–112.

Duellman WE, Trueb L (1994) Biology of Amphibians. Baltimore: Johns Hopkins.

Dunia R, Narins PM (1989) Temporal resolution in frog auditory-nerve fibers. J Acoust Soc Am 85:1630–1638.

Dunn RF (1978) Nerve fibers of the eighth nerve and their distribution to the sensory nerves of the inner ear in the bullfrog. J Comp Neurol 182:621–636.

Dunn RF (1980) Reciprocal synapses between hair cells and first order afferent dendrites in the crista ampullaris of the bullfrog. J Comp Neurol 193:255–264.

Eatock RA, Corey DP, Hudspeth AJ (1987) Adaptation of mechanoelectrical transduction in hair cells of the bullfrog's sacculus. J Neurosci 7:2821–2836.

Egert D (1993) The Physiological Basis of Tuning in the Bullfrog Sacculus (doctoral dissertation). Berkeley: University of California.

Egert D, Lewis ER (1995) Temperature-dependence of saccular nerve fiber response in the North American bullfrog. Hear Res 84:72–80.

Eggermont JJ (1988) Mechanisms of sound localization in anurans. In: Fritzsch B, Ryan MJ, Wilczynski W, Hetherington TE, Walkowiak W (eds) The Evolution of the Amphibian Auditory System. New York: Wiley, pp. 307–336.

Ehret G, Capranica RR (1980) Masking patterns and filter characteristics of auditory nerve fibers in the green treefrog (*Hyla cinerea*). J Comp Physiol 141:1–12.

Ehret G, Moffat AJM, Capranica RR (1983) Two-tone suppression in auditory nerve fibers of the green treefrog (*Hyla cinerea*). J Acoust Soc Am 73:2093–2095.

Ehret G, Tautz J, Schmitz B, Narins PM (1990) Hearing through the lungs: lung-eardrum transmission of sound in the frog *Eleutherodactylus coqui*. Naturwissenschaften 77:192–194.

Ehret G, Keilworth E, Kamada T (1994) The lung-eardrum pathway in three treefrog and four dendrobatid frog species: some properties of sound transmission. J Exp Biol 195:329–343.

Evans BN (1988) Motile Response Patterns and Ultrastructural Observations of the Isolated Outer Hair Cell (doctoral dissertation). Houston: University of Texas Health Science Center.

Evans EF (1977) Frequency selectivity at high signal levels of single units in cochlear nerve and nucleus. In: Evans EF, Wilson JP (eds) Psychophysics and Physiology of Hearing. London: Academic Press, pp. 185–196.

Feng AS (1980) Directional characteristics of the acoustic receiver of the leopard frog (*Rana pipiens*): a study of the eighth nerve auditory responses. J Acoust Soc Am 68:1107–1114.

Feng AS, Gerhardt HC, Capranica RR (1976) Sound localization behavior of the green treefrog (*Hyla cinerea*) and barking treefrog (*H. gratiosa*). J Comp Physiol 107:241–252.

Feng AS, Narins PM, Capranica RR (1975) Three populations of primary auditory fibers in the bullfrog (*Rana catesbeiana*): their peripheral origins and frequency sensitivities. J Comp Physiol 100:221–229.

Feng AS, Shofner WP (1981) Peripheral basis of sound localization in anurans. Acoustic properties of the frog's ear. Hear Res 5:201–216.

Fermin CD, Lovett AE (1989) The sensory epithelia and gelatinous membranes of the chick inner ear. Assoc Res Otolaryngol Abstr 12:324.

Fermin CD, Igarashi M, Yoshihara T (1987) Ultrastructural changes of statoconia after segmentation of the otolithic membrane. Hear Res 28:23–34.

Flock Å, Flock B (1966) Ultrastructure of the amphibian papilla in the bullfrog. J Acoust Soc Am 40:1262.

Frishkopf LS, Capranica RR, Goldstein MH Jr (1968) Neural coding in the bullfrog's auditory system—a teleological approach. Proc Inst Elec Electron Engr 56:969–980.

Frishkopf LS, Flock Å (1974) Ultrastructure of the basilar papilla, an auditory organ in the bullfrog. Acta Otolaryngol 77:176–184.

Frishkopf LS, Geisler CD (1966) Peripheral origin of auditory responses recorded from the eighth nerve of the bullfrog. J Acoust Soc Am 40:469–472.

Frishkopf LS, Goldstein MH Jr (1963) Responses to acoustic stimuli from single units in the eighth nerve of the bullfrog. J Acoust Soc Am 35:1219–1228.

Fritzsch B (1987) The inner ear of the ceolacanth fish *Latimeria* has tetrapod affinities. Nature 327:331–333.

Fritzsch B (1992) The water-to-land transition: evolution of the tetrapod basilar papilla, middle ear, and auditory nuclei. In: Webster DB, Fay RR, Popper AN (eds) The Evolutionary Biology of Hearing. New York: Springer-Verlag, pp. 351–375.

Fritzsch B, Wake MH (1988) The inner ear of gymnophione amphibians and its nerve supply: a comparative study of regressive events in a complex sensory system. Zoomorphology 108:210–217.

Geisler CD, Bergeijk WA van, Frishkopf LS (1964) The inner ear of the bullfrog. J Morphol 114:43–58.

Gerhardt HC, Rheinlaender J (1980) Accuracy of sound localization in a miniature dendrobatid frog. Naturwissenschaften 67:362–363.

Gerhardt HC, Rheinlaender J (1982) Localization of an elevated sound source by the green tree frog. Science 217:663–664.

Gold T (1948) Hearing II: the physical basis of the action of the cochlea. Proc Roy Soc B 135:492–498.

Goldberg JM, Brown PB (1969) Response of binaural neurons of dog superior olivary complex to dichotic tonal stimuli: some physiological mechanisms of sound localization. J Neurophysiol 32:613–636.

Goldstein JL, Kiang NYS (1968) Neural correlates of the aural combination tone $2f_1-f_2$. Proc Inst Elec Electron Engr 56:981–992.

Goldstein MH, Frishkopf LS, Geisler CD (1962) Representation of sounds by response of single units in the eighth nerve of the bullfrog. J Acoust Soc Am 34:734.

Harrison HS (1902) On the perilymphatic spaces of the amphibian ear. Int Meschr Anat Physiol 19:222–261.

Henry KR, Lewis ER (1992) One-tone suppression in the cochlear nerve of the gerbil. Hear Res 63:1–6.

Henson OW Jr (1974) Comparative anatomy of the middle ear. In: Keidel WD, Neff WD (eds) Handbook of Sensory Physiology, vol. 1: Auditory System. Berlin: Springer-Verlag, pp. 39–110.

Hetherington TE (1988) Metamorphic changes in the middle ear. In: Fritzsch B, Ryan MJ, Wilczynski W, Hetherington TE, Walkowiak W (eds) The Evolution of the Amphibian Auditory System. New York: Wiley-Interscience, pp. 339–357.

Hetherington TE, Jaslow AP, Lombard RE (1986) Comparative morphology of the amphibian opercularis system, I. General design features and functional interpretation. J Morphol 190:43–61.

Hillery CM, Narins PM (1984) Neurophysiological evidence for a traveling wave in the amphibian ear. Science 225:1037–1039.

Hillery CM, Narins PM (1987) Frequency and time domain comparison of low-frequency auditory fiber responses in two anuran amphibians. Hear Res 25:233–248.

Hillman DE (1969) New ultrastructural findings regarding a vestibular ciliary apparatus and its possible functional signficance. Brain Res 13:407–412.

Hillman DE (1976) Morphology of the peripheral and central vestibular systems. In: Llinás R, Precht W (eds) Frog Neurobiology. Berlin: Springer-Verlag, pp. 452–480.

Hillman DE, Lewis ER (1971) Morphological basis for a mechanical linkage in otolithic receptor transduction. Science 174:416–419.

Honrubia V, Strelioff D, Sitko S (1976) Physiological basis of cochlear transduction and sensitivity. Ann Otol Rhinol Laryngol 85:697–710.

Houseley GD, Norris CH, Guth PS (1989) Electrophysiological properties and morphology of hair cells isolated from the semicircular canal of the frog. Hear Res 38:259–276.

Howard J, Hudspeth AJ (1987) Mechanical relaxation of the hair bundle mediates adaptation in mechanoelectrical transduction by the bullfrog's saccular hair cell. Proc Natl Acad Sci USA 84:3064–3068.

Hudspeth AJ (1982) Extracellular current flow and the site of transduction by vertebrate hair cells. J Neurosci 2:1–10.

Hudspeth AJ, Corey DP (1977) Sensitivity, polarity, and conductance change in the response of vertebrate hair cells to controlled mechanical stimuli. Proc Natl Acad Sci USA 74:2407–2411.

Hudspeth AJ, Jacobs R (1979) Stereocilia mediate transduction in vertebrate hair cells. Proc Natl Acad Sci USA 76:1506–1509.

Hudspeth AJ, Lewis RS (1988) Kinetic analysis of voltage- and ion-dependent conductances in saccular hair cells of the bullfrog, *Rana catesbeiana*. J Physiol 400:237–274.

Iurato S (1962) Submicroscopic structure of the membranous labyrinth III. The supporting structure of Corti's organ (basilar membrane, limbus spiralis and spiral ligament). Z Zellforsch 56:40–96.

Jaramillo F, Hudspeth AJ (1991) Localization of the hair cell's transduction channels at the hair bundle's top by iontophoretic application of a channel blocker. Neuron 7:409–420.

Jaslow AP, Hetherington TE, Lombard RE (1988) Structure and function of the amphibian middle ear. In: Fritzsch B, Ryan MJ, Wilczynski W, Hetherington TE, Walkowiak W (eds) The Evolution of the Amphibian Auditory System. New York: Wiley-Interscience, pp. 69–91.

Jørgensen JM (1981) On a possible hair cell turnover in the inner ear of the caecilian *Ichthyophis glutinosus* (Amphibia: Gymnophiona). Acta Zool 62:171–186.

Jørgensen MB, Christensen-Dalsgaard J (1991) Peripheral origins and functional characteristics of vibration-sensitive VIIIth nerve fibers in the frog *Rana temporaria*. J Comp Physiol 169:341–347.

Jørgensen MB, Gerhardt HC (1991) Directional hearing in the gray tree frog Hyla versicolor: eardrum vibrations and phonotaxis. J Comp Physiol 169:177–183.

Jørgensen MB, Schmitz B, Christensen-Dalsgaard J (1991) Biophysics of directional hearing in the frog *Eleutherodactylus coqui*. J Comp Physiol 168:223–232.

Katsuki Y, Suga N, Kanno Y (1962) Neural mechanism of the peripheral and central auditory system in monkeys. J Acoust Soc Am 34:1396–1410.

Kemp DT (1979) Evidence of mechanical nonlinearity and frequency-selective wave amplification in the cochlea. Arch Otorhinolaryngol 244:37–45.

Kemp DT, Martin JA (1976) Active resonance systems in audition. Int Congr Audiol Abstr 13:64–65.

Kiang NYS, Watanabe T, Thomas EC, Clark LF (1965) Discharge patterns of single fibers in the cat's auditory nerve. Massachusetts Inst Tech Res Monog 35.

Kingsbury BD, Reed HD (1909) The columella auris in amphibia. J Morphol 20:549–628.

Klump G, Gerhardt HC (1989) Sound localization in the barking treefrog. Naturwissenschaften 76:35–37.

Koyama H, Lewis ER, Leverenz EL, Baird RA (1982) Acute seismic sensitivity in the bullfrog ear. Brain Res 250:168–172.

Kronester-Frei A (1978) Ultrastructure of the different zones of the tectorial membrane. Cell Tissue Res 193:11–23.

Lawson DP (1993) The reptiles and amphibians of the Korup National Park Project, Cameroon. Herp Nat Hist 1:27–90.

Lewis ER (1976) Surface morphology of the bullfrog amphibian papilla. Brain Behav Evol 13:196–215.

Lewis ER (1977) Comparative scanning electron microscopy study of the anuran basilar papilla. Ann Proc Electron Microsc Soc Am 35:632–633.

Lewis ER (1978) Comparative studies of the anuran auditory papillae. Scan Electr Microsc 1978(II):633–642.

Lewis ER (1981a) Evolution of inner-ear auditory apparatus in the frog. Brain Res 219:149–155.

Lewis ER (1981b) Suggested evolution of tonotopic organization in the frog amphibian papilla. Neurosci Lett 21:131–136.

Lewis ER (1984) On the frog amphibian papilla. Scan Electr Microsc 1984(IV):1899–1913.

Lewis ER (1986) Adaptation, suppression and tuning in amphibian acoustical fibers. In: Moore BCJ, Patterson RD (eds) Auditory Frequency Selectivity. New York: Plenum, pp. 129–136.

Lewis ER (1988) Tuning in the bullfrog ear. Biophys J 53:441–447.

Lewis ER (1992) Convergence of design in vertebrate acoustic sensors. In: Webster DB, Fay RR, Popper AN (eds) The Evolutionary Biology of Hearing. New York: Springer-Verlag, pp. 163–184.

Lewis ER (1996) A brief introduction to network theory. In: Berger SA, Goldsmith W, Lewis ER (eds) Introduction to Bioengineering. Oxford: Oxford University Press, pp. 261–338.

Lewis ER, Henry KR (1994) Dynamic changes in tuning in the gerbil cochlea. Hear Res 79:183–189.

Lewis ER, Henry KR (1995) Nonlinear effects of noise on phase-locked cochlear-nerve responses to sinusoidal stimuli. Hear Res 92:1–16.

Lewis ER, Leverenz EL (1979) Direct evidence for an auditory place mechanism in the frog amphibian papilla. Soc Neurosci Abstr 5:25.

Lewis ER, Leverenz EL (1983) Morphological basis for tonotopy in the anuran amphibian papilla. Scan Electr Microsc 1983:189–200.

Lewis ER, Li CW (1973) Evidence concerning the morphogenesis of saccular receptors in the bullfrog (Rana catesbeiana). J Morphol 139:351–361.

Lewis ER, Li CW (1975) Hair cell types and distributions in the otolithic and auditory organs of the bullfrog. Brain Res 83:35–50.

Lewis ER, Lombard RE (1988) The amphibian inner ear. In: Fritzsch B, Ryan MJ, Wilczynski W, Hetherington TE, Walkowiak W (eds) The Evolution of the Amphibian Auditory System. New York: Wiley, pp. 93–123.

Lewis ER, Nemanic P (1972) Scanning electron microscope observations of saccular ultrastructure in the mudpuppy (Necturus maculosus). Z Zellforsch 123:441–457.

Lewis ER, Baird RA, Leverenz EL, Koyama H (1982) Inner ear: dye injection reveals peripheral origins of specific sensitivities. Science 215:1641–1643.

Lewis ER, Leverenz EL, Koyama H (1982) The tonotopic organization of the bullfrog amphibian papilla, an auditory organ lacking a basilar membrane. J Comp Physiol 145:437–445.

Lewis ER, Leverenz EL, Bialek WS (1985) The Vertebrate Inner Ear. Boca Raton, FL: CRC Press.

Lewis ER, Sneary M, Yu XY (1990) Further evidence for tuning mechanisms of high dynamic order in lower vertebrates. In: Dallos P, Geisler CD, Matthews JW, Ruggero MA, Steele CR (eds) The Mechanics and Biophysics of Hearing. Berlin: Springer-Verlag, pp. 139–147.

Lewis ER, Hecht EI, Narins PM (1992) Diversity of form in the amphibian papilla of Puerto Rican frogs. J Comp Physiol 171:412–435.

Lewis RS, Hudspeth AJ (1983) Frequency tuning and ionic conductance in hair cells of the bullfrog sacculus. In: Klinke R, Hartmann R (eds) Hearing—Physiological Bases and Psychophysics. Berlin: Springer-Verlag, pp. 17–24.

Li CW, Lewis ER (1974) Morphogenesis of auditory receptor epithelia in the bullfrog. In: Johari O, Corvin I (eds) Scanning Electron Microscopy, vol. III. Chicago: IIT Research Institute, pp. 791–798.

Li CW, Lewis ER (1979) Structure and development of vestibular hair cells in the larval bullfrog. Ann Otol Rhinol Laryngol 88:427–437.

Liff H (1969) Responses from single auditory units in the eighth nerve of the leopard frog. J Acoust Soc Am 45:512–513.

Liff H, Goldstein MH Jr (1970) Peripheral inhibition in auditory nerve fibers in the frog. J Acoust Soc Am 47:1538–1547.

Lim DJ (1972) Fine morphology of the tectorial membrane. Its relationship to the organ of Corti. Arch Otolaryngol 96:199–215.

Loftus-Hills JJ (1973) Comparative aspects of auditory function in Australian anurans. Aust J Zool 21:353–367.

Loftus-Hills JJ, Johnstone BM (1970) Auditory function, communication and the brain-evoked response in anuran amphibians. J Acoust Soc Am 47:1131–1138.

Lombard RE (1970) A Comparative Morphological Analysis of the Inner Ear of Salamanders (doctoral dissertation). Chicago: University of Chicago.

Lombard RE (1977) Comparative morphology of the inner ear in salamanders (Caudata; Amphibia). Contrib Vert Evol 2:1–140.

Lombard RE (1980) The structure of the amphibian auditory periphery: a unique experiment in terrestrial hearing. In: Popper AN, Fay RR (eds) Comparative Studies of Hearing in Vertebrates. New York: Springer-Verlag, pp. 121–138.

Lombard RE, Bolt JR (1979) Evolution of the tetrapod ear: an analysis and reinterpretation. Biol J Linn Soc 11:19–76.

Lombard RE, Straughan IR (1974) Functional aspects of anuran middle ear structures. J Exp Biol 61:71–93.

Long GR, van Dijk P, Wit H (1996) Temperature dependence of spontaneous otoacoustic emissions in the edible frog (*Rana esculenta*). Hear Res 98:22–28.

Lowenstein O, Wersäll J (1959) Functional interpretation of the electron microscopic structure of the sensory hairs in the cristae of the elasmobranch *Raja clavata* in terms of directional sensitivity. Nature (Lond) 184:1807–1808.

Marcus H (1935) Zur Entstehung der stapesplatte bei *Hypogeophis*, Anat Anz 80:142–146.

Megela AL (1984) Diversity of adaptation patterns in responses of eighth nerve fibers in the bullfrog, *Rana catesbeiana*. J Acoust Soc Am 75:1155–1162.

Megela AL, Capranica RR (1981) Response patterns to tone bursts in peripheral auditory systems of anurans. J Neurophysiol 46:465–478.

Marmo F, Franco E, Balsamo G (1981) Scanning electron microscopy and x-ray diffraction studies of otoconia in the lizard *Podarcus s. sicula*. Cell Tiss Res 218:265–270.

Marmo F, Balsamo G, Franco E (1983) Calcite in the statoconia of amphibians: a detailed analysis in the frog *Rana esculenta*. Cell Tiss Res 233:35–43.

McNally WJ, Tait J (1925) Ablation experiments on the labyrinth of the frog. Am J Physiol 75:155–179.

Michelsen A, Jorgensen M, Christensen-Dalsgaard J, Capranica RR (1986) Directional hearing of awake, unrestrained treefrogs. Naturwissenschaften 73:682–683.

Millot J, Anthony J (1965) Anatomie de *Latimeria chalumnae*, vol. II. Paris: CNRS.

Moffat AJM, Capranica RR (1974) Sensory processing in the peripheral auditory system of treefrogs (*Hyla*). J Acoust Soc Am 55:480.

Moffat AJM, Capranica RR (1976) Auditory sensitivity of the saccule in the American toad (*Bufo americanus*). J Comp Physiol 105:1–8.

Møller AR (1977) Frequency selectivity of single auditory-nerve fibers in response to broad band noise stimuli. J Acoust Soc Am 62:136–142.

Møller AR (1978) Frequency selectivity of the peripheral auditory analyzer studied using broadband noise. Acta Physiol Scand 104:24–32.

Monath T (1965) The opercular apparatus of salamanders. J Morphol 116:149–170.

Mullinger AM, Smith JJB (1969) Some aspects of the gross and fine structure of the amphibian papilla in the labyrinth of the newt, *Triturus cristatus*. Tissue Cell 1:403–416.

Narins PM (1975) Electrophysiological determination of the function of the lagena in terrestrial amphibians. Biol Bull 149:438.

Narins PM (1987) Coding of signals in noise by amphibian auditory nerve fibers. Hear Res 26:145–154.

Narins PM, Capranica RR (1976) Sexual differences in the auditory system of the treefrog, *Eleutherodactylus coqui*. Science 192:378–380.

Narins PM, Ehret G, Tautz J (1988) Accessory pathway for sound transfer in a neotropical frog. Proc Natl Acad Sci USA 85:1508–1512.

Narins PM, Hillery CM (1983) Frequency coding in the inner ear of anuran amphibians. In: Klinke R, Hartmann R (eds) Hearing—Physiological Bases and Psychophysics. Berlin: Springer-Verlag, pp. 70–76.

Narins PM, Lewis ER (1984) The vertebrate ear as an exquisite seismic sensor. J Acoust Soc Am 76:1384–1387.

Narins PM, Wagner I (1989) Noise susceptibility and immunity of phase locking in amphibian auditory-nerve fibers. J Acoust Soc Am 85:1255–1265.

Narins PM, Zelick R (1988) The effects of noise on auditory processing and behavior in amphibians. In: Fritzsch B, Ryan MJ, Wilczynski W, Hetherington TE, Walkowiak W (eds) The Evolution of the Amphibian Auditory System. New York: Wiley-Interscience, pp. 511–536.

Palmer AR, Pinder AC (1984) The directionality of the frog ear described by a mechanical model. J Theor Biol 110:205–215.

Palmer AR, Wilson JP (1982) Spontaneous and evoked acoustic emissions in the frog *Rana esculenta*. J Physiol (Lond) 324:66P.

Passmore NI, Capranica RR, Telford SR, Bishop PJ (1984) Phonotaxis in the painted reed frog (*Hyperolius marmoratus*). J Comp Physiol 154:189–197.

Penna M, Narins PM (1989) Effects of acoustic overstimulation on spectral and temporal processing in the amphibian auditory nerve. J Acoust Soc Am 85:1617–1629.

Pickles JO (1993) A model for the mechanics of the stereociliar bundle on acousticolateral hair cells. Hear Res 68:159–172.

Pickles JO, Comis SD, Osborne MP (1984) Cross links between stereocilia in the guinea pig organ of Corti, and their possible relation to sensory transduction. Hear Res 15:103–112.

Pinder AC, Palmer AR (1983) Mechanical properties of the frog ear: vibration measurements under free- and closed-field acoustic conditions. Proc R Soc Lond [B] 219:371–396.

Pitchford S, Ashmore JF (1987) An electrical resonance in hair cells of the amphibian papilla of the frog, *Rana temporaria*. Hear Res 27:75–84.

Platt C (1977) Hair cell distribution and orientation in goldfish otolithic organs. J Comp Neurol 172:283–287.

Platt C (1983) Retention of generalized hair cell patterns in the inner ear of the primitive flatfish *Psettodes*. Anat Rec 207:503–508.

Pote KG, Hauer CR, Shabanowitz J, Hunt DF, Kretsinger RH (1993) The major protein of frog otoconia is a homolog of phospholipase A2. Biochemistry 32:5017–5024.

Pote KG, Ross MD (1991) Each otoconia polymorph has a protein unique to that polymorph, Comp Biochem Physiol [B] 98:287–295.

Pote KG, Weber CH, Kretsinger RH (1993) Inferred protein content and distribution from density measurements of calcitic and aragonitic otoconia. Hear Res 66:225–232.

Purgue AP (1997) Tympanic sound radiation in the bullfrog Rana catesbeiana. J Comp Physiol A 181:438–445.

Purgue AP, Narins PM (1997a) Morphology and wave transmission properties of the perilymphatic duct in the bullfrog, *Rana catesbeiana*. Assoc Res Otolaryngol Abstr 20:140.

Purgue AP, Narins PM (1997b) The vibration patterns of the contact membrane: the key to the stimulation of the tectorial membrane in the bullfrog, *Rana catesbeiana*? Assoc Res Otolaryngol Abstr 20:140.

Rauch S, Rauch I (1974) Physico-chemical properties of the inner ear especially ion transport. In: Keidel WD, Neff WD (eds) Handbook of Sensory Physiology, vol. 1: Auditory System. Berlin: Springer-Verlag, pp. 647–682.

Retzius G (1881) Das Gehörorgan der Wirbelthiere. I. Gehörorgander Fische und Amphibien. Stockholm: Samson and Wallin.

Rheinlaender J, Gerhardt HC, Yager DD, Capranica RR (1979) Accuracy of phonotaxis by the green treefrog (*Hyla cinerea*). J Comp Physiol 133:247–255.

Rheinlaender J, Klump G (1988) Behavioral aspects of sound localization. In: Fritzsch B, Ryan MJ, Wilczynski W, Hetherington TE, Walkowiak W (eds) The Evolution of the Amphibian Auditory System. New York: Wiley, pp. 297–305.

Rheinlaender J, Walkowiak W, Gerhardt HC (1981) Directional hearing in the green treefrog: a variable mechanism? Naturwissenschaften 68:430–431.

Robbins RG, Bauknight BS, Honrubia MD (1967) Anatomical distribution of efferent fibers in the VIIIth cranial nerve of the bullfrog (*Rana catesbeiana*). Acta Otolaryngol 64:436–448.

Ronken DA (1991) Spike discharge properties that are related to the characteristic frequency of single units in the frog auditory nerve. J Acoust Soc Am 90:2428–2440.

Rose GJ, Capranica RR (1985) Sensitivity to amplitude modulated sounds in the anuran auditory nervous system. J Neurophysiol 53:446–465.

Ross RJ, Smith JJB (1977) Detection of substrate vibrations by salamanders: eighth cranial nerve activity. Can J Zool 57:368–374.

Ross RJ, Smith JJB (1978) Detection of substrate vibrations by salamanders: inner ear sense organ activity. Can J Zool 56:1156–1162.

Ross RJ, Smith JJB (1980) Detection of substrate vibrations by salamanders: frequency sensitivity of the ear. Comp Biochem Physiol 65A:167–172.

Sachs MB (1964) Responses to acoustic stimuli from single units in the eighth nerve of the green frog. J Acoust Soc Am 36:1956–1958.

Sachs MB, Kiang NYS (1968) Two-tone inhibition in auditory nerve fibers. J Acoust Soc Am 43:1120–1128.

Sarasin P, Sarasin F (1890) Zur Entwicklungsgeschichte und Anatomie der ceylonesischen Blindwühle *Ichthyophis glutinosus*. Das Gehörorgan. Erg Naturwiss Forsch auf Ceylon (Wiesbaden) 2:207–222.

Sarasin P, Sarasin F (1892) Über das Gehörorgan der Ceaciliiden. Anat Anz 7:812–815.

Schmitz B, White TD, Narins PM (1992) Directionality of phase locking in auditory nerve fibers of the leopard frog *Rana pipiens pipiens*. J Comp Physiol 170:589–604.

Schwartz JJ, Simmons AM (1990) Encoding of a spectrally-complex communication sound in the bullfrog's auditory nerve. J Comp Physiol 166:489–499.

Shepherd GMG, Corey DP (1994) The extent of adaptation in bullfrog saccular hair cells. J Neurosci 14:6217–6229.

Shofner WP, Feng AS (1983) A quantitative light microscopic study of the bullfrog amphibian papilla tectorium: correlation with the tonotopic organization. Hear Res 11:103–116.

Shotwell SL, Jacobs R, Hudspeth AJ (1981) Directional sensitivity of individual vertebrate hair cells to controlled deflection of their hair bundles. Ann NY Acad Sci 374:1–10.

Simmons AM, Ferragamo M (1993) Periodicity extraction in the anuran auditory nerve I: "pitch-shift" effects. J Comp Physiol 172:57–69.

Simmons AM, Shen Y, Sanderson MI (1996) Neural and computational basis for periodicity extraction in frog peripheral auditory system. Aud Neurosci 2:109–133.

Simmons DD, Bertolotto C, Narins PM (1992) Innervation of the amphibian and basilar papillae in the leopard frog: reconstruction of single labeled fibers. J Comp Neurol 322:191–200.

Simmons DD, Bertolotto C, Narins PM (1994) Morphological gradients in sensory hair cells of the amphibian papilla of the frog, *Rana pipiens pipiens*. Hear Res 80:71–78.

Simon EJ, Hilding DA, Kashgarian M (1973) Micropuncture study of the mechanism of endolymph production in the frog. Am J Physiol 225:114–118.

Smith JB (1968) Hearing in terrestrial urodeles: a vibration sensitive mechanism in the ear. J Exp Biol 48:191–205.

Smotherman MS, Narins PM (1997) Variations of the electrical properties of hair cells isolated from the amphibian papilla of the leopard frog. Assoc Res Otolaryngol Abstr 20:141.

Smotherman MS, Narins PM (1998) The effect of temperature on electrical resonance in leopard frog saccular hair cells. J Neurophysiol 79:312–321.

Steel KP (1985) Composition and properties of the mammalian tectorial membrane. In: Drescher D (ed) Auditory Biochemistry. Springfield, IL: CC Thomas, pp. 351–365.

Steyger PS, Wiederhold ML, Batten J (1995) The morphogenic features of otoconia during larval development of *Cynops pyrrhogaster*, the Japanese red-bellied newt. Hear Res 84:61–71.

Stiebler IB, Narins PM (1990) Temperature-dependence of auditory nerve responses properties in the frog. Hear Res 46:63–82.

Strelioff D, Haas G, Honrubia V (1972) Sound-induced electrical impedance changes in the guinea pig cochlea. J Acoust Soc Am 51:617–620.

Tachibana M, Saito H, Machino M (1973) Sulfated acid mucopolysaccharides in the tectorial membrane. Acta Otolaryngol 76:37–46.

Taylor EH (1969) Skulls of Gymnophiona and their significance in the taxonomy of the group. Univ Kansas Sci Bull 48:585–687.

Villiers CGS de (1932) Über das Gehörskelett der aglossen Anuren. Anat Anz 71:305–331.

Villiers CGS de (1934) Studies of the cranial anatomy of *Ascaphus truei* Stejneger. Bull Mus Comp Zool Harvard Coll 77:1–38.

Vlaming MSMG, Aertsen AMHJ, Epping WJM (1984) Directional hearing in the grass frog (*Rana temporaria* L.): I. Mechanical vibrations of tympanic membrane. Hear Res 14:191–201.

Wang J, Narins PM (1996) Directional masking of phase locking in the amphibian auditory nerve. J Acoust Soc Am 99:1611–1620.

Wang J, Ludwig TA, Narins PM (1996) Spatial and spectral dependence of the auditory periphery in the northern leopard frog. J Comp Physiol 178:159–172.

Wever EG (1973) The ear and hearing in the frog, *Rana pipiens*. J Morphol 141:461–478.

Wever EG (1974) The evolution of vertebrate hearing. In: Keidel MD, Neff WD (eds) Handbook of Sensory Physiology, vol. 1: Auditory System. Berlin: Springer-Verlag, pp. 423–454.

Wever EG (1975) The caecilian ear. J Exp Biol 191:63–72.

Wever EG (1976) Origin and evolution of the ear of vertebrates. In: Masterton RB, Hodos W, Jerison H (eds) Evolution of Brain and Behavior in Vertebrates. Hillsdale, NJ: Lawrence Erlbaum, pp. 89–105.

Wever EG (1979) Middle ear muscles of the frog. Proc Natl Acad Sci USA 76:3031–3033.

Wever EG (1985) The Amphibian Ear. Princeton, NJ: Princeton University Press.

Wever EG, Gans C (1976) The caecilian ear: further observations. Proc Natl Acad Sci USA 73:3744–3746.

White JS (1978) A light and scanning electron microscopic study of the basilar papilla in the salamander, *Ambystoma tigrinum*. Am J Anat 151:437–452.

White JS (1986) Comparative features of the surface morphology of the basilar papilla in five families of Salamanders (Amphibia; Caudata). J Morphol 187:201–217.

White JS, Baird IL (1982) Comparative morphological features of the caecilian inner ear with comments on the evolution of amphibian auditory structures. Scan Electr Microsc 1982(3):1301–1312.

White TD, Schmitz B, Narins PM (1992) Directional dependence of auditory sensitivity and frequency selectivity in the leopard frog. J Acoust Soc Am 92:1953–1961.

Whitehead ML, Wilson JP, Baker RJ (1986) The effects of temperature on otoacoustic emission tuning properties. In: Moore BCJ, Patterson RD (eds) Auditory Frequency Selectivity. New York: Plenum, pp. 39–48.

Wiederhold ML, Yamashita M, Larsen KA, Batten JS, Koike H, Asashima M (1995) Development of the otolith organs and semicircular canals in the Japanese red-bellied newt, *Cynops pyrrhogaster*. Hear Res 84:41–51.

Wilczynski W, Zakon HH, Brenowitz EA (1983) A sex difference in basilar papilla tuning in the *Hyla crucifer* auditory system and its behavioral significance. Soc Neurosci Abstr 9:531.

Wilczynski W, Zakon HH, Brenowitz EA (1984) Acoustic communication in spring peepers. Call characteristics and neurophysiological aspects. J Comp Physiol 155:577–584.

Wilczynski W, Resler C, Capranica RR (1987) Tympanic and extratympanic sound transmission in the leopard frog. J Comp Physiol 161:659–669.

Wislocki GB, Ladman AJ (1955) Selective and histochemical staining of the otolithic membranes, cupulae and tectorial membrane of inner ear. J Anat (Lond) 89:3–12.

Wit HP, van Dijk P, Segenhout JM (1994) Wiener kernel analysis of the inner ear function in the American bullfrog. J Acoust Soc Am 95:904–919.

Witschi E (1949) The larval ear of the frog and its transformation during metamorphosis. Z Natur 4(b):230–242.

Witschi E (1955) The bronchial columella of the ear of larval Ranidae. J Morphol 96:497–512.

Wolodkin GJ, Yamada WM, Lewis ER, Henry KR (1997) Spike rate models for auditory fibers. In: Lewis ER, Long GR, Lyon RF, Narins PM, Steele CR, Poinar E (eds) Diversity in Auditory Mechanics. Singapore: World Scientific Press, pp. 104–110.

Yamada WM, Wolodkin GJ, Lewis ER, Henry KR (1997) Wiener kernel analysis and the singular value decomposition. In: Lewis ER, Long GR, Lyon RF, Narins PM, Steele CR, Poinar E (eds) Diversity in Auditory Mechanics. Singapore: World Scientific Press, pp. 111–118.

Yano J, Sugai T, Sugitani M, Ooyama H (1990) Observations of the sensing and tectorial membrane in bullfrog amphibian papilla: their possible functional roles. Hear Res 50:237–244.

Yu XL (1991) Signal Processing Mechanics in Bullfrog Ear Inferred from Neural Spike Trains (doctoral dissertation). Berkeley: University of California.

Yu XL, Lewis ER, Feld D (1991) Seismic and auditory tuning curves from bullfrog saccular and amphibian papillar axons. J Comp Physiol 169:241–248.

Zakon HH, Wilczynski W (1988) Anuran eighth nerve physiology. In: Fritzsch B, Ryan MJ, Wilczynski W, Hetherington TE, Walkowiak W (eds) The Evolution of the Amphibian Auditory System. New York: Wiley-Interscience, pp. 125–155.

Zelick R, Narins PM (1985) Temporary threshold shift, adaptation, are recovery characteristics of frog auditory nerve fibers. Hear Res 17:161–176.

5
Anatomy of the Central Auditory Pathways of Fish and Amphibians

CATHERINE A. MCCORMICK

1. Introduction

1.1 Evolutionary Considerations

The vertebrate inner ear is phylogenetically ancient. The inner ear of both extinct and extant species of the earliest vertebrates, the Agnatha (jawless fishes), has a variety of anatomical characteristics that were not retained in later vertebrates, and it is uncertain whether extant agnathans can hear (Popper and Fay, Chapter 3). This chapter describes the acoustic circuits of the three classes of jawed anamniotes: the Chondrichthyes (cartilaginous fish), the Osteichthyes (bony fish), and the Amphibia.

The anamniotes constitute well over half of all living vertebrate species, and of the anamniotes, the bony fishes are by far the largest group. As is the case in all vertebrates, the extant species of each class are the descendants of ancestral populations that are now extinct. Any given species will retain some characteristics that were present in these ancestral populations— primitive features—but will have other characteristics that are derived relative to the ancestral condition. Recognizing this is crucial to understanding the organizational variations that appear to be present in the acoustic circuits of anamniotes, and underscores why these circuits must be studied in more than one species within each class.

This chapter first discusses the acoustic circuits of cartilaginous and bony fishes, and then those of the amphibians. Each subsection is preceded by an overview of the acoustic portions of the inner ear. Readers should refer to Popper and Fay (Chapter 3) and Lewis and Narins (Chapter 4) for more extensive discussions of the auditory periphery. Gaps in our knowledge and suggestions for future research are described throughout the chapter. The summary presents a discussion of the neuroanatomy of the auditory system of anamniotes in a broader context, highlighting unresolved anatomical, evolutionary, and functional issues.

1.2 Acoustic Receptors in the Jawed Fishes

The inner ear of the jawed fishes primitively contains seven end organs. The three cristae of the semicircular canals universally function as accelerometers and hence are vestibular. The remaining four end organs—the macula neglecta, saccule, lagena, and utricle—have all been implicated in hearing in various species.

The hair cells of the macula neglecta are covered by a cupula. The function of this end organ is unknown in bony fishes, but in chondrichthyans, the macula neglecta is responsive to vibration or to sound (Corwin 1981). The saccule, lagena, and utricle are each covered by an otolithic membrane and a calcareous otolith. Although the saccule and lagena are commonly said to be the main auditory end organs of fishes, each of the otolithic end organs can potentially encode the particle motion waveform of sound via an inertial mode of sound reception. The saccule, lagena, and utricle of goldfish can all be stimulated by whole-body acceleration, and are directionally sensitive to the axis of particle motion (Fay 1984). In elasmobranchs, vibration/acoustic sensitivity characterizes portions of the utricle and saccule (Lowenstein and Roberts 1951; Budelli and Macadar 1979; Corwin 1981). However, a given otolithic end organ may be bifunctional, that is, it may have gravistatic and auditory roles (Platt 1983). Unfortunately, the distribution of these functions among the three otolithic end organs has rarely been studied in a single species, and what has been learned from the existing studies cannot a priori be generalized to all other members of the class. For example, we do not know whether the particle motion sensitivity of all three of the goldfish otolithic end organs is a primitive or a derived feature, or if the lagena of all chondrichthyans is solely gravistatic. In future studies, the functional characteristics of the otolithic end organs need to be examined in species chosen on the basis of taxonomic status. Collectively, such studies could provide a basis for reasonable speculation on how these functions were initially distributed, and how this compares to otolith end organ functions in later-derived species.

The oscillations of a gas-filled structure, such as the swim bladder, in a sound field provide an "indirect mode" of acoustic stimulation of the inner ear (Fay and Popper 1975), and allow detection of the pressure waveform of sound. Species in which the gas-filled structure is physically coupled to the inner ear are particularly sensitive to sound pressure. Such coupling occurs only in certain species belonging to the most recently derived division of the bony fishes, the teleosts. Examples include mormyrids, otophysans, and clupeids (Section 2.2.2).

Whether or not the lateral line neuromasts respond to sound has been a matter of debate for close to a century. It is possible that the lateral line mechanoreceptors of a fish close to the source of a near-field sound can be

stimulated (Sand 1981; Kalmijn 1988), but whether such stimuli are interpreted in the same manner as sound detected by the inner ear is unknown.
Sound pressure detection by the cephalic lateral line of clupeids is a consequence of the specialized anatomical relationship between the lateral line
and extensions of the swim bladder known as auditory bullae (Allen et al.
1976; Denton and Blaxter 1976). Other specialized anatomical relationships
between the lateral line and the swim bladder or the inner ear have been
described, but their functional meaning has not been explored (Webb and
Blum 1990; Bleckmann et al. 1991). The central pathways of the lateral line
system are largely separate from those of the inner ear, at least below the
level of the forebrain, but there are points of convergence at every level
(reviewed in McCormick 1992). Therefore, relevant information about
mechanosensory lateral line circuits will be provided. For a more complete
discussion, see McCormick (1989) and Will (1989a).

1.3 Criteria for Recognizing the Central Acoustic Circuits of Fishes

Fish do not have an inner ear end organ homologous to the organ of Corti.
Hearing is subserved by some subset of the three otolithic end organs and,
in cartilaginous fish, the macula neglecta. As will become apparent in the
following sections, the central inputs of these end organs are complexly
organized within a series of first-order octaval nuclei, presumably reflecting
the fact that at least some of these end organs have dual functions as both
auditory and vestibular receptors. Fish do not have discrete auditory nuclei,
like the dorsolateral nucleus of frogs and toads (Section 3.4) or the cochlear
nuclei of amniotes. How, then, can the acoustic circuits of fish be
recognized?

One strategy has been to assume that vertebrate auditory circuits are
organized according to a common plan. Within the brain stem, for example,
first-order auditory populations would be expected to have a major projection to a division of the midbrain tectum different from that supplied by the
optic nerve; the mammalian example is the projection of the cochlear nuclei
to the inferior colliculus. Using normal (nonexperimental) cell and fiber
stains, comparative neuroanatomists working in the first half of the 20th
century discovered such connections between the "acousticolateral area" of
the medulla and the torus semicircularis of the midbrain (Ariens Kappers
et al. 1967). Experimental tract-tracing studies that followed refined our
understanding of these connections. A deficiency of this strategy is that it
only allows us to confirm or refute the presence of connections that we
expect based on what we know in other vertebrates. To discover any
additional auditory areas that do not conform to expected patterns of
connectivity, a second strategy—functional analysis—must be employed.

Electrophysiological and other techniques (such as metabolic labeling studies) can discover new acoustic populations, confirm whether or not suspected acoustic areas are in fact auditory, and can uncover the functional characteristics of auditory processing at all levels of the neuraxis. There are relatively few such studies in fish, but those that are available confirm the acoustic nature of some of the structures that have been identified using neuroanatomical techniques.

Therefore, the following description of acoustic circuits in fish is a conservative one. Neuroanatomy, however, does have the power to reveal new avenues to pursue. For example, the anatomical relationship between cells in first- and second-order acoustic populations and the cerebellar crest (Sections 2.2.2 and 2.4.2) may indicate that the cerebellum of fish plays a unique role in modulating the activity of these neurons.

2. The Auditory Pathways of Fishes

2.1 The Octavolateralis Area

The nuclei that receive direct input from the inner ear—the octaval nuclei—are constituents of the octavolateralis area of the medulla (Figs. 5.1, 5.2, and 5.3). The octavolateralis area also includes the first-order nuclei of the mechanosensory lateral line system, which appear to be ubiquitous among jawed fishes, and the variably present nuclei of the electrosensory system. The octavolateralis area is covered dorsally for most of its rostrocaudal extent by a fiber layer, the cerebellar crest. The cerebellar crest is composed primarily of axons arising at least from populations of cerebellar granule cells. The portion of the cerebellar crest that overlies the lateral line mechanosensory and electrosensory nuclei is sometimes referred to as the molecular layer of these nuclei. In some species, the cerebellar crest is also intimately associated with one of the first-order octaval nuclei—the descending nucleus (Section 2.2.1).

2.2 First-Order Octaval Nuclei

Among the nuclei that receive all or most of their primary input from the inner ear, four appear to be universally present in cartilaginous and bony fishes (reviewed in McCormick 1992). These four nuclei are arranged in a roughly rostral to caudal sequence through the medulla as follows: the anterior nucleus, nucleus magnocellularis, the descending nucleus, and the posterior nucleus (Figs. 5.1, 5.2, and 5.3). All of the inner ear end organs, including the cristae of the semicircular canals, project to each of these nuclei, indicating that each octaval nucleus contributes at least in part to vestibular circuits. Therefore, there is no first-order nucleus that is dedicated to acoustic processing in the fishes. However, neurons located within the

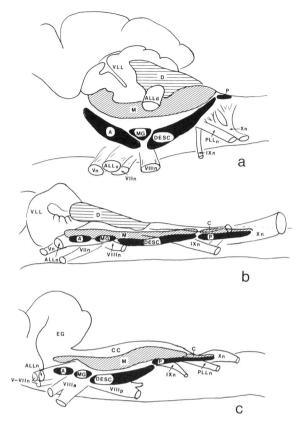

FIGURE 5.1. Lateral views of the hindbrains of (A) the skate *Raja eglanteria* (after Barry 1987), (B) the sturgeon *Scaphyrinchus platorynchus* (after New and Northcutt 1984), and (C) the bowfin *Amia calva* (after McCormick 1981, © John Wiley & Sons, Inc.) showing the nuclei of the octavolateralis area of the medulla. Horizontal lines, electrosensory nucleus dorsalis; diagonal lines, mechanosensory nucleus medialis; darkened areas, octaval nuclei; A, anterior octaval nucleus; ALLn, anterior lateral line nerve; ALLd, dorsal root of the anterior lateral line nerve; ALLv, ventral root of the anterior lateral line nerve; C, nucleus caudalis; CC, cerebellar crest; D, dorsal nucleus; DESC, descending octaval nucleus; EG eminentia granularis; IX n, glossopharyngeal nerve; M, nucleus medialis; MG, nucleus magnocellularis; P, posterior octaval nucleus; PLLn, posterior lateral line nerve; Vn, trigeminal nerve; V-VIIn, trigeminal and facial nerves; VIIn, facial nerve; VIIIn, octaval nerve; VIIIa, anterior ramus of the octaval nerve; VIIIp, posterior ramus of the octaval nerve; VLL, vestibulolateral lobe of the cerebellum; Xn, vagus nerve.

descending nucleus and, in many species, within the anterior nucleus give rise to projections that ascend to the midbrain area known or suspected to be acoustic. These neurons are located primarily in dorsomedial zones of the descending and anterior nuclei that receive otolithic input and, additionally

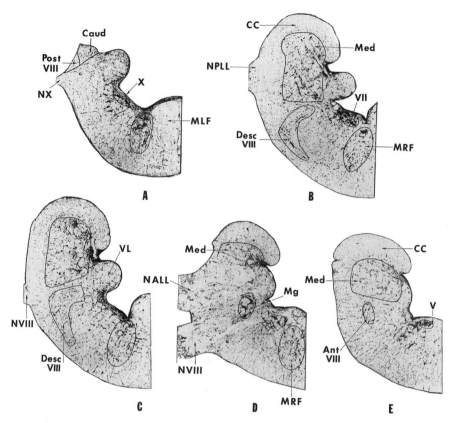

FIGURE 5.2. Transverse cresyl violet–stained sections illustrating the octavolateralis area of the medulla of a primitive ray-finned fish, *Amia calva*. (McCormick 1981, © John Wiley & Sons, Inc.) Levels A through E are in a caudal to rostral order. Ant VIII, anterior octaval nucleus; Caud, nucleus caudalis; CC, cerebellar crest; Desc VIII, descending octaval nucleus; Med, nucleus medialis; Mg, nucleus magnocellularis; MRF, medial reticular formation; NALL, anterior lateral line nerve; NPLL, posterior lateral line nerve; NVIII, octaval nerve; NX, vagus nerve; V, trigeminal motor nucleus; VL, vagal lobe; X; motor nucleus of the vagus nerve.

in chondrichthyans, input from the macula neglecta (Fig. 5.4). Thus, on the basis of their primary inputs and their outputs, dorsal zones within the descending and anterior nuclei likely process acoustic information. It is possible that other regions within these two nuclei, and perhaps others, are also auditory but have not been identified because they do not have direct connections to the auditory midbrain. Such neurons might project to other higher-order acoustic structures, or might have intrinsic connections.

The following sections discuss characteristics of the anterior and descending nuclei in greater detail, and provide information about other first-order populations that may contribute to sound processing.

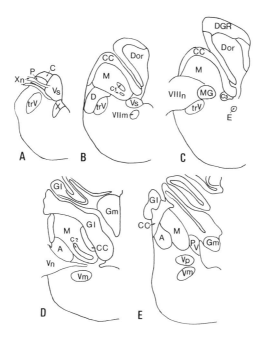

FIGURE 5.3. Transverse sections through the medulla of a cartilaginous fish, *Dasyatis sabina*, illustrating the octavolateralis nuclei. Levels A through E are in a caudal to rostral order. (Redrawn and modified after Puzdrowski and Leonard 1993.) A, anterior octaval nucleus; C, nucleus caudalis; C1, C2, C3, cell plates 1, 2, and 3; CC, cerebellar crest; D, descending octaval nucleus; DGR, dorsal granular ridge; Dor, nucleus dorsalis; E; efferent octaval nucleus; Gl; lateral granular mass; Gm, medial granular mass; M, nucleus medialis; MG, nucleus magnocellularis; P, posterior octaval nucleus; PV, periventricular nucleus; trV, descending trigeminal tract; VIIm, facial motor nucleus; VIIIn, octaval nerve; Vm, trigeminal motor nucleus; Vn, trigeminal nerve; Vp, principal trigeminal nucleus; Vs, vagal lobe; X, motor nucleus of the vagus nerve; Xn, vagus nerve.

2.2.1 Primitive and Derived Organization of the Anterior and Descending Nuclei

It is probable that the primitive condition in jawed fishes is that both the anterior and the descending nuclei give rise to ascending auditory projections, because this is the case in both a chondrichthyan (*Raja eglanteria*; Barry 1987) and a primitive osteichthyan (*Amia calva*; Braford and McCormick 1979). This is also the case in most teleosts that have been studied (Bell 1981a; Echteler 1984; Finger and Tong 1984; Braford et al. 1993). The goldfish *Carassius* is the only species in which there may be no direct projections from the anterior nucleus to the acoustic midbrain (McCormick and Hernandez 1996), although such a projection is present in other otophysans (Echteler 1984; Finger and Tong 1984).

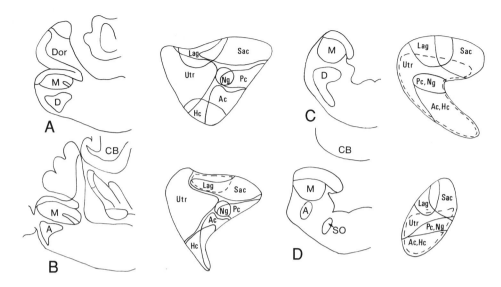

FIGURE 5.4. Transverse hemisections through the medulla of the skate *Raja eglanteria*, a cartilaginous fish (after Barry 1987) and the bowfin *Amia calva*, a primitive bony fish (after McCormick and Braford 1988) showing the termination sites of the inner ear end organs in the descending octaval nucleus (A,C) and the anterior octaval nucleus (B,D). A, anterior octaval nucleus; Ac, terminals of anterior semicircular canal nerve; CB, cerebellum; D, descending octaval nucleus; Dor, nucleus dorsalis; Hc, terminals of horizontal semicircular canal nerve; Lag, terminals of lagenar nerve; M, nucleus medialis; Ng, terminals of macula neglecta nerve; Pc, terminals of posterior semicircular canal nerve; Sac, terminals of saccular nerve; SO, secondary octaval population; Utr, terminals of utricular nerve.

Both the morphology of and input to the dorsal portion of the descending nucleus in *Raja* and *Amia* may reflect the primitive condition for anamniotes (McCormick 1992). In these species, the dorsal portion of the descending nucleus is positioned ventral to nucleus medialis—the main first-order nucleus of the mechanosensory lateral line (see Figs. 5.2B, 5.3B, and 5.7A). Saccular fibers terminate most medially within the dorsal descending nucleus, while lagenar fibers terminate in a laterally adjacent, but overlapping area (Fig. 5.4A,C). Utricular fibers terminate laterally and ventrally in the descending nucleus in *Raja*, partially overlapping the lagenar and semicircular canal terminal areas (Fig. 5.4A; Barry 1987); in *Amia*, utricular fibers terminate ventral to the saccular and lagenar zones, overlapping both of them as well as the projection areas of the semicircular canals (Fig. 5.4C; McCormick 1983a). Although there are minor differences, the overall pattern of inputs is quite similar in the two species. Moreover, neurons that project to the area of the midbrain that is presumed to be acoustic are located primarily in the dorsomedial portion of the descending nucleus (as well as the anterior nucleus) in both species.

In some species of teleosts, the descending nucleus retains all or some of these primitive characteristics (see Fig. 5.7A, E—pattern A). For example, the dorsal portion of the descending nucleus is located ventral to nucleus medialis in *Gillicthyes* (goby; Northcutt 1981), *Crenicichla* (pike cichlid; McCormick 1983b, *Anguilla* (European eel; Meredith et al. 1987), and *Opsanus* (toadfish; Highstein et al. 1992). In some of these species, however, there are differences in the patterns of octaval input. For example, the descending nucleus of *Opsanus* has a unique rostromedial division that receives heavy input from the lagena and the horizontal semicircular canal and lighter input from all other otic end organs.

In contrast, the dorsal portion of the descending nucleus of certain other teleosts has a derived morphology and may in addition have derived inputs. This is known or appears to be the case in species of otophysans (Fig. 5.5A,B; also see Fig. 5.9A; Carr and Matsubara 1981; McCormick and Braford 1993, 1994), clupeids (Fig. 5.5D,E; McCormick 1997), and osteoglossomorphs (Fig. 5.6; McCormick 1992; Braford et al. 1993). In all of these species the descending nucleus has a dorsomedial extension that lies medial to nucleus medialis, and ventral to the most medial portion of the cerebellar crest. Based on its afferent inputs, the dorsomedial extension may in some species result from displacement and possibly hypertrophy primarily of the saccular termination zone, or, in other species, may represent a displacement of a larger portion of the dorsal descending nucleus. This morphological specialization is represented in Figure 5.7 as pattern B, in which the dorsomedial extension is relatively small, and as pattern C, in which the dorsomedial extension is more substantial.

The dorsomedial extension has been referred to as the pars dorsalis of the descending nucleus (Finger and Tong 1984) or the dorsomedial zone of the descending nucleus (McCormick and Braford 1993, 1994) in the otophysans *Ictalurus* and *Carassius*. McCormick and Braford subdivided the remainder of the descending nucleus in these two species into intermediate and ventral zones on morphological and connectional criteria (Fig. 5.5A; also see Fig. 5.9A). The dorsomedial zone and the most medial portion of the intermediate zone receive the bulk of their primary input from the saccule, although the lagena and, particularly at caudal levels, the utricle also project to these areas. In *Carassius*, the dorsomedial zone and the medial portion of the intermediate zone are the only regions of the descending nucleus from which axons to the auditory midbrain originate. Within the intermediate zone, neurons that are lateral to the saccular projection area receive input from the lagena and the utricle. Thus, in otophysans, the dorsomedial and intermediate zones may together represent the dorsal portion of the descending nucleus present in *Amia* and *Raja*.

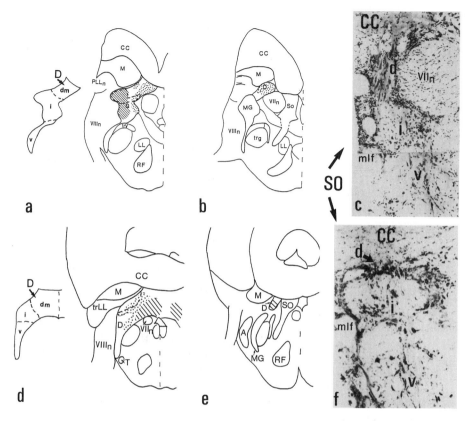

FIGURE 5.5. Summary diagram of the morphological characteristics of the descending octaval nucleus and the secondary octaval population in the goldfish, an otophysan (A,B,C) and the gizzard shad, a clupeid (D,E,F). The dorsomedial (dm), intermediate (i), and ventral (v) zones of the descending nucleus are shown in levels A and D. Photomicrographs C and F are cresyl violet–strained close-ups of the SO showing its dorsal (d), intermediate (i), and ventral (v) subdivisions; they show the SO on the side of the brain contralateral to levels B and E. Floating v's, termination site of the saccular nerve; diagonal lines, termination site of the utricular nerve; area enclosed by dashed line, termination site of the lagenar nerve; A, anterior octaval nucleus; CC, cerebellar crest; D, descending octaval nucleus; LL, lateral lemniscus; M, nucleus medialis; MG, nucleus magnocellularis; mlf, medial longitudinal fasciculus; PLLn, posterior lateral line nerve; RF, reticular formation; SO, secondary octaval population; T, nucleus tangentialis; trg, secondary gustatory tract; tr LL, primary lateral line tracts; trV, descending trigeminal tract; VIIn, facial nerve; VIIIn, octaval nerve.

Among osteoglossomorph fishes, *Pantodon* (butterfly fish) and *Osteoglossum* (arawana) have a descending nucleus that has a small dorsomedial extension in which saccular and lagenar fibers terminate dorsally and utricular fibers terminate more ventrally (Figs. 5.6C,D and 5.7E;

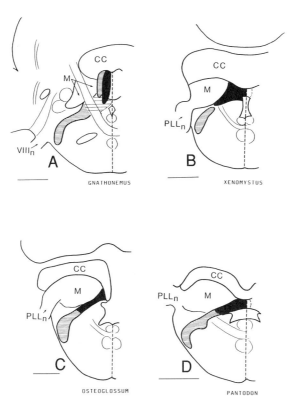

FIGURE 5.6. Hemisections through the medulla of four osteoglossomorphs, (A) the mormyrid *Gnathonemus*, (B) the knifefish *Xenomystus*, (C) the arawana *Osteoglossum*; (D) the butterfly-fish *Pantodon*. In these species, a dorsal or dorsomedial zone of the descending nucleus lies medial to the lateral line column(s). Projection zones of the saccular and lagenar nerves are shown in black, while those of the utricular and semicircular canal nerves are indicated by the horizontal lines. Areas of overlap at the borders are present but not shown. A is based upon the data of Bell (1981b). CC, cerebellar crest; M, nucleus medialis; PLLn, posterior lateral line nerve; VIIIn, eighth nerve.

McCormick 1992). A similarly located cell zone that receives otolithic input is present in *Xenomystus* (African knifefish) and mormyrids such as *Gnathonemus* (elephant-nose fish) and is most likely also a dorsal portion of the descending nucleus (Figs. 5.6A,B and 5.7E; McCormick 1992). The organization of the connections to this cell zone is most completely known in mormyrids, in which it was referred to as the medial portion of the anterior lateral line lobe (Bell 1981b). The inputs to the dorsomedial extension are organized in roughly medial to lateral columns, such that the saccule terminates most medially, lagenar fibers are laterally adjacent to saccular fibers, and utricular fibers are most lateral (Fig. 5.6A). Since all the otolithic end organs project to the dorsomedial extension, it may represent

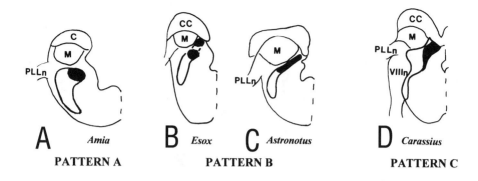

PATTERN A **PATTERN B** **PATTERN C**

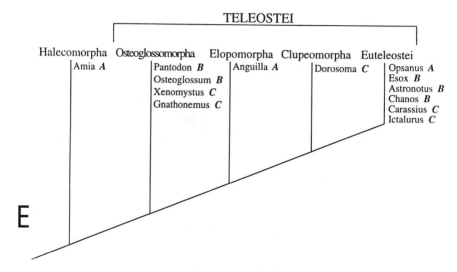

FIGURE 5.7. Different morphologies of the descending octaval nucleus are shown in level a (pattern A), levels b and c (pattern B), and level d (pattern C). The darkened area within the descending nucleus represents the location of neurons that receive otolithic input and that project to the area of the torus semicircularis known or presumed to be acoustic. The cladogram in level e shows the distribution of patterns A, B, and C among ray-finned fishes. Abbreviations as in Figure 5.6.

all or most of the dorsal portion of the descending nucleus. Consistent with this interpretation is the fact that the remainder of the descending nucleus receives input almost exclusively from the utricle and semicircular canal cristae, and thus likely represents the ventral portion of the descending nucleus. A further specialization of the dorsomedial zone of mormyrids and *Xenomystus* is the bilateral projection it receives at least from the saccule; in mormyrids, the lagena provides a bilateral input as well (Bell 1981b; Braford et al. 1993).

Two studies have examined the octavolateralis area of clupeids (Meredith 1985; McCormick 1997). In *Dorosoma cepedianum* (gizzard shad) the descending nucleus is divided into dorsomedial, intermediate, and ventral zones similar to those recognized in otophysans (McCormick 1997). The pattern of projections from the otolithic end organs to the dorsomedial zone is unique in that fibers from the utricle, rather than the saccule, terminate bilaterally upon its most medial cells (Fig. 5.5D). The majority of the saccular input to the dorsomedial zone lies laterally adjacent to that of the utricle, while lagenar fibers supply the most laterally positioned cells in this zone. The dorsomedial location of the utricular terminal field may be related to the fact that in clupeids, the utricle is a sound pressure detector by virtue of its coupling to gas-filled auditory bullae (Allen et al. 1976; Denton and Blaxter 1976). In *Clupea* (herring), Meredith (1985) similarly described a projection of the utricle to the most medial region of the descending nucleus. However, additional utricular afferents terminate bilaterally in the medial part of nucleus medialis. This latter input may possibly be supplying an unrecognized dorsomedial zone of the descending nucleus, like that present in *Dorosoma* (McCormick 1997).

2.2.2 The Dorsomedial Zone of the Descending Nucleus of Teleosts: Functional and Evolutionary Considerations

As described in Section 2.2.1, a specialized dorsomedial zone of the descending nucleus is present in certain teleosts (Fig. 5.7B,C,D). This zone is characterized by its topographic relationships: it lies medial to nucleus medialis, and ventral to the cerebellar crest. In some species the dorsomedial zone is known to project to the presumed acoustic area of the midbrain (as do dorsomedially located neurons in the primitively organized descending nucleus of other species; Fig. 5.7A), but beyond this, the functional significance of the specialized dorsomedial zone is unknown. However, two features of this zone—its prominence in species in which the inner ear is coupled to a gas-filled structure, and the close relationship of some of its neurons to the cerebellar crest—may with further study provide important insights into acoustic processing in these, and perhaps all, species of fish.

The dorsomedial zone was first recognized in species in which the saccule is specialized for sound pressure detection by virtue of its physical connection to the swim bladder—otophysans and certain species of osteoglossomorphs. This led to the initial hypothesis that the dorsomedial zone is a central correlate of such coupling, functionally related to specialized acoustic capacities in such species (McCormick 1989). While this zone was subsequently recognized in a clupeid species (McCormick 1997) in which, along with all other Clupeidae, the utricle is specialized for sound pressure detection, it was also identified in two species of osteoglossomorphs that lack inner ear–swim bladder coupling (McCormick

1992). Thus, the presence of the dorsomedial zone cannot be attributed to the presence of an otophysic ear. It appears however, that the dorsomedial zone is cytoarchitectonically more distinct and is larger in species in which sound pressure stimuli are channeled directly to the inner ear (Fig. 5.7D—pattern C; McCormick 1992).

Preliminary studies indicate that the descending nucleus may have a specialized morphology, relative to that of *Amia*, in many, if not all, teleosts. This was not recognized previously because the most dorsomedial cells of the descending nucleus—those that are proximate to the cerebellar crest—are not always directly contiguous with the cells in the remainder of the descending nucleus (Fig. 5.7B). In some species, the terminal fields of afferents from the otolithic end organs do not overlie the somata of the most dorsomedial cells. Thus, the euteleosts *Esox* (tiger muskie), *Chanos* (milkfish), and *Astronotus* (oscar) appear, on the basis of normal cell stains, to have a descending nucleus morphology like that of *Amia* (Fig. 5.7A; Meredith and Butler 1983). Moreover, the pattern of otolithic end organ inputs to the descending nucleus in these species is like that of *Amia* (Fig. 5.4C,D), again suggesting that a specialized dorsomedial zone is absent. However, additional experimental studies (McCormick, 1998; O'Marra and McCormick, personal observations) indicate that *Esox*, *Chanos*, and *Astronotus* possess a small group of fusiform cells that underlie the most medial portion of the cerebellar crest and that project, along with dorsal neurons in the underlying "descending nucleus proper," to the nucleus centralis of the torus semicircularis (the presumed auditory area; Fig. 5.7B,C). The fusiform cells extend ventral dendrites into the terminal field of at least the saccule and thus likely receive first-order input. Like neurons in the specialized dorsomedial zone of other teleosts, the fusiform cells also extend dorsal dendrites into the cerebellar crest. It is possible that the primitive level of organization for the descending nucleus in teleosts is one in which the majority of its neurons are located ventral to nucleus medialis, as in *Amia*, but in addition includes a small dorsomedial zone in association with the cerebellar crest (Fig. 5.7B,C—pattern B). The hypertrophy of this dorsomedial zone would produce the specialized descending nucleus morphology present in otophysans, osteoglossomorphs, and clupeids (Fig. 5.7D—pattern C). Future comparative studies are needed to test this hypothesis, as well as to determine whether a small dorsomedial zone is also present in nonteleost bony fishes.

Whether the dorsomedial zone is small or large, some of its neurons have a dorsal dendrite that penetrates the overlying cerebellar crest (Bell 1981a; Finger and Tong 1984; McCormick 1992; McCormick and Hernandez 1996; McCormick, 1998; O'Marra and McCormick, personal observations). This raises the possibility that cerebellar crest axons may modulate the activity of these neurons. This has been shown to be the case for similarly located neurons of the electrosensory and mechanosensory lateral line nucleus, which, like the dorsal neurons of the dorsomedial zone, extend a dendrite into the cerebellar crest (reviewed by Montgomery et al. 1995). In addition to addressing this possibility with physiological studies, anatomical studies are required to determine the location·(s) of the somata of the cerebellar crest axons that overlie the dorsomedial zone, whether these somata constitute populations that are distinct from those involved in modulation of the lateral line nuclei, and what the inputs to these somata are.

2.2.3 The Posterior and Magnocellular Nuclei

The functions of the posterior nucleus (Figs. 5.1, 5.2A, and 5.3A) and of nucleus magnocellularis (Figs. 5.1, 5.2D, and 5.3C) are unknown. The posterior nucleus does not project directly to either the spinal cord or to the midbrain. However, the magnocellular nucleus, which receives input from the mechanosensory lateral line as well as from all or many of the inner ear end organs (e.g., McCormick 1981, 1983a,b; Meredith and Butler 1983; New and Northcutt 1984; Highstein et al. 1992), does project directly to the spinal cord (Barry 1987; Prasada-Rao et al. 1987; McCormick personal observations). Such a projection could indicate either its role in vestibular processing, or alternatively its participation in a startle reflex like that driven by the Mauthner cell (Larsell 1967; Barry 1987; Puzdrowski and Leonard 1993). The Mauthner cell is described in Section 2.3.1.

2.2.4 The Periventricular Nucleus and Cell Plate C3 of Chondrichthyans

In the chondrichthyan *Dasyatis* (stingray), two additional areas of the medulla receive inner ear input (Puzdrowski and Leonard 1993): the periventricular nucleus (Fig. 5.3E) and cell plate C3 (Fig. 5.3C). It is not known which otic end organs project to the periventricular nucleus in *Dasyatis*; the octaval input to this nucleus as well as the input from the anterior lateral line are light. A periventricular nucleus that receives input from all otic end organs is present in the skate *Raja* (Barry 1987), but it is unclear whether it is the same nucleus as that described in *Dasyatis* (Puzdrowski and Leonard 1993). Barry (1987) speculates that the periventricular nucleus in *Raja* is acoustic based on a 2-deoxyglucose study in another chondrichthyan, the guitarfish *Platyrhinoides* (Corwin and Northcutt 1982).

Cell plate C3 is one of three aggregations of neurons—C1 (Fig. 5.3B), C2 (Fig. 5.3D), and C3 (Fig. 5.3C)—associated with the octavolateralis area in chondrichthyans (Smeets et al. 1983; Puzdrowski and Leonard 1993). C3 is a collection of spindle and stellate neurons that lie ventral to a medial region of the cerebellar crest in *Dasyatis* and *Raja* (Puzdrowski and Leonard 1993). In *Dasyatis*, very few octaval fibers, and no lateral line fibers, reach C3, although C3 projects to the mechanosensory area of the midbrain (Puzdrowski and Leonard 1991, 1993). C3 may be therefore be a higher-order population associated with the auditory system, lateral line, or both (Puzdrowski and Leonard 1993). Cell plates C1 and C2, also referred to as nucleus X (Northcutt 1978) were hypothesized to be a higher-order acoustic population, such as the superior olive (Corwin and Northcutt 1982). However, C1 and C2 receive input from mechanosensory anterior lateral line fibers but not the eighth nerve (Bodznick and Schmidt 1984; Puzdrowski and Leonard 1993). Therefore, whether C1 and C2 are acoustic or mechanosensory, or are integration areas is uncertain.

2.2.5 The Nucleus Tangentialis of Teleosts

Among fishes, a tangential nucleus (Fig. 5.5D; also see Fig. 5.9A) is present only in the teleosts. The tangential nucleus is likely vestibular as most of its primary input arises from the three semicircular canals; in some species, the utricle may also project here (Bell 1981b; Meredith and Butler 1983; Highstein et al. 1992). Its known outputs to the spinal cord and oculomotor nucleus (Bell 1981a; Torres et al. 1992) likewise suggest a vestibular role.

2.3 Inner Ear Projections to Other Brain Stem Areas

In addition to the octaval nuclei, the octaval nerve projects to several other hindbrain structures: the Mauthner cell, the reticular formation, the nucleus medialis, and the cerebellum.

2.3.1 The Mauthner Cell and the Reticular Formation

The Mauthner (M) cell may be involved in acoustic detection and localization. Under certain circumstances a sound stimulus, and perhaps other hydrodynamic stimuli (Eaton and Popper 1995) to the M cell initiates a fast startle response called the c-start. This response initially orients the fish away from the stimulus and is thought to help the fish escape predation. The M cell is theorized to determine sound direction by comparing particle motion inputs arising from the otolithic end organs and pressure input arising mainly from the saccule (Eaton et al. 1995; Fay 1995). The sources of eighth input to the M cell are incompletely known. In the goldfish (Zottoli et al. 1995) and possibly in other species (Meredith and Butler 1983), fibers from the saccule and utricle synapse directly on the lateral dendrite of the M cell, but whether the lagena also contributes an input has not been investigated. Other, indirect inputs to the goldfish M cell arise from the inner ear, the trunk neuromasts, the optic tectum, and the somatosensory system (Zottoli and van Horne 1983; reviewed in Zottoli et al. 1995). The relatively few studies in non-otophysans indicate or suggest that the M cell varies in structure and function (reviewed in Zottoli et al. 1995). The complement of otolithic end organs that project to the M cell may also vary, particularly among species that have an inner ear coupled to a gas-filed structure versus species that lack such coupling (Popper and Edds-Walton 1995). Studies done in a comparative framework are needed.

Eaton and colleagues (1991) review evidence that the c-start is controlled by a network of reticulospinal neurons that includes the Mauthner cell, rather than by the Mauthner cell alone. In the absence of the Mauthner cell, these reticulospinal neurons can effect the c-start, apparently in response to acoustic stimuli. Octaval nerve projections to the reticular formation have been reported in many species, although the specific sources of the input are known only for *Astronotus* (utricle and semicircular canals; Meredith and Butler 1983) and *Opsanus* (utricle, lagena, semicircular canals; Highstein et al. 1992). These is no evidence concerning the modality of the otolithic input in either of these species. Direct input from inner ear acoustic receptors to reticular neurons is therefore a possibility that should be explored.

2.3.2 Nucleus Medialis

Nucleus medialis (Figs. 5.1, 5.2B,C,D,E, 5.3B,C,D, and 5.5; also see Fig. 5.9A,B) is the main termination site of mechanosensory lateral line fibers (reviewed in McCormick 1989). The octaval nerve also projects to nucleus medialis in most species, although the intensity of such projections appears to vary. In chondrichthyans and certain teleosts (Barry 1987; Dunn and Koester 1987; Meredith et al. 1987; Puzdrowski and Leonard 1993), the inner ear may have more substantial projections to nucleus medialis than those reported in *Amia* and in other teleosts (McCormick 1981, 1997; Meredith and Butler 1983; Finger and Tong 1984; McCormick and Braford 1993, 1994). In some teleosts, there appears to be no octaval input to nucleus medialis (Northcutt 1981; Highstein et al. 1992). Interestingly, the scant octaval input to nucleus medialis in *Amia* and the heavier octaval input in *Raja* (a chondricthyan) and *Anguilla* (a teleost) arise from the semicircular canal cristae as well as from the otolithic end organs, suggesting that integration within nucleus medialis may involve either vestibular information, or both vestibular and auditory information. Overlap in lateral line and octaval connections also occurs in other areas within the medulla. Lateral line mechanosensory projections to nucleus magnocellularis, a probable vestibular area, were mentioned previously (Section 2.2.3) Lateral line input also reaches auditory areas. The dorsomedial zone of the descending nucleus of otophysans (Puzdrowski 1989; New and Singh 1994) and clupeids (Meredith 1985; McCormick 1997) receives light input from one or both of the lateral line nerves. The significance of overlap among lateral line, auditory, and vestibular inputs at the level of the medulla is unknown.

2.3.3 Cerebellum

First-order octaval fibers, usually originating from each of the inner ear end organs, reach granule cell populations and, in some species, other neurons, within the vestibulolateral lobe of the cerebellum (Fig. 5.1) in cartilaginous and bony fishes. These granule cell populations—the main octaval nerve terminal areas—are the medial and lateral granular masses in chondrichthyans (Fig. 5.3D,E), and the eminentia granularis in osteichthyans (Fig. 5.5E).

In some species, the octaval nerve projects diffusely throughout the vestibulolateral lobe, presumably overlapping the termination sites of first-order mechanosensory lateral line fibers, whereas in other species, octaval and lateral line inputs are more segregated. The locations of the terminal zones of fibers from individual otic end organs have not been studied in detail. However, it appears that not all parts of the inner ear project to the vestibulocerebellum with equal intensity. For example, in *Raja* and in certain teleosts, saccular projections to the cerebellum are minimal compared to the projections from other otic end organs (Bell 1981b; Barry 1987; Meredith and Butler 1983; Highstein et al. 1992).

It is generally assumed that first-order octaval fibers provide the cerebellum with vestibular information. It is unknown whether first-order auditory input also reaches the cerebellum. However, acoustic processing of some type likely occurs; Echteler (1985) recorded acoustic responses from the eminentia granularis in the carp. One rationale for investigating the possibility that the vestibulolateral lobe contributes to acoustic processing is the fact that its granule cells are a major source of the axons that form the cerebellar crest. It is known that these axons influence the processing of mechanosensory input by nucleus medialis (Montgomery et al. 1995). Similarly, any acoustic neurons in primary (or higher-order, Section 2.4.2) nuclei upon which cerebellar crest axons terminate would likewise be modulated by the vestibulolateral lobe; neurons in the dorsomedial zone of the descending nucleus are an obvious candidate.

2.4 Higher-Order Acoustic Areas in the Medulla and Midbrain

The efferent connections of the first-order acoustic areas of the medulla are incompletely known. Current information has been derived primarily from studies in which neuronal tract-tracers have been introduced into the midbrain or into a limited number of areas in the medulla. These and other studies collectively indicate that axons from dorsal regions of the descending and anterior nuclei and a secondary octaval population form a lateral lemniscus that innervates one or more midbrain structures. A population located at the level of the isthmus may also contribute to this projection.

2.4.1 The Octavolateralis Area of the Midbrain: Organization and Hindbrain Inputs

The midbrain area that receives input from first-order acoustic populations is also the terminus of axons from first-order mechanosensory lateral line and electrosensory populations. An ascending lemniscal pathway originates from nuclei associated with each of these divisions of the octavolateralis system. In many species, each of these lemnisci terminates in a relatively modality-specific region or nucleus of the midbrain. In others, the degree of segregation is unclear. The presence of modality-specific areas does not rule out the existence of bi- and multimodal areas. For example, the octavolateralis area of the midbrain of the trout contains regions where acoustic and lateral line units overlap. Electrophysiological studies have also categorized a variety of bimodal units, for example visual-auditory and visual-lateral line (Schellart 1983; Schellart and Kroese 1989).

In chondrichthyans the octavolateralis area of the midbrain is called the lateral mesencephalic complex (LMC), whereas in most osteichthyans it is called the torus semicircularis (TS). The LMC and TS are believed to be homologous to one another as well as to the octavolateralis midbrain area of amphibians (TS), reptiles (TS), birds (nucleus mesencephalicus lateralis dorsalis), and mammals (inferior colliculus) (Wilczynski 1988). In each case, the octavolateralis area of the midbrain is further subdivided into regions or nuclei.

The LMC of chondrichthyans varies in its level of differentiation. The boundaries of its constitutent nuclei are unclear in the (primitive) sharks *Squalus* and *Cephalloscyllum*, hence the degree to which acoustic, lateral line, and electrosensory areas are segregated has been difficult to determine (Boord and Northcutt 1988; Boord and Montgomery 1989). In at least some skates and rays, however, there are four distinct nuclei within the LMC: lateral, mediodorsal, medioventral, and anterior (Fig. 5.8A; Boord and Northcutt 1982). In *Raja*, the lateral and mediodorsal nuclei are terminal areas of the electrosensory and lateral line mechanosensory lemnisci, respectively (Boord and Northcutt 1982). It is not definitively known which modalities are processed in the medioventral nucleus. In *Raja*, bilateral input to the medioventral nucleus may arise from portions of the descending and anterior nuclei that are supplied by the saccule, utricle, and macula neglecta (Barry 1987). This finding, as well as the results of a physiological and metabolic study in the guitarfish *Platyrhinoides* (Corwin and Northcutt 1982), suggests that the medioventral nucleus is an acoustic structure. However, it is also possible that the medioventral nucleus receives input from nucleus medialis, cell plates C1 and C2 of the medulla (nucleus X of Corwin and Northcutt 1982), and from the nucleus of the lateral lemniscus (Barry 1987). There is contradictory or inconclusive information concerning whether these latter nuclei process acoustic information, lateral line information, or both (Boord and Northcutt 1982; Corwin and Northcutt 1982; Bodznick and Schmidt 1984; Puzdrowski and Leonard 1993). Thus, further investigations are needed to establish the existance of a discrete acoustic population within the LMC of chondrichthyans.

Detailed studies of the TS—the midbrain octavolateralis area in osteichthyans—are available for only a few teleosts. Teleosts that possess

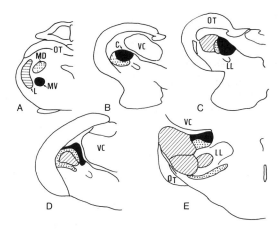

FIGURE 5.8. Functional zones within the lateral mesencephalic complex of the chondrichthyan *Raja* (A) and the torus semicircularis of various ray-finned fishes: the otophysan *Carassius* (B), the osteoglossomorph *Xenomystus* (C), and the mormyrid *Gnathonemus* (D). Dots, known or presumed mechanosensory lateral line area; darkened area, known or presumed acoustic area; horizontal lines, primitive electrosensory midbrain area; diagonal lines, various independently evolved electrosensory midbrain areas; C, nucleus centralis; L, lateral nucleus; LL, lateral lemniscus; MD, mediodorsal nucleus; MV, medioventral nucleus; OT, optic tectum; VC, valvula cerebelli.

an electrosensory system have a region within the TS dedicated to such processing (Bullock and Heiligenberg 1986). The location of this region relative to those of the lateral line mechanosensory and acoustic systems varies (Fig. 5.8B,C,D), as does its nomenclature. In contrast, the mechanosensory lateral line region—nucleus ventralis or nucleus ventrolateralis—is positioned lateral or ventrolateral to the main acoustic region—nucleus centralis (Fig. 5.8B,C,D). In mormyrids (Osteoglossomorpha), nucleus centralis and all or a part of nucleus ventrolateralis appear to be included within a larger structure, the mediodorsal nucleus (Fig. 5.8D; Bell 1981a; Haugede-Carre 1983; Crawford 1993). The posteroventral nucleus is a second lateral line mechanosensory region in mormyrids. Similarly, the mechanosensory region of the TS of another osteoglossomorph, *Xenomystus*, has two separate subdivisions (Fig. 5.8C; Braford 1982, 1986). In otophysans, an additional acoustic area, the medial pretoral nucleus (Fig. 5.9C), is also present (see below).

A similar subset of nuclei or neuronal populations are known or suspected to project to nucleus centralis (or to the equivalent cells in the mediodorsal nucleus of mormyrids) in species of osteoglossomorphs—*Gnathonemus* and *Xenomystus*—and otophysans—*Cyprinus, Ictalurus*, and *Carassius* (Bell 1981a; Echteler 1984; Finger and Tong 1984; Braford et al. 1993; McCormick and Hernandez 1996). As in *Raja* and *Amia*, two of these inputs arise from first-order acoustic populations in the descending and anterior nuclei (Figs. 5.10 and 5.11; Section 2.2.1); a possible exception is the apparent absence of such an input from the anterior nucleus in *Carassius*. Within these nuclei, dorsal or dorsomedially located neurons that receive input from the otolithic end organs project bilaterally, with contralateral predominance, to nucleus centralis. The saccule provides the predominant input to the dorsal areas from which the acoustic lemniscus originates, which is consistent with the saccule's specialized capacity for sound pressure detection in these species.

Other inputs to nucleus centralis arise from higher-order structures—the secondary octaval population and the paralemniscal nucleus—that are presumably acoustic in whole or in part (Figs. 5.10 and 5.11). However, functional studies of the modalities processed by these structures are lacking. The secondary octaval population/superior olive (SO) is located in the medulla. The inputs to the SO have not been fully studied, but in otophysans they appear to include the dorsal descending nucleus (Finger and Tong 1984; McCormick and Hernandez 1996). The SO is known to project bilaterally to the nucleus centralis or to its homologue in a number of species (Braford and McCormick 1979; Bell 1981a; Echteler 1984; Finger and Tong 1984; Braford et al. 1993; McCormick and Hernandez 1996). Other inputs to the SO, its structure, and its distribution among fishes are discussed in Section 2.4.2. The paralemniscal, or isthmoreticular, nucleus is located at the level of the isthmus within and surrounding the lateral lemniscus (Braford and McCormick 1979; Bell 1981a; Echteler 1984; Finger and Tong 1984; McCormick and Hernandez 1996). This nucleus projects bilaterally to nucleus centralis. Its inputs are unknown.

In mormyrids, one or more populations within the cerebellum appear to project to the mediodorsal nucleus (Bell 1981a), but it is unclear whether these projections terminate in its acoustic portion, its lateral line mechanosensory portion, or both.

Neurons within nucleus centralis/the mediodorsal nucleus participate in commissural connections in some species (Bell 1981a; Echteler 1984; Finger and Tong 1984; Braford et al. 1993). In mormyrids and otophysans, these connections appear to be organized homotopically. The functional significance of these connections is unknown. Reciprocal connections between the torus and the optic tectum have also been observed in many species (Bell 1981a; Grover and Sharma 1981; Luiten 1981; Echteler 1984; Finger and Tong 1984; McCormick and Hernandez 1996). In some cases, these projections involve both nucleus centralis and the ventrolateral nucleus. Toral inputs to the optic tectum may contribute to various orientation responses (Echteler 1984).

As mentioned above, the medial pretoral nucleus (Fig. 5.9C) is an additional acoustic structure in the midbrain of otophysans. It receives input from nucleus centralis (Fig. 5.10; Finger and Tong 1984; Striedter 1991, 1992) and from the central posterior nucleus of the diencephalon (Striedter

1991; Section 2.5.1), and gives rise to fibers that descend to the secondary octaval population and possibly the dorsal portion of the descending nucleus (Finger and Tong 1984; Section 2.4.2). The medial pretoral nucleus varies in size among otophysans; relative to the catfish *Ictalurus*, for example, the medial pretoral nucleus is small in the goldfish and minuscule in the gymnotid *Apteronotus* (Striedter 1991). Additional studies are

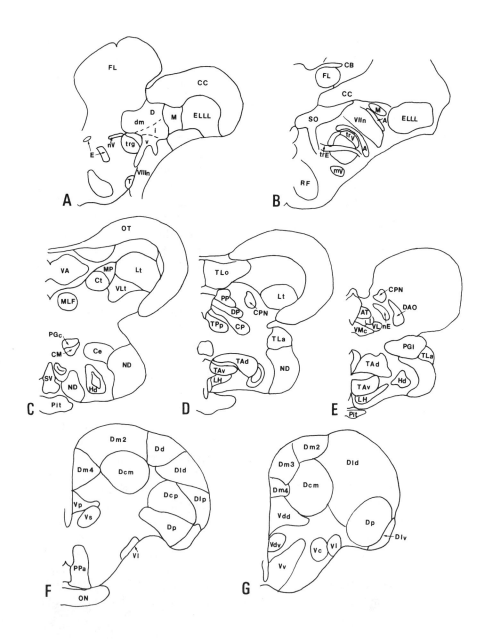

required to determine whether the medial pretoral nucleus is present in non-otophysans.

2.4.2 The Secondary Octaval Population/Superior Olive (SO) of Osteichthyes

As mentioned above, a number of cell populations in the hindbrain potentially contribute to higher-order acoustic processing. These include cell plates C1–C3 and the nucleus of the lateral lemnsicus of chondrichthyans, and the secondary octaval population (SO) and paralemniscal nucleus of osteichthyans (Section 2.4.1). All these populations require further study to determine their connections and/or functional characteristics; some may be

◄──

FIGURE 5.9. Caudal (A) through rostral (G) hemisections through the brain of the catfish, *Ictalurus punctatus*. (Levels A and B are after McCormick and Braford 1993; levels C–G are redrawn and modified from Striedter 1991.) A, anterior octaval nucleus; AT, anterior thalamic nucleus; CB, cerebellum; CC, cerebellar crest; Ce, central nucleus of the inferior lobe; CM, mammillary body of preglomerular complex; CP, central posterior nucleus; CPN, central pretectal nucleus; Ct, nucleus centralis of torus semicircularis; D, descending octaval nucleus; DAO, dorsal accessory optic nucleus; Dcm, central nucleus of area dorsalis, medial division; Dcp, central nucleus of area dorsalis, posterior division; Dld, laterodorsal nucleus of area dorsalis; Dlp, lateroposterior nucleus of area dorsalis; Dlv, ventrolateral nucleus of area dorsalis; dm, dorsomedial zone of descending octaval nucleus; Dm 2, Dm 3, Dm 4, medial nucleus of area dorsalis, subdivisions 2 through 4; DP, dorsal posterior thalamic nucleus; Dp, posterior nucleus of area dorsalis; E, efferent lateral line and octaval nuclei; ELLL, electrosensory lateral line lobe; FL, facial lobe; Hd, dorsal periventricular hypothalamic nucleus; I, intermediate nucleus; i, intermediate zone of descending octaval nucleus; LH, lateral nucleus of the hypothalamus; Lt, lateral nucleus of the torus semicircularis; M, nucleus medialis; MLF, medial longitudinal fasciculus; MP, medial pretoral nucleus; mV, trigeminal motor nucleus; ND, nucleus diffusus of the inferior lobe; nE, nucleus electrosensorius; nV, descending trigeminal nucleus; ON, optic nerve; OT, optic tectum; PGc, commissural nucleus of preglomerular complex; PGl, lateral nucleus of preglomerular complex; Pit, pituitary gland; PP, periventricular pretectum; PPa, anterior parvocellular preoptic area; RF, reticular formation; SO, secondary octaval population; SV, saccus vasculosus; T, nucleus tangentialis; TAd, dorsal subdivision of anterior tuberal nucleus; TAv, ventral subdivision of anterior tuberal nucleus; TLa, torus lateralis; TLo, torus longitudinalis; TPp, periventricular nucleus of posterior tuberculum; trE, efferent octavolateralis tract; trg, secondary gustatory tract; trV, descending trigeminal tract; v, ventral zone of descending octaval nucleus; VA, valvula cerebelli; Vc, central nucleus of area ventralis; Vdd, dorsal part of dorsal nucleus of area ventralis; Vdv, ventral part of dorsal nucleus of area ventralis; VIIn, facial nerve; VIIIn, octaval nerve; Vl, lateral nucleus of area ventralis; VL, ventrolateral nucleus of ventral thalamus; VLt, ventrolateral nucleus of the torus semicircularis; VMc, caudal subdivision of ventromedial nucleus; Vp, posterior nucleus of area ventralis; Vs supracommissural nucleus of area ventralis.

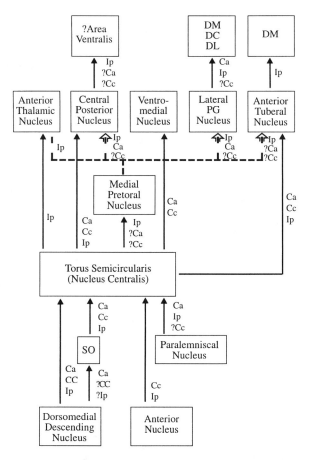

FIGURE 5.10. Simplified summary of possible acoustic circuits in three otophysans: *Carassius auratus* (Ca), *Cyprinus carpio* (Cc), and *Ictalurus punctatus* (Ip). A question mark before a species initials indicates that the projection has not been investigated; the question mark before Area Ventralis indicates that this projection requires further study. The laterality of the connections is not indicated. DC, area dorsalis pars centralis; DL, area dorsalis pars lateralis; DM, area dorsalis pars medialis, PG, preglomerular; SO, secondary octaval nucleus.

found to be components of the lateral line mechanosensory circuits rather than, or in addition to, those of the acoustic system. Among these populations, however, the secondary octaval population will be discussed further because its connectivity and organizational characteristics strongly suggest an acoustic role.

From one to three regions have been identified within the SO. The first region that was identified contains the ventral, fusiform cells described below. These neurons were hypothesized to be comparable to the SO of land vertebrates on the basis of their location (Fig. 5.4D) and their output

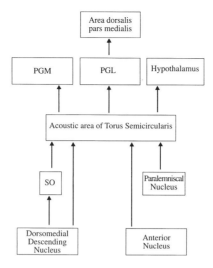

FIGURE 5.11. Simplified summary of possible acoustic circuits in an osteo-glossomorph, *Xenomystus nigri*. The laterality of the connections is not indicated. SO, secondary octaval nucleus; PGM, medial preglomerular nucleus; PGL, lateral preglomerular nucleus.

to the torus semicircularis, and hence are labeled SO in many studies (e.g., Braford and McCormick 1979; Bell 1981a; Echteler 1984; McCormick 1989, 1992; McCormick and Braford 1993, 1994). However, there is insufficient evidence that either the ventral fusiform neurons, or any other higher-order acoustic neurons in bony or cartilaginous fishes, are homologous or functionally analogous to the SO of tetrapods. The term *secondary octaval population* was therefore introduced to avoid implications about the evolutionary relationship or function of this group of neurons (McCormick and Hernandez 1996).

Most information about the SO comes from studies of otophysans. The otophysan SO has dorsal, intermediate, and ventral regions (Fig. 5.5B,C; McCormick and Hernandez 1996). Fusiform cells with a marked vertical orientation constitute the dorsal region. These cells have a dorsal dendrite that penetrates the cerebellar crest and a ventral dendrite that is a constituent of a neuropil ventral to the fusiform somata. The intermediate region of the SO contains spherical cells that lie largely ventral to the neuropil. Fusiform cells make up the SO's ventral region, most of which is located ventral to the internal arcuate tract. The ventral fusiform neurons were referred to as the SO in the carp *Cyprinus*, a species closely related to the goldfish (Echteler 1984). A previous study in the catfish *Ictalurus* (Finger and Tong 1984) described a medial auditory nucleus that includes the dorsal and spherical regions of the SO as well as a portion of the dorsomedial zone of the descending nucleus (McCormick and Hernandez 1996). At least some of the perilemniscal reticular cells identified by Finger and Tong (1984) constitute the ventral, fusiform cell region of the SO.

The connections of the otophysan SO have not been systematically studied, but certain aspects of its circuitry are known (Finger and Tong 1984; McCormick and Hernandez 1996). It is presumably a second-order acoustic population because it receives bilateral input from dorsomedial regions of the descending nucleus. In the goldfish, the SO projects bilaterally to nucleus centralis of the torus semicircularis but not likely to the ventrolateral (mechanosensory lateral line) nucleus (Fig. 5.10). However, in the catfish, some perilemniscal reticular neurons receive input from nucleus medialis and project to nucleus ventrolateralis (Finger and Tong 1984). It is unclear whether these perilemniscal neurons belong to the SO, or whether they are a topographically separate population; there is therefore a possibility that a portion of the SO is either mechanosensory or is an acoustic-lateral line integration center. In the goldfish, there is preliminary evidence that some of the spherical and ventral fusiform cells of the SO may be involved in descending acoustic pathways (McCormick and Hernandez 1996). Nucleus centralis projects bilaterally to some of the spherical cells, and some spherical and ventral fusiform cells project ipsilaterally to dorsomedial regions of the descending nucleus. In the catfish, descending axons from three locations project to the medial auditory nucleus and may therefore be contacting the portion of this structure that is homologous to the SO of the goldfish: nucleus centralis, the medial pretoral nucleus, and granule cells associated with the medial auditory nucleus (Finger and Tong 1984). Finally, the cerebellar crest may provide input to the SO via the dorsal dendrites of the dorsal fusiform cells. Study of the origins and functional significance of these inputs can the potentially contribute to our understanding of how sound is processed at lower levels of the neuraxis.

At least two of the three regions of the SO have been identified in other species of bony fishes. In *Porichthyes*, the midshipman, at least one dorsal region is present in addition to the ventral-most SO population (Bass et al. 1994). The milkfish *Chanos*, which is a member of the group most closely related to otophysans, has all three regions of the SO (McCormick, 1998). All three regions appear to be present in the clupeid *Dorosoma* (Fig. 5.5E,F; McCormick 1997), but experimental studies are needed to confirm that this structure projects to the auditory midbrain. Studies in the osteoglossomorphs *Xenomystus* and *Gnathonemus* have identified the ventral fusiform region (Bell 1981a; Braford et al. 1993) and the dorsal fusiform region (Braford et al. 1993; McCormick, unpublished observations), but the spherical cell region may be absent. A studies in the primitive euteleost *Esox* has likewise identified only the dorsal and ventral fusiform regions (McCormick, 1998). Taken together, these studies suggest that the SO of teleosts is composed of at least the dorsal and ventral fusiform cells. Only the ventral fusiform cells of the SO have been identified in the bowfin *Amia*

(Fig. 5.4D; Braford and McCormick 1979; McCormick 1989). *Amia*, as well as other nonteleosts, should be reexamined to determine whether or not the dorsal fusiform cells of the SO are present, as may generally be the case in teleosts.

2.5 Acoustic Areas of the Forebrain

The diencephalic and telencephalic auditory areas are incompletely known. In chondrichthyans there are no anatomical studies of forebrain auditory pathways at all, although evoked potentials to sound (as well as lateral line and electrosensory stimuli) have been recorded from the telencephalon (Bullock and Corwin 1979). Mechanosensory lateral line pathways from the midbrain have been traced to two diencephalic nuclei: the posterior central and lateral posterior nuclei (Boord and Montgomery 1989).

In bony fishes, comparative data on auditory areas of the forebrain are very limited. Connectivity patterns may vary among taxa. Pooling information from all studies in bony fishes, nuclei in each of the four regions of the diencephalon receive input from the acoustic midbrain in some or all species examined to date: (1) the dorsal thalamus—central posterior and anterior nuclei, (2) the ventral thalamus—ventromedial nucleus, (3) the posterior tuberculum (preglomerular complex), and (4) the hypothalamus. Each of these nuclei/regions receives a complex array of additional sensory inputs, and it is unclear how these nonacoustic inputs influence auditory processing.

2.5.1 Dorsal Thalamus: Connections of the Central Posterior and Anterior Nuclei

The central posterior nucleus receives input from the torus semicircularis in two nonteleosts—*Amia* and *Lepisosteus* (Braford and McCormick 1979; Striedter 1991)—but the modalities being conveyed are unknown. There is more detailed information on the connections of the torus to the dorsal thalamus in otophysans (Fig. 5.10), as is discussed below. Based on studies in mormyrids and *Xenomystus*, the torus of osteoglossomorphs may lack connections with the dorsal thalamus (Fig. 5.11); further studies are needed to confirm this possibility (Braford et al. 1993; reviewed in Braford and McCormick 1993).

The central posterior nucleus of the dorsal thalamus is a major target of the auditory regions of the torus semicircularis in otophysans (Fig. 5.9D). This nucleus receives a substantial ipsilateral input from nucleus centralis in the carp (Echteler 1984), catfish (Striedter 1991), and goldfish (Rosenberger and McCormick, unpublished observations), and is acoustically responsive (Echteler 1985). Magnocellular neurons in the lateral portion of the central posterior nucleus project back to nucleus centralis in all of the above-mentioned species. At least in the catfish, the medial pretoral nucleus (discussed above) and the central posterior nucleus are also reciprocally connected (Striedter 1991).

In addition to its prominant acoustic input from nucleus centralis and the medial pretoral nucleus, the central posterior nucleus receives other projections. Not all of the latter inputs are functionally categorized. For example, the central posterior nucleus of catfish is reciprocally connected with the anterior tuberal nucleus (Fig. 5.9D,E) of the hypothalamus (Striedter 1991). Because the anterior tuberal nucleus receives mechanosensory as well as auditory efferents from the torus semicircularis, Striedter (1991) suggests that the central posterior nucleus may not be wholly acoustic. Echteler's (1985) electrophysiological study, however, found no evidence for mechanosensory lateral line processing in the central posterior nucleus of the carp. An electrosensory input to the central posterior nucleus originates from the nucleus electrosensorius in gymnotids (Keller et al. 1990).

The central posterior nucleus has interconnections with additional midbrain and diencephalic structures. Striedter (1990) reported that in catfish, a discrete ventromedial area of central posterior nucleus projects to the optic tectum. Interestingly, there is variation among otophysans in connections between the central posterior nucleus and the lateral preglomerular nucleus (Fig. 5.9E; Striedter 1992), another recipient of toral efferents (see Section 2.5.3).

The central posterior nucleus projects to the telencephalon at least in catfish, but the location of the projection is not precisely known. Labeling studies suggest that the area ventralis of the telencephalon is one possible target (Striedter 1991). Area ventralis is in the subpallium (Section 2.5.5); its dorsal portion may be homologous to the striatum (Nieuwenhuys 1963, Braford 1995).

The catfish is the only species in which connections from nucleus centralis and the medial pretoral nucleus to the anterior nucleus of the dorsal thalamus (Fig. 5.9E) have been reported (Fig. 5.10; Striedter 1991). These connections are ipsilateral. The anterior nucleus also receives input from the retina and has a substantial output to the optic tectum (Striedter 1990). In the goldfish and the pacu, the anterior nucleus projects to the lateral preglomerular nucleus (Section 2.5.3) but surprisingly this projection is absent in catfish and gymnotids.

2.5.2 Ventral Thalamus: Connections of the Ventromedial Nucleus

Connections between the auditory midbrain and the ventromedial nucleus (Fig. 5.9E) are variably present among otophysans. Echteler (1984) reported a sparse ipsilateral projection of nucleus centralis to the ventromedial nucleus in the carp. Although this projection is present in the goldfish (Rosenberger and McCormick, personal observations), it is apparently absent in the catfish (Striedter 1991). However, the ventromedial nucleus of catfish projects to both nucleus centralis and to the medial pretoral nucleus (Striedter 1991). It is not known whether this projection is present in other otophysans.

Striedter (1992) demonstrated in species from four otophysan subgroups that the ventromedial nucleus projects to the lateral preglomerular nucleus. As is discussed below, the lateral preglomerular nucleus has complex connections with a number of nuclei within the octavolateralis system.

In the scorpaenid *Sebasticus*, the torus as a whole projects to the "nucleus ventromedialis thalami" of Schnitzlein (Murukami et al. 1986); the latter is a composite structure that probably includes the ventromedial nucleus recognized in otophysans and other species (Braford 1995).

2.5.3 Posterior Tuberculum: Connections of the Preglomerular Complex

The posterior tuberculum of bony fishes is a complex and poorly understood diencephalic region. It contains a periventricular area that may be present in land vertebrates (Braford and Northcutt 1983; Neary and Northcutt 1983) and uniquely in ray-finned fishes, a well-developed group of migrated nuclei that are difficult to compare to structures in land vertebrates. There is uncertainty as to whether all of the migrated nuclei are actually developmentally related to the posterior tuberculum, instead having migrated to this area secondarily after originating in some other diencephalic region (Braford 1995). Whatever their origins, the migrated nuclei are the major diencephalic source of input to the telencephalon. At least one of these groups, the lateral preglomerular nucleus of the preglomerular complex (Fig. 5.9E), is involved in acoustic processing in at least some species.

The lateral preglomerular nucleus is present in both nonteleosts (*Amia*) and various teleosts. In *Amia*, the torus semicircularis projects ipsilaterally to the lateral preglomerular nucleus (as well as to other divisions of the preglomerular complex), but it is not known whether this input is carrying auditory, mechanosensory, or both classes of input (Braford and McCormick 1979). There is more specific information in other species, however.

In the osteoglossomorph *Xenomystus*, toral input to the lateral preglomerular nucleus apparently originates only from its acoustic portion (Fig. 5.11; Braford et al. 1993). In mormyrids, the mediodorsal nucleus of the torus semicircularis, which has auditory and mechanosensory divisions (Section 2.4.1), also projects heavily to the lateral preglomerular nucleus (Bell 1981a), but whether one or both modalities are conveyed is unknown. Braford and McCormick (1993) speculate that all osteoglossomorph fishes may lack toro-diencephalic lateral line connections; if this is the case, then the lateral preglomerular nucleus of osteoglossomophs may be solely or primarily involved with acoustic processing. Recall that the torus semicircularis does not appear to provide input to any region of the dorsal or ventral thalamus in osteoglossomorphs (Section 2.5.1 and 2.5.2). As will be discussed below, there is, however, a toral projection to the hypothalamus, at least in *Xenomystus* (Section 2.5.4). In *Xenomystus*, there is also a small projection from nucleus centralis of the torus to the medial preglomerular nucleus (Fig. 5.11; Braford et al. 1993); the further connections of this nucleus are unknown.

The connections of the lateral preglomerular nucleus in otophysans are somewhat different from those in osteoglossomorphs, and apparently vary among different otophysan species (Striedter 1991, 1992). Considering first the differences between osteoglossomorphs and otophysans, the otophysan lateral preglomerular nucleus receives a small input from the ventrolateral (mechanosensory lateral line) nucleus of the torus semicircularis, but no direct projection from nucleus centralis (acoustic). Rather, the lateral preglomerular nucleus receives efferents from the medial pretoral nucleus, which itself receives a major, presumably acoustic, input from nucleus centralis (Fig. 5.10; Striedter 1991, 1992). In addition, the telencephalic outputs of the lateral preglomerular nucleus are not identical (see Section 2.5.5).

Striedter (1991, 1992) demonstrated that the lateral preglomerular nucleus has different patterns of connectivity with other diencephalic structures in otophysan species from four subgroups: cyprinids (*Carassius*, goldfish), characins (*Colossoma*, pacu), catfishes (*Ictalurus*, channel catfish), and gymnotoids (*Apteronotus*, chocolate ghost knife fish).

The central posterior nucleus of the dorsal thalamus (Section 2.5.1) is one structure that has variable connections with the lateral preglomerular nucleus. The two nuclei are reciprocally connected in *Carassius* and *Colossoma*. In *Apteronotus*, there is only a projection from the central posterior nucleus to the lateral preglomerular nucleus, while in *Ictalurus*, the are no interconnections between the two nuclei. Striedter (1992) suggests that the loss of reciprocal connections is related to the increased size of the central posterior nucleus in catfishes and gymnotids compared to the other two species.

The ventromedial nucleus of the ventral thalamus (Section 2.5.2) provides input to the lateral preglomerular nucleus in all four species examined. However the interconnections between the lateral preglomerular nucleus and the anterior tuberal nucleus of the hypothalamus (Section 2.5.4) vary. The lateral preglomerular nucleus provides bilateral input to the anterior tuberal nucleus in all four species. A reciprocal connection is present in *Carassius*, *Colossoma*, and *Apteronotus*, but is lost in *Ictalurus*.

The otophysan lateral preglomerular nucleus also projects, with some intergroup variability, to other populations within the diencephalon and pretectum (Striedter 1991, 1992). Although future studies may prove otherwise, these populations are not known to be involved with the auditory or lateral line systems, and so will not be discussed here.

2.5.4 Hypothalamus

The hypothalamus and acoustic areas in the torus semicircularis are bilaterally and reciprocally interconnected in otophysans (Echteler 1984; Striedter 1991). Nucleus centralis of the torus and the medial pretoral nucleus specifically supply the dorsal and ventral divisions of the anterior tuberal nucleus (Figs. 5.9D,E and 5.10). In catfish, the dorsal portion of the anterior tuberal nucleus projects back to the medial pretoral nucleus. According to Echteler (1984), the anterior tuberal nucleus projects to nucleus centralis in *Cyprinus*, the carp. In contrast, the bilateral hypothalamic projection to nucleus centralis in catfish originates from the dorsal periventricular hypothalamus that lies just medial to the dorsal anterior tuberal nucleus (Striedter 1991).

An apparent acoustic input to the hypothalamus is also present in the osteoglossomorph *Xenomystus* (Fig. 5.11; Braford et al. 1993). The sparse, bilateral projection from the acoustic portion of the torus supplies neurons that surround the lateral portion of the hypothalamic ventricles. It is not known whether this area is comparable in any way to the otophysan anterior tuberal nucleus.

The otophysan anterior tuberal nucleus also likely receives input from the mechanosensory (Finger and Bullock 1982; Rosenberger and McCormick, personal observations) and electrosensory lateral line systems (Keller et al. 1990). As noted above (Section 2.5.3), the anterior tuberal nucleus receives bilateral input from the lateral preglomerular nucleus in four otophysan species, and, with the exception of catfish, reciprocal connections are present. Such connections could conceivably convey information about one or more of the octavolateralis modalities.

Echtler (1984) suggested that the projection of nucleus centralis to the anterior tuberal nucleus might indicate that sound plays a role in reproductive and/or aggressive behavior in otophysans. This suggestion may also apply to input from other sensory systems, and may extend to species in other taxa (i.e., osteoglossomorphs). Such inputs are reminiscent of, though not identical to, those present in anurans (Section 3.6.5).

2.5.5 Telencephalon

The telencephalon of bony fishes is divided into an area dorsalis and an area ventralis. According to Nieuwenhuys (1963) and Braford (1995), the area dorsalis is the homologue of the pallium of other vertebrates, while the area ventralis is the subpallium (other hypotheses are reviewed in Northcutt and Braford 1980).

The area dorsalis (Fig. 5.9F,G) includes lateral (DL), dorsal (DD), medial (DM), central (DC), and posterior (DP) subdivisions. Some of these subdivisions have in turn been further divided in various species. Thus, DM may contain a number of zones, as is the case in otophysans (DM-1 through DM-4; Bass 1981). Similarly, DC has zones that are closely associated with

DM, DP, etc., and these zones are called, respectively, DC-m, DC-p, etc. Area dorsalis as a whole potentially includes structures that are homologous to the pallial amygdala and to cortical areas of land vertebrates. In mammals, the auditory system sends projections to regions within both of these structures (Winer 1992).

The area ventralis also contains a number of subdivisions. The area ventralis potentially includes structures that are the homologues of the striatum, subpallial amygdala, septal area, etc., of land vertebrates. The dorsal portion of the area ventralis may be the homologue of the striatum (Braford 1995). In mammals, the striatum is another telencephalic area that receives auditory input (reviewed in Winer 1992).

Two acoustically responsive telencephalic areas have been identified in an otophysan, the carp *Cyprinus* (Echteler 1985). They lie within a caudal portion of DM and a caudal portion of DC.

In otophysans, DM and the lateral preglomerular nucleus (Section 2.5.3) are reciprocally connected (Striedter 1991, 1992). In species containing a DM with highly differentiated zones, the region of DM that receives input from the lateral preglomerular nucleus is the most discrete; that is, a small area of DM-3 in *Apteronotus* (a gymnotid) as opposed to large portions of a combined DM2-3 in *Carassius* (a cyprinid; Striedter 1992). Catfish have a unique projection from the anterior tuberal nucleus (Section 2.5.4) to DM, specifically to DM-2 and DM-4 (Finger 1980; Striedter 1991). In contrast, the input from the lateral preglomerular nucleus supplies DM-3, suggesting functional differences (Striedter 1991).

In the osteoglossomorph *Xenomystus*, a caudal area of DM is apparently acoustic (Braford et al. 1993). This area receives input from a region of the lateral preglomerular nucleus that is in turn supplied by the auditory area of the torus semicircularis (Fig. 5.11).

As noted above, Echteler (1984) also recorded acoustic activity from the DC in *Cyprinus*. In otophysans, at least a portion of DC is reciprocally connected with the lateral preglomerular nucleus. As is the case for DM, in species with a greater number of subdivisions in DC, such as the gymnotid *Apteronotus*, the lateral preglomerular nucleus has the most discrete projection (Striedter 1992). In otophysans and in the osteoglossomorph *Xenomystus*, DC also provides a descending input to the nucleus centralis of the torus semicircularis (Echteler 1984; Braford et al. 1993).

In otophysans, other structures in the area dorsalis receive input from the lateral preglomerular nucleus; whether or not these projections convey acoustic information is unknown. For example, the lateral preglomerular nucleus projects to the dorsal part of the DL in *Carassius*, *Colossoma*, *Ictalurus*, and *Apteronotus*, to the ventral part of DL only in *Apteronotus*, and to DD in *Colossoma*, *Ictalurus*, and *Apteronotus* (but not *Carassius*) (Striedter 1991, 1992). The meaning of these variations is unknown. Finally, projections from DP and DC-p to the lateral preglomerular nucleus of otophysans presumably convey information about olfaction. Such connections underscore the need to look, in a more detailed way, for possible functional subdivisions within the lateral preglomerular nucleus.

Little is known about possible auditory regions within the area ventralis. Striedter (1991) hypothesized that the central posterior nucleus (which appears to be an important acoustic structure in the dorsal thalamus; Section 2.5.1) may project within the area ventralis. Further experiments are needed both to confirm this and to localize the specific target site.

3. The Acoustic Pathways of Amphibians

3.1 Acoustic Receptors in Amphibians

Whereas the inner ear of jawed fish shows little, if any, variability in its complement of sensory end organs, the inner ear of extant amphibians—the Lissamphibia—in this regard is quite variable. The complex distribution of inner ear receptors among amphibians is discussed by Lewis and Narins (Chapter 4). I will make only some general points below.

All amphibians retain at least some of the otic end organs present in fish and, at least in many species, the otolith-covered saccule and lagena remain acoustic (i.e., responding at least to vibratory/seismic stimuli). All amphibians have an additional, membrane-covered acoustic receptor, the amphibian papilla, which may have evolved from the macula neglecta (White and Baird 1982; Fritzsch and Wake 1988). The amphibian papilla is not present in reptiles, birds, and mammals. A second membrane-covered acoustic receptor, the basilar papilla, is also a primitive feature of the lissamphibia, but is lost secondarily in some species. The basilar papilla may have originated in amphibians, or, alternatively, may have originated in sarcopterygian fishes (Fritzsch 1987). Whether the basilar papilla of amphibians is (Fritzsch 1992) or is not (Lombard and Bolt 1979; Wever 1985) homologous to the basilar papilla/organ of Corti of amniotes is also a matter of debate.

A given amphibian species, therefore, may have up to four otic end organs that are responsive to sound. In species lacking a tympanic ear, as is the case for all urodeles and apodans and for certain anurans, vibratory/seismic stimuli may constitute the dominant acoustic signal. Some interpretations of the paleontological data support the notion that this was the primitive mode of hearing in the Amphibia (e.g., Bolt and Lombard 1992). Airborne hearing requires a tympanum as well as additional structures; the tympanic ear of anurans is generally considered to be independently derived from that of later vertebrate radiations (e.g., Lombard 1980; Lombard and Bolt 1988).

As will be described below, the octavolateralis area of anurans is organized differently from that of urodeles and apodans. These differences might be related to the evolution of airborne hearing and/or requirements at the first-order level necessary for processing vocalization signals.

3.2 The Octavolateralis Area

As is the case in fish, the octavolateralis area of the medulla contains the first-order nuclei of the auditory, vestibular, mechanosensory lateral line, and electrosensory systems. The mechanosensory and electrosensory systems and their first-order nuclei are variably present. Unlike fish, a cerebellar crest is not associated with any of these nuclei.

There are two general patterns of organization of the amphibian octavolateralis area—that of urodeles and apodans, and that of anurans (McCormick 1988; Will 1988). The patterns are distinguished from one another in the degree to which nuclei have migrated away from their embryonic positions at the ventricle, and by the presence or absence of a discrete acoustic nucleus.

In urodeles and apodans, most neurons of the medulla, including those of the octavolateralis area, occupy a largely periventricular position (Fig. 5.12A,B). The result is that the alar plate consists of a series of cell populations along the ventricle having relatively poorly differentiated boundaries. In contrast, some of the octavolateralis nuclei in anurans have migrated away from the ventricle and these nuclei are generally easier to distinguish (Fig. 5.12C,D).

The second difference between nonanurans and anurans involves the constituents of the octaval column. Anurans have evolved an auditory nucleus, the dorsal or dorsolateral nucleus (Fig. 5.12C,D), which is not present in urodeles and apodans. The dorsolateral nucleus has sometimes been homologized with the cochlear nuclei of amniotes (Larsell 1934;

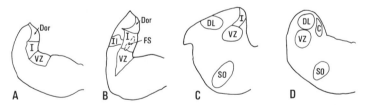

FIGURE 5.12. Hemisections through the medulla of four amphibians illustrating the constituents of the octavolateralis area. A, the urodele *Ambystoma mexicanum*; B, the apodan *Ichthyophus kohtaoensis*; C, the primitive anuran *Xenopus laevis*; D, the derived anuran *Rana catesbeiana*. (A, B, and C are redrawn from Will 1988; D is redrawn from Wilczynski 1988, © John Wiley & Sons, Inc. Reprinted by permission of John Wiley & Sons, Inc.) C, nucleus caudalis; DL, dorsolateral (acoustic) nucleus; Dor, dorsal (electrosensory) nucleus; FS, fasciculus solitarius; I, nucleus intermedius (mechanosensory lateral line); Il, nucleus intermedius pars lateralis; SO, superior olive; VZ, ventral octaval zone.

Ariens Kappers et al. 1967; Matesz 1979), but others consider it more likely that the dorsolateral nucleus and the amniote cochlear nuclei have evolved independently (Will et al. 1985a; McCormick 1988; Will 1988).

The following sections discuss the nuclear constituents of the octavolateralis area in the three amphibian orders and their inputs from the octaval nerve.

3.3 The Octavolateralis Area in Urodeles and Apodans

The octavolateralis area in nonanurans is divided into a dorsal nucleus, an intermediate nucleus, and a ventral zone (Fig. 5.12A,B). The dorsal nucleus is likely homologous to the dorsal electrosensory nucleus of lampreys, chondrichthyans, and primitive bony fishes. The intermediate nucleus is for the most part homologous to the nucleus medialis (mechanosensory lateral line) of fish. In apodans with free-swimming larvae, an additional, migrated group lateral to intermediate nucleus is present and is called the intermediate nucleus pars lateralis (Fig. 5.12B; Will and Fritzsch 1988; Will 1989); it will be discussed further below. The ventral zone, along with a saccular termination area lateral to the intermediate nucleus (see below), may be homologous to the octaval nuclei of fish (Fig. 5.13A,B). In urodeles and apodans, Will (1989) recognized two regions within the ventral zone that he homologized to the caudal and magnocellular nuclei of anurans. Braford and McCormick (in press) and McCormick (1988) provisionally recognize four divisions of the octaval column in the apodan *Typhlonectes* and in urodeles that are potential homologues of the four octaval nuclei in nonteleost fish.

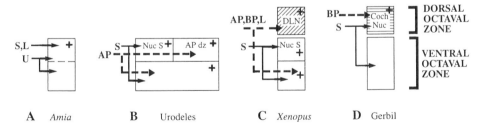

FIGURE 5.13. Schematized summary of inner ear inputs to the dorsal and ventral octaval columns. The ventral octaval column is unshaded. The dorsal octaval column is indicated by diagonal lines in anurans, and by horizontal lines in amniotes to reflect its probable nonhomology. Arrows entering these columns indicate inputs. Plus marks indicate a projection from at least one subdivision of a column to the acoustic midbrain. AP, amphibian papilla; AP dz, dorsal terminal zone of amphibian papilla; BP, basilar papilla, Coch Nuc, cochlear nuclei; DLN, dorsolateral nucleus, L, lagena; Nuc S, nucleus saccularis of Matesz; U, utriculus.

Lateral to each periventricular zone is a neuropil in which the afferents of various cranial nerves terminate. In the urodeles (Will and Fritzsch 1988) electrosensory afferents terminate in the neuropil associated with the dorsal nucleus, and mechanosensory lateral line afferents terminate in the dorsal portion of the neuropil associated with the intermediate nucleus. Afferents from the saccule and from the amphibian papilla have dorsal and ventral terminal fields (Fig. 5.13B). The dorsal terminal field of the saccule occupies the ventral portion of the neuropil adjacent to the caudal region of the intermediate nucleus (Fig. 5.12A), whereas the dorsal terminal field of the amphibian papilla is located within the ventral zone neuropil immediately ventral to the saccular termination zone. The ventral terminal fields of the saccule and amphibian papilla overlap within a portion of the ventral zone—the nucleus magnocellularis and its neuropil. Saccular fibers also reach the reticular formation.

Afferents from end organs that are totally vestibular have a somewhat different projection pattern. Fibers from two of the semicircular canal cristae project largely to the neuropil of the entire ventral zone, the magnocellular nucleus, the reticular formation, nucleus cerebelli, and the eminentia granularis of the cerebellum (Will and Fritzsch 1988).

According to Will and Fritzsch (1988), the projections of the octaval nerve in apodans are generally like those of urodeles. One difference concerns the location of one of the two saccular terminal fields; the more dorsal field is in the ventral portion of the intermediate nucleus pars lateralis (Fig. 5.12B) rather than in the neuropil of the ventral intermediate nucleus. The other terminal field is in a portion of the ventral zone neuropil near the fasciculus solitarius. The intermediate nucleus pars lateralis may not be present in apodans that lack a free-swimming larval stage, such as *Typhlonectes*. In this species, which lacks both the mechanosensory lateral line system and the intermediate nucleus, the octaval nerve terminates in the four divisions of the octaval column discussed above (Braford and McCormick, in press).

Although the afferent terminal fields of the lateral line and octaval nerves may be confined to areas of neuropil specific to a given nucleus, convergence of sensory information within those nuclei is possible. Whereas the dendrites of some neurons extend laterally into the neuropil of its nucleus, other neurons have dendrites that ramify within the neuropil of other nuclei (Gomez Segade 1980; Gomez Segade and Carrato Ibanez 1981; Will 1989). This factor may result in categories of neurons that receive only one modality, and other categories that receive information from two or more sensory systems. The functional significance of such cross-talk is unknown. The neurons in the ventral intermediate nucleus are one population that receives a discrete input, in that they extend their dendrites only into the saccular projection zone.

3.4 The Octavolateralis Area in Anurans

The octaval nuclei constitute the entire octavolateralis area of most adult anurans. In tadpoles and in the few species that retain a mechanosensory lateral line system postmetamorphically, the intermediate nucleus is present and is located in a periventricular position like that of nonanuran amphibians (Fig. 5.12C). An electrosensory system and the dorsal electrosensory nucleus are absent in both larval and adult life stages (Fritzsch et al. 1984).

The anuran octaval nuclei form a dorsal zone, which is absent in nonanurans, and a ventral zone, which may be homologous to the ventral zone of nonanurans (Fig. 5.13C; Will et al. 1985b). The dorsolateral, or dorsal acoustic, nucleus, is the sole component of the dorsal zone (Figs. 5.12C,D, 5.13C, and 5.14A). Two to four nuclei have been recognized in

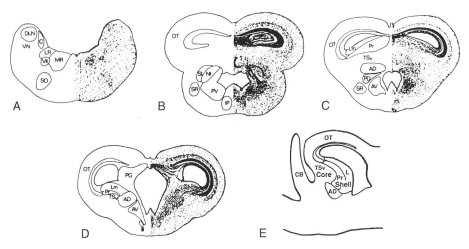

FIGURE 5.14. Auditory nuclei in the brain stem of the bullfrog, *Rana catesbeiana*. In levels A–D, the boundaries of nuclei in the photomicrographs of cresyl violet–stained sections are shown in the drawings of the contralateral side of the brain. E is a line drawing of a sagittal section through the medial torus semicircularis, illustrating the shell and core arrangement of the torus. Rostral is to the right. (From Wilczynski 1988, © John Wiley & Sons, Inc. Reprinted by permission of John Wiley & Sons, Inc.) AD, anterodorsal tegmentum; AV, anteroventral tegmentum; CB, cerebellum; DLN, dorsolateral nucleus; III, oculomotor nucleus; IP, interpeduncular nucleus; L, laminar nucleus of torus semicircularis; LR, lateral reticular formation; MR, medial reticular formation; NC, nucleus caudalis; NI, nucleus isthmi; OT, optic tectum; PG, pretectal gray; PV, posteroventral tegmentum; Pr, principal nucleus of torus semicircularis; SI, secondary isthmal nucleus; SO, superior olive; SR, superficial reticular nucleus; TSv, ventral toral zone; VII, facial motor nucleus; VN, ventral vestibular nuclei (ventral octaval zone).

the ventral zone (Opdam et al. 1976; Matesz 1979; Nikundiwe and Nieuwenhuys 1983; Will et al. 1985a). In some cases this may reflect species differences but in other cases different criteria have been used for recognizing nuclear subdivisions. I will use the nomenclature of Will and colleagues (1985a) to facilitate discussion of their experimental data on a number of anuran species. This scheme recognizes the following four nuclei in the ventral zone: anterior nucleus, lateral octavus nucleus, medial vestibular nucleus, caudal nucleus.

Despite differences in nomenclature, the experimental data of Will and coworkers (1985a) and of Matesz (1979) converge in their key findings. This is particularly significant in light of the fact that the nine species studied collectively span a broad taxonomic range and include both primitive and derived genera. The authors conclude that the dorsolateral nucleus is the main first-order acoustic nucleus, but there are additional acoustic areas within the ventral zone that may be homologous to the acoustic areas in nonanuran amphibians and, perhaps, fish.

The dorsolateral nucleus (Figs. 5.12C,D and 5.14A) receives input from the basilar papilla, the amphibian papilla, and from the lagena (Matesz 1979; Fuzessary and Feng 1981; Will et al. 1985a). Furthermore, it is associated with a ventral neuropil (the nucleus saccularis of Matesz) supplied by saccular fibers (Matesz 1979; Will et al. 1985a; Will and Fritzsch 1988). Therefore, all four acoustic end organs project within, or near, the dorsal acoustic zone in at least some anurans (Fig. 5.13C). These studies expand on previous studies that concluded that the dorsolateral nucleus is the exclusive terminus of fibers from the amphibian and basilar papillae (Ariens-Kappers et al. 1967; Gregory 1972).

The dorsolateral nucleus is tonotopically organized (Fuzessery and Feng 1981, 1983; Feng 1986a) and is homotopically connected to the contralateral dorsolateral nucleus (Feng 1986a). Its dorsomedial portion receives fibers from the basilar papilla, which encodes high-frequency sound. The amphibian papilla, a mid- to low-frequency receptor, projects to the remainder of the dorsolateral nucleus (Lewis et al. 1980; Fuzessery and Feng 1981; Will et al. 1985a). The lowest frequencies are represented in the ventral part of the dorsolateral nucleus. The lagena has a small projection to the lateral or ventrolateral part of the dorsolateral nucleus, and the saccule projects to the neuropil ventral to the dorsolateral nucleus.

Three areas within the ventral octaval zone (Fig. 5.12C,D) are supplied by all four acoustic end organs in at least some anurans: the caudal nucleus, regions within the lateral octaval nucleus, and the dorsal portion of the medial vestibular nucleus (Will and Fritzsch 1988). The caudal nucleus (Figs. 5.12D and 5.14A) receives the majority of the acoustic input supplying the ventral zone. The amphibian and basilar papillae both project here. The saccule and lagena supply the ventral portion of the caudal nucleus. The dorsal portion of the caudal nucleus likely also receives saccular input because of its association with a neuropil that contains saccular afferents.

According to Will and Fritzsch (1988), this is the same neuropil that lies ventral to the dorsolateral nucleus at more rostral levels. They consider this neuropil, and the dorsal portion of the caudal nucleus, to be homologous to the neuropil and associated ventral portion of nucleus intermedius in nonanuran amphibians (Fig. 5.12A,B).

The amphibian and basilar papillae provide a small input to dorsal portions of the lateral octaval and medial vestibular nuclei. The saccule also projects to these areas, but the lagena projects more ventrally (Will et al. 1985a; Will and Fritzsch 1988).

In general, the amphibian and basilar papillae do not project to other nuclei within the octavolateralis area or to the cerebellum. According to Matesz (1979) they project to the superior olivary nucleus (Section 3.6) in *Rana esculenta*, but this projection is absent or minimal in other anurans (Will and Fritzsch 1988).

The saccule and lagena have additional projections that include the anterior nucleus of the ventral octaval zone, the reticular formation, the nucleus cerebelli, the cerebellar auricle, and areas within the cerebellar corpus (Will 1988). The significance of these structures to auditory processing is unknown. Will (1988) reports that many of the octaval nuclei have complex interconnections within the octavolateralis area. For example, the anterior nucleus projects to the dorsolateral and caudal nuclei (perhaps relaying saccular and lagenar information) as well as to the oculomotor complex. The caudal nucleus, which receives input from all four acoustic end organs, is reciprocally interconnected with the dorsolateral nucleus but also projects to all of the other octavolateralis nuclei. As was pointed out in the discussion of primary acoustic areas in fish (Section 2.2) there are likely unrecognized acoustic areas within the primary octaval nuclei of anurans that do not project to higher levels of the neuraxis, but instead influence auditory processing at the level of the medulla.

3.5 Inner Ear Projections to Other Brain Stem Areas: The Mauthner Cell

The Mauthner cell is variably present among amphibians. Interestingly, its distribution is not correlated with the degree to which the postmetamorphic adult is terrestrial (Will 1991).

In urodeles and anurans, some Mauthner cell dendrites extend into the terminal zones of the octaval nerve and, when present, the mechanosensory lateral line nerve (Will 1986). Inputs to these dendrites have been demonstrated in a number of cases (Cochran et al. 1980; Will 1986). At least some larval anurans that possess Mauthner cells show a startle response that initiates with a C-bend resembling the C-start of fish (Rock et al. 1981; Will 1991). An electrophysiological study suggests that this response is driven by the Mauthner cell, which itself can be driven by electrically stimulating the octaval nerve or by delivering vibratory stimuli to the otic capsule (Rock et al. 1981). Adult anurans that possess Mauthner cells have a startle response, but it does not involve a C-bend (Will 1991). In *Xenopus*, this response is elicited by tapping on the rim of a water-filled aquarium, suggesting either sound or lateral line activation (Will 1991). While there is as yet no evidence that this input specifically drives the Mauthner cell (as opposed to other reticulospinal cells), similarity among characteristics of the startle responses of fish and *Xenopus* support the notion that this is the case.

Functional studies of the Mauthner cell in fish are available for only a few species, and thus may be limited in their generality. Similarly, comparative studies of Mauthner cell function that include species from the three extant amphibian groups are needed before we can determine whether the amphibian Mauthner cell localizes sound in the manner suggested for fish.

3.6 Higher-Order Acoustic Areas in the Medulla and Midbrain

3.6.1 Urodeles and Apodans

There is essentially no information on how the auditory circuits are organized beyond the first-order level in nonanuran amphibians. In urodeles, neurons in the dorsal electrosensory nucleus, the intermediate nucleus, and the magnocellular nucleus appear to project to the midbrain, but the details of the connections are unknown (Will and Fritzsch 1988).

3.6.2 Anurans

There is a good deal of information about the auditory brain-stem structures of anurans (Fritzsch et al. 1988). The brain-stem circuits that originate from the main first-order acoustic population, the dorsolateral nucleus, parallel those from the amniote cochlear nuclei (Fig. 5.15). Thus there are anuran counterparts to the amniote superior olive, nuclei of the lateral lemniscus, and inferior colliculus.

The auditory midbrain tectum in anurans, the torus semicircularis (Fig. 5.14B,C), is considered homologous both to the torus semicircularis of fish and to the mammalian inferior colliculus (and its homologues in reptiles and birds). However, just as the dorsolateral nucleus may be nonhomologous to the cochlear nuclei, the anuran counterparts of the superior olive and the nuclei of the lateral lemniscus—the superficial reticular nucleus of anurans—may likewise be examples of independent evolution. Structures are presumed homologous if they are present in a common ancestor. Neither a superior olive nor a superficial reticular nucleus have been identified in urodeles and apodans; if these structures are in fact absent, this could be evidence for their absence in the common ancestor of amphibians and reptiles. It is also possible that these two structures originated in bony fishes (Section 2.4.2), were present in the early amphibian radiations, and were retained in anurans and amniotes but not in

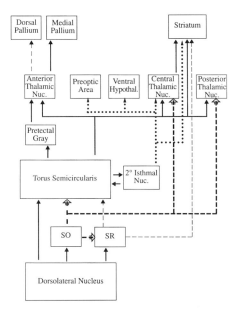

FIGURE 5.15. Simplified summary of possible acoustic circuits in ranid anurans. The laterality of the connections is not indicated. 2° Isthmal Nuc., secondary isthmal nucleus; SO, superior olive, SR, superficial reticular nucleus.

nonanurans. Therefore, comparative studies in primitive species of bony fish, urodeles, and apodans are needed to establish whether or not these structures share a common evolutionary source, or whether they evolved independently several times.

3.6.3 The Superior Olivary Nucleus and the Superficial Reticular Nucleus of Anurans

The anuran superior olive (Fig. 5.14A) is a single, tonotopically organized nucleus. The superficial reticular nucleus is positioned lateral to the reticular formation (Fig. 5.14B). Its caudal portion, which lies partly within the lateral lemniscus, appears to be the region of the nucleus that is part of the auditory brain-stem circuitry (Wilczynski 1988).

The superior olive receives bilateral input from the dorsolateral and caudal nuclei and from the contralateral superior olive (Fuller and Ebbesson 1973; Feng 1986a,b; Will 1988). The superficial reticular nucleus receives contralateral input from the dorsolateral nucleus and ipsilateral input from the superior olive (Feng 1986a,b). Below the level of the torus, the superior olive projects to the dorsolateral nucleus (bilaterally) and to the superficial reticular nucleus (contralaterally; Wilczynski 1981; Feng 1986b).

The torus semicircularis receives bilateral ascending projections from the superior olive and the superficial reticular nucleus and provides ipsilateral descending connections to them (e.g., Rubinson and Skiles 1975; Wilczynski 1981; Feng 1986b). There is a small, direct ipsilateral projection of the superior olive to the posterior thalamic nucleus and to a restricted caudal area of the central thalamic nucleus (Fig. 5.15; Rubinson and Skiles 1975; Feng 1986b). The superficial reticular nucleus projects directly to the striatum (Fig. 5.15; Marin et al. 1997). The striatum receives acoustic input from midbrain and diencephalic structures (discussed below), but whether the projection from the superficial reticular nucleus conveys auditory information is unknown.

3.6.4 The Torus Semicircularis of Anurans

The anuran torus semicularis is a complexly organized structure that has five subdivisions (Potter 1965). According to Wilczynski (1988), some of these subdivisions contribute to an overall organization consisting of a shell of neurons, which receive a variety of inputs, surrounding a central core, which receives most of the auditory information and in addition has multimodal regions (Fig. 5.14E). The following description of the torus utilizes Wilczynski's concept and interpretation of intrinsic toral boundaries, and notes differences between this and other interpretations.

The subependymal and commissural midline nuclei are functionally uncharacterized toral subdivisions. The other three subdivisions—the laminar nucleus, principal nucleus, and ventral toral zone—are arranged, respectively, from dorsal to ventral within the torus (Fig. 5.14C–E). These subdivisions receive auditory and/or other ascending and descending inputs.

The laminar and principal nuclei form the dorsal boundary of the toral shell. These nuclei are continuous with a group of neurons, the antrodorsal tegmental field, that forms the ventral boundary of the torus (Grover and Grusser-Cornehls 1985), thereby forming the ventral portion of the toral shell (Fig. 5.14E; Wilczynski 1988). Some neurons in the laminar nucleus extend dendrites into the principal nucleus and, to a lesser extent, into the ventral zone. Similarly, the dendrites of some principal nucleus neurons extend into the ventral zone. Auditory input may therefore reach laminar and principal neurons via these dendrites as well as through projections directly into these nuclei, described below.

The ventral toral zone (Fig. 5.14C–E) is the main recipient of ascending auditory projections. This zone, which forms the central core of the torus (Wilczynski 1988), has had its boundaries defined in a variety of ways. Its lateral portion includes magnocellular cells and this portion alone has sometimes been celled nucleus magnocellularis. Some investigators consider its medial region, which has fewer magnocellular cells, to be a ventral extension of the overlying principal nucleus (Wilczynski 1988). However, using connectional and other criteria, Wilczynski (1988) concludes that the lateral and medial portions of the ventral toral zone are a single structure and are a key site for acoustic processing within the midbrain.

In ranids, the dorsolateral nucleus and superior olive project bilaterally (Fig. 5.15) to the ventral toral zone (i.e., the nucleus magnocellularis/ventral principal nucleus of some investigators). The input from the superior olive is predominantly ipsilateral. In *Xenopus*, the ventral toral zone includes a lateral line mechanosensory area (termed the lateral magnocellular nucleus by Will et al. 1985b; Will 1988) that receives input from nucleus intermedius. This lateral line area lies lateral to the auditory area (termed the medial magnocellular nucleus by Will et al. 1985b; Will 1988); this lateral/mechanosensory-to-medial/auditory toral organization is also present in fish (Section 2.4.1). The superficial reticular nucleus projects to the torus, but the exact area of termination is uncertain.

Ascending auditory inputs also reach other toral areas. Like the ventral zone, the ventral part of the principal nucleus and the lateral part of the laminar nucleus receive bilateral input from the dorsolateral nucleus and the superior olive. As described above, these nuclei may also receive auditory input via their ventral dendrites.

The ventral toral zone receives a substantial projection from the secondary isthmal nucleus (Fig. 5.14B), which overlaps the inputs of the ascending auditory system (Neary and Wilczynski 1986; Neary 1988). The secondary isthmal nucleus, which also receives input from the torus (Fig. 5.15), may be involved in circuitry that links the auditory and endocrine systems (discussed below). According to Wilczynski (1988) and Neary (1988), the secondary isthmal nucleus (which does not appear to be a gustatory structure) probably corresponds to the nucleus of the lateral lemniscus of Hall and Feng (1987).

Finally, nonauditory input to the torus includes ascending somatosensory afferents that supply the ventral toral zone, the laminar and the principal nuclei, and input from the optic tectum (Wilczynski 1988).

3.6.5 Forebrain Auditory Regions of Anurans

The forebrain efferents of the torus course to most of the nuclei of the dorsal and ventral thalamus and to the striatum (Neary 1988; Feng and Lin 1991). Thalamic toral recipients themselves have a variety of connections. Only the forebrain structures and circuitry that appear to be most directly associated with hearing are discussed below and illustrated in Figures 5.15 and 5.16. The review chapters of Neary (1988) and Wilczynski (1988) should be consulted for a more complete discussion of structures that receive input from the torus and their further connections.

The caudal portion of the central thalamic nucleus is one region that may be essentially only concerned with acoustic processing (Neary 1988; Marin et al. 1997). The rostral portion of the central thalamic nucleus receives

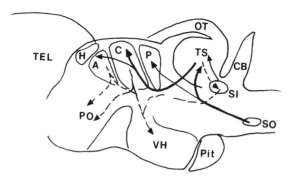

FIGURE 5.16. Schematized summary of possible diencephalic acoustic connections in ranid anurans. Nuclei have been projected onto a sagittal view of the brain; rostral is to the left. (Redrawn and modified from Neary 1988.) A, anterior thalamic nucleus; C, central thalamic nucleus; CB, cerebellum; H, habenula; P, posterior thalamic nucleus; Pit, pituitary; PO, preoptic area; SI, secondary isthmal nucleus; SO, superior olive; TS, torus semicircularis; TEL, telencephalon, VH, ventral hypothalamus.

somatosensory input (Munoz et al. 1994, 1995). The central nucleus is reciprocally connected with the toral magnocellular nucleus—that is, the ventral toral zone—and with the secondary isthmal nucleus described above (Hall and Feng 1987; Neary 1988; Feng and Lin 1991). The central thalamic nucleus also receives ascending auditory input from the ipsilateral superior olive (Hall and Feng 1987). Hall and Feng (1987) demonstrated that the central thalamic nucleus is involved in time domain analysis of complex auditory stimuli. They consider the central thalamic nucleus and the posterior thalamic nucleus (discussed below) to be components of parallel circuits through the diencephalon that process different aspects of an acoustic signal.

Within the telencephalon, the central thalamic nucleus apparently projects only to the striatum (Fig. 5.15; Vesselkin et al. 1980; Wilczynski and Northcutt 1983; Neary 1988; Marin et al. 1997). Interestingly, the secondary isthmal nucleus, like the torus semicircularis, also projects to the striatum. In addition to this presumed auditory input, the striatum receives a wide variety of other sensory and nonsensory inputs (e.g., Neary 1988, 1995; Marin et al. 1997). There is evidence that the striatum is anatomically heterogeneous in that inputs from different sources terminate in different areas (Marin et al. 1997), but no area specific to the auditory system is known. The striatum projects back indirectly to two acoustic structures via the entopeduncular nucleus. This basal forebrain structure is either closely associated with the striatum (Wilczynski 1988) or is its caudal continuation (Marin et al. 1997). Some striatal efferents course to the posterior entopeduncular nucleus, which in turn projects to the central thalamic nucleus (Neary 1988). Another striatal pathway supplies the anterior entopeduncular nucleus, which projects to the laminar and principal nuclei of the torus semicircularis (Wilczynski 1981).

The central thalamic nucleus has two additional targets: the anterior preoptic area and the ventral hypothalamus. The secondary isthmal nucleus as well as other nuclei associated with the auditory system (discussed below) also project to these structures (Fig. 5.16). The ventral hypothalamus and anterior preoptic areas are both gonadotropin-releasing hormone (GnRH) control centers and contain neurons that are responsive to acoustic stimuli, including conspecific mating calls (reviewed in Wilczynski et al. 1993). These projections may provide an anatomical substrate through which acoustic input influences vocalization and the hormonal control of reproductive behavior (Neary 1988; Wilczynski et al. 1993).

Toral input also reaches the anterior nucleus of the thalamus (Figs. 5.15, 5.16), which has been shown to be acoustically responsive (Megala and Capranica 1983). Some of this input is direct, although most is relayed through the pretectal gray (Wilczynski 1978; Neary and Wilczynski 1979; Neary 1988; Feng and Lin 1991). Because it also receives inputs from the visual and somatosensory system, the anterior nucleus as a whole is probably multimodal (Neary 1988). The anterior thalamic nucleus has two telencephalic targets—the dorsal and the medial pallium. The projection to the dorsal pallium, a possible isocortical homologue (Striedter 1997), is light; it is not known whether it conveys acoustic information. On the other hand, the medial pallium is acoustically responsive, although it is not devoted exclusively to auditory processing (Vesselkin and Ermakova 1978; Mudry and Capranica 1980). The medial pallium may be associated with the limbic system (Northcutt and Ronan 1992; Bruce and Neary 1995; Striedter 1997) and projects back upon the anterior thalamic nucleus along with other limbic structures. Like nucleus centralis and the secondary isthmal nucleus, the anterior nucleus projects to the ventral hypothalamus and to the anterior preoptic area (Allison and Wilczynski 1991); it also receives connections from these structures (Neary and Wilczynski 1977).

The posterior thalamic nucleus (Figs. 5.15 and 5.16) is another example of a diencephalic structure that receives substantial, but not exclusive, sensory information from the auditory system. Its direct input from the superior olive is joined by input from the central thalamic nucleus and the torus (Feng 1986b; Hall and Feng 1987; Neary 1988; Feng and Lin 1991). The caudal portion of this nucleus is reciprocally connected to a tonotopically organized area of the torus. Feng and Lin (1991) suggest that this connection may underlie the responsiveness of posterior thalamic neurons to the spectral components of sound (Hall and Feng 1978). Nuclei in receipt of input from the eye, from the optic tectum, and possibly from the somatosensory system also project to the posterior thalamic nucleus (reviewed in Neary 1988). Descending projections to this nucleus originate from the ventral hypothalamus and the striatum (Neary and Wilczynski 1977).

The torus also provides a light input to the lateral thalamic nucleus, and to two ventral thalamic nuclei—the ventromedial thalamic nucleus and the ventral part of the ventrolateral thalamic nucleus (Neary 1988; Feng and Lin 1991). At least some of this input is acoustic based on electrophysiological evidence (Megala and Capranica 1983), but Neary (1988, 1995) characterizes them as multimodal structures based on their inputs from other sensory systems. The lateral thalamic nucleus and the ventral part of the ventrolateral nucleus project to the striatum, but the ventromedial nucleus projects to nucleus accumbens (Marin et al. 1997).

In summary, ascending auditory information appears to be channeled to the medial pallium, and to a lesser extent the dorsal pallium, via the anterior thalamic nucleus, and to the striatum via the torus semicircularis, the central thalamic nucleus, the secondary isthmal nucleus, and the superficial reticular nucleus. Other categories of input reach these telencephalic areas in addition to this auditory input. All of the thalamic nuclei that receive auditory/toral input also receive other categories of sensory input. The central thalamic nucleus is one forebrain nucleus that is known to contain a localized population of neurons dedicated to auditory processing (Marin et al. 1997). The posterior thalamic nucleus may be another such nucleus (Feng and Lin 1991). More detailed analyses are needed to determine whether or not similar populations are present in other thalamic nuclei that have acoustically responsive neurons, such as the anterior nucleus. It is noteworthy that almost all midbrain and forebrain areas that are, or that are potentially, auditory (at least in part) project to the ventral hypothalamus and/or the preoptic area. In addition to the inputs shown in Figure 5.15, the medial pallium projects to the ventral hypothalamus, the medial pallium and the preoptic area project to the striatum, and the striatum is reciprocally connected with the hypothalamus (e.g., Neary 1995). Auditory information from both midbrain and forebrain structures appears to be channeled via multiple and complicated routes to basal forebrain structures that play key roles in the control of reproductive and social behavior of anurans. Wilczynski and colleagues (1993) provide a broader discussion of the relationship between sensory input and control of the endocrine system, pointing out that further work is necessary before we can conclude to what extent such connections are unique to anurans.

4. Summary and Suggestions for Future Studies

Acoustic receptors within the inner ear evolved early in vertebrates. It is reasonable to assume that this peripheral evolution was accompanied by the establishment of central circuits for processing sound information. Even though the acoustic circuits of fish and amphibians have not been investigated as thoroughly as those of amniotes, particularly mammals, it is clear that the flow of information from the periphery to the telencephalon follows the same general pattern in all classes of jawed vertebrates (Fig. 5.17) that have been sufficiently studied. Assuming that jawless fish possess an auditory sense, study of their acoustic circuits is an area for future research. Further study of the ascending auditory pathways in cartilaginous fish is also needed to elucidate brain-stem connections and to determine which areas of the forebrain, if any, are involved in sound processing.

The common pattern illustrated in Figure 5.17 represents a sequence of information flow. This sequence can be defined as primitive, because it evolved early in evolutionary time, at least as early as the evolution of bony fish. During the course of evolution, each vertebrate class has superimposed modifications/specializations upon this pattern, some of which characterize all species in the class, and others that are unique to a subset of those

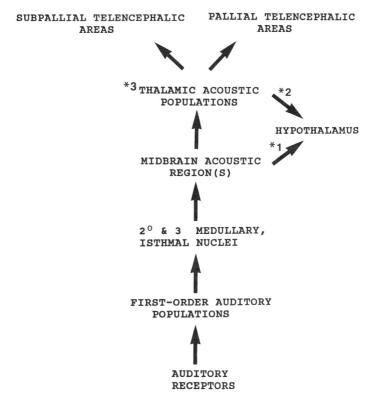

FIGURE 5.17. Sequence of information flow through the acoustic circuits of jawed vertebrates. Further study is required to determine whether the dorsomedial area of the telencephalon in ray-finned fishes is part of the pallium. *1, Hypothalamic pathways in otophysans; *2, hypothalamic pathways in anurans; *3, it is unknown whether the lateral preglomerular nucleus of teleosts is part of the thalamus.

species. For example, even though fish do not have separate nuclei dedicated to auditory processing at the first-order level, that is, they lack homologues of the cochlear nuclei, they nevertheless (1) have first-order populations that receive acoustic input, and (2) have a lateral lemniscus that arises from these populations to supply the acoustic midbrain as well as other brain-stem structures.

Another way of expressing this is to state, or at least hypothesize, that the auditory circuits of amniotes, including mammals, very likely originated as a result of modifications in the auditory circuits of the amphibians and bony fishes that were ancestral to them. This hypothesis has a number of implications. One is, that we cannot compare the acoustic brain areas of amniotes to those of a species of amphibian or fish without knowing the degree to which the acoustic structures of that species conform to the primitive state for their class. For example, comparisons that imply homology between the dorsolateral and superior olivary nuclei of anurans and the cochlear and superior olivary nuclei of amniotes are currently without foundation. Based on analysis of living amphibians, we have no evidence that these nuclei existed in the labyrinthodont ancestor of anurans and amniotes. That is not to say that there is nothing to be gained from comparisons of nonhomologous structures. Study of the independent evolution of common functional features, as well as the evolution of diverse functions, may reveal some of the constraints and possibilities for the neural machinery that processes acoustic information.

Another implication of the above hypothesis is that the primitive organization of acoustic circuits can help us to uncover features of the acoustic circuits of amniotes that may have been overlooked as a result of historical factors and/or their lack of prominence relative to other features.

I will futher explore these two implications below, pointing out issues that might be pursued in future studies.

4.1 First-Order Acoustic Populations: Evolution and Directions for Future Studies

In his review of brain-stem auditory structures in amphibians, Wilczynski (1988) hypothesized that "evolutionary changes in the tetrapod auditory system have not involved fundamental changes in the pattern of brainstem pathways such as the addition of major new auditory centers or significant changes in long connecting pathways, but rather have involved a reorganization of nuclei within each station of a preset chain of brainstem auditory centers." This hypothesis implies that as structural reorganization occurs at a given level of the neuraxis, one or more mechanisms function to keep the sequence of flow through the circuitry as a whole intact.

The first-order acoustic populations of fish and nonanuran amphibians (i.e., urodeles and apodans) reside within a ventral octaval column that primitively contains four octaval nuclei. In other words, fish and nonanurans lack a dorsal octavus column containing a discrete auditory nucleus. Three questions arise from this. First, how do acoustic afferents from the otolithic end organs of fish and amphibians centrally segregate themselves from vestibular afferents arising from those same otolithic end organs or from other structures of the ear? Furthermore, how do afferents from the basilar and amphibian papillae of nonanuran amphibians find their way to the area supplied by otolithic acoustic fibers? Second, is there any evolutionary/developmental relationship between the ventral octavus column of fish and nonanuran amphibians, and the additional, dorsal octavus column of anurans and amniotes? How and why did a dorsal octavus column originate, and how do acoustic afferents find their way to this column? Third, what is the evolutionary fate of the primitive acoustic populations within the ventral octavus column? Might these populations be retained in amniotes?

There are no definitive answers to these three questions. However, some experimental data and hypotheses that address aspects of these questions are summarized below.

All investigations in bony fish that have specified the origins of the lateral lemnsicus find that the cell bodies of these axons are part of a dorsomedial population within the descending octaval nucleus and, in many species, a dorsal population within the anterior octaval nucleus. It is possible that this organization may be the result of at least a crude tonotopic organization within these nuclei. Recalling the dorsolateral nucleus of anurans (but not implying homology with the descending nucleus), the inputs of the two papillar end organs and the lagena are ordered such that the nucleus has dorsomedially to ventrolaterally a high- to low-frequency tonotopy. Interestingly, input from the anuran saccule extends the pattern beyond the boundaries of the dorsal octaval column into the ventral octaval column. Based on patterns of inner ear end-organ input in fish, future physiological studies may reveal that responses to vibratory stimuli are ordered such that there is a dorsomedial/high-frequency to ventrolateral/low-frequency gradient in the descending and anterior nuclei. Although the developmental origins of this ordering would not be revealed, its existence might help to explain not only why particular areas in the descending and anterior nuclei are auditory, but also variations seen in the end-organ input patterns. For example, utricular afferents in clupeid fish have a unique projection to the dorsomedial zone of the descending nucleus that supplants the usual input of the saccule and lagena. These utricular fibers, which are very likely the afferents from expanded, apparently specialized acoustic populations within the utricular maculae, may be directed to the primitive acoustic area within the descending nucleus by a "functional sorting" mechanism (Wilczynski 1984). Similarly, the high-frequency afferents of the amphibian and basilar papillae of amphibians may have originally tapped into such high-frequency populations within the ventral octaval column, thus coming into proximity with the acoustic projections of otolithic end organs. Such overlap occurs in nonanurans (Section 3.3) as well as certain anurans (Section 3.4).

The second question concerns the origin of the dorsal octavus column in anurans and amniotes. A dorsal octavus column is not present in fish. Rather, the dorsal or dorsomedial portion of the descending nucleus appears to have undergone hypertrophy and/or dorsomedial displacement in a number of fish taxa, presumably independently. Therefore, a portion of the ventral octavus column—the dorsal portion of the descending nucleus—comes to lie in a dorsal location within the medulla, and it is this portion that gives rise to a substantial component of the lateral lemniscus (in species that have been studied). Although such dorsally located acoustic neurons have been interpreted as additional, or new, nuclei homologous to those of the tetrapod dorsal octaval column (Fritzsch et al. 1990), this interpretation is in my view not supported on morphological or connectional grounds (McCormick 1992; McCormick and Braford 1993), and, from an evolutionary standpoint, is not parsimonious in view of what we know about the organization of the octaval nuclei in fish.

As discussed in Section 2.2.2, recent data suggest that primitively, the descending nucleus of teleosts includes a small number of dorsally located neurons that are segregated to some degree from those in the main portion of the nucleus. These dorsal-most neurons receive otolithic input via long ventral dendrites that extend into the terminal field of the eighth nerve, and probably cerebellar input via dorsal dendrites that extend into the overlying cerebellar crest. A possible selective pressure for the hypertrophy/displacement of the dorsomedial descending nuclei in other teleosts (i.e., otophysans, certain osteoglossomophs, and clupeids) may be to increase the number of neurons in the acoustic portion of the descending nucleus that are (presumably) contacted by cerebellar crest axons. These questions then arise: (1) Do cerebellar crest fibers in fact synapse on the dendrites of these dorsally located acoustic neurons? (2) What is the functional significance of this contact? and (3) What are the sources of the cerebellar crest axons that potentially modulate the activity of these neurons? Investigations of these questions in both first- and second-order (see below) structures may reveal important and possibly unique features of sound processing in bony fish. Physiological studies of cerebellar crest fiber modulation on neurons in the first-order mechanosensory and electrosensory unclei indicate that such input is part of a mechanism to eliminate reafference and in addition influences the temporal characteristics of neurons in these nuclei (reviewed by Montgomery et al. 1995).

The origin of the dorsal octavus column of anurans is unknown. In *Xenopus*, one of the more primitive anurans, the papillar and otolithic acoustic end organs have dual projections; that is, they project to the ventral octavus column, as occurs in nonanurans, and to the dorsal octavus column—the dorsolateral nucleus. This pattern of dual projections may indicate that a portion of the ventral octavus column gave rise to the dorsolateral nucleus either through hypertrophy and migration during development, or through duplication (McCormick and Braford 1988). Fritzsch and colleagues (1984) provide some evidence for this hypothesis in a study that found that portions of the anuran dorsolateral acoustic nucleus arise from the same embryonic source as neurons in the ventral octavus column. However, the authors also found an additional, dorsal source for this nucleus. One possibility is that the dorsal embryonic field gives rise to the primitive electrosensory nucleus in nonanurans and primitive fish (Fritzsch et al. 1984; Fritzsch 1988); another is that it represents a newly evolved embryonic field (McCormick and Braford 1988). Interestingly, at least the dorsal cochlear nucleus of mammals may have multiple embryonic sources (Altman and Bayer 1980; Willard and Martin 1986).

The evolutionary origins of the cochlear nuclei of amniotes are likewise speculative. As discussed above, the cochlear nuclei (and its amniote homologues) and the anuran dorsolateral nucleus probably arose independently. Prior to experimental methodologies that allowed the bulk tracing of nerves from specific inner ear end organs, it was generally concluded that the cochlear nuclei (or their homologues) are the exclusive domain of afferents from the basilar papilla, or organ of Corti. Note that this same assumption, which turned out to be false, was made about the inputs to the dorsolateral nucleus of anurans; only the amphibian papilla and basilar papilla were thought to supply these neurons. Therefore, the traditionally accepted dichotomy between dorsally located (i.e., dorsal octavus column) acoustic nuclei versus ventrally located (i.e., ventral octavus column) vestibular nuclei needs to be reevaluated.

There are three general categories of experiments that bear upon such a reevaluation. First, neuroanatomical studies in reptiles, birds, and mammals can determine whether afferents from the basilar papilla are confined to the dorsal octavus column, and whether otolithic end-organ afferents are confined to the ventral octavus column. Studies of saccular projections in mammals indicate that unexpected projections to the cochlear nuclei exist, although their functional significance is unknown (Kevetter and Perachio 1989). Second, neuroanatomical studies can also determine whether there are populations within the ventral octavus column (i.e., the vestibular nuclei) of amniotes that project to the auditory midbrain. Third, electrophysiological studies can evaluate whether unorthodox sites in the ventral octavus column that are suspected to be acoustic are in fact so, as well as reevaluate response properties of the otolithic end organs. One recent study, for example, describes acoustically responsive fibers from the saccule in the cat (McCue and Guinan 1994). Collectively, studies that span the three amniote classes and that attempt to define the primitive condition of inner ear projections for each class may, along with comparative developmental studies, lend some insight into the evolutionary origins of the cochlear nuclei.

An additional issue concerns the significance of bilateral projections from the inner ear to the dorsomedial zone of the descending nucleus (Section 2.2.1). This specialized projection has been documented in only two of the osteoglossomorph taxa that have been studied—mormyrids and xenomystines—and in clupeids. The sources of the projection are the saccule (mormyrids and xenomystines), lagena (mormyrids), and the utricle (clupeids). The common features among these species are that (1) the bilateral input originates in whole or in part from the otolithic end organ that is at the point of coupling between the inner ear and a gas-filled structure, and (2) this otophysic coupling is paired. In other words, each inner ear is contacted by one of a pair of diverticula of the swim bladder, or one of a pair of gas-filled pouches developmentally originating from the swim bladder (Stipetić 1939; Dehadrai 1957; Allen et al. 1976). This anatomical arrangement contrasts with a single, midline otophysic coupling such as occurs in otophysans. It is possible that bilateral acoustic projections and paired otophysic contacts are functionally related.

4.2 Higher-Order Brain Stem and Forebrain Auditory Centers

Higher-order nuclei below the level of the midbrain that are known or presumed to be acoustic are present in bony fish and at least anuran amphibians. Comparative neuroanatomical studies in sarcopterygian and nonteleost bony fish and in nonanuran amphibians are needed to establish whether any of these structures are homologous to the superior olivary nuclei and nuclei of the lateral lemnsicus of amniotes. The apparent elaboration of the secondary octaval population in certain teleosts is an unexpected finding of unknown significance (Section 2.4.2). Like certain acoustic neurons in the dorsomedial descending nucleus, neurons in the dorsal secondary octaval population may be influenced by cerebellar crest input. Other populations within the secondary octaval population may be part of descending auditory pathways. One set of future studies might explore the connections and functions of the secondary octaval population in species in which it is well developed, such as otophysans. A second set of studies might investigate whether dorsally located components of the secondary octaval population are common to all teleosts. Just as the descending nucleus of some species may have a dorsally located component that is not easily recognized using normal cell stains (Section 2.2.2), the secondary octaval population may likewise have a more complex structure in teleosts than has previously been recognized.

The isthmal reticular nucleus of anurans, a tegmental nucleus that relays acoustic information from the torus semicircularis to the hypothalamus, occupies an unusual, and possibly unique, position in the ascending acoustic circuitry (Section 3.6.4). The function of this nucleus in anurans and its possible homologue in other vertebrates are unknown.

Neuroanatomical and functional data concerning acoustic areas within the midbrain torus semicircularis are more complete for anuran amphibians, particularly ranids, than for any other anamniote group. As in amniotes, the anuran torus semicircularis is composed of regions that are concerned primarily with processing and relaying acoustic information, as well as regions that receive a variety of sensory and other inputs in addition to those concerned with hearing. This feature of the auditory midbrain— that it is a multimodal integration area as well as an area concerned with audition—may well be common to all vertebrates. A better understanding of the location of functional subpopulations within the torus of fish and nonanuran amphibians would likely lead to a better understanding, from both functional and evolutionary standpoints, of toral inputs to the diencephalon and the further projections of these diencephalic structures to the telencephalon.

The anuran torus, through its diencephalic projections, influences four regions—the hypothalamus/preoptic area, the striatum, the medial pallium (a possible hippocampal homologue), and the dorsal pallium (a possible isocortical homologue). Whether any of these pathways have evolutionary counterparts in amniotes is unknown; further studies in both nonanurans and bony fish are needed in order to have the data necessary to construct a well-formulated hypothesis.

Braford (1995) has pointed out that studies of ascending sensory pathways in nonmammals often attempt to homologize the diencephalon-to-telencephalon components of these pathways to the well-known thalamocortical pathways of mammals. Thus, presumed auditory pathways from midbrain to diencephalon to telencephalon in anamniotes are sometimes assumed to be the primitive homologue of the lemniscal pathways ascending from the central nucleus of the inferior colliculus to the ventral division of the medial geniculate body to primary auditory isocortex. However, mammals have a variety of extralemniscal auditory pathways that arise from areas of the inferior colliculus other than the central nucleus. Each of these areas has specific projections to one or more multimodal areas within the medial geniculate body, which in turn may project to a variety of cortical areas other than the primary auditory cortex, to the amygdala, and/or to the striatum (for a review, see Oliver and Huerta 1992, Winer 1992). Understanding which, if any, of these lemniscal and extralemniscal pathways are present in anamniotes has been a challenging task. Particularly in fish, interpreting the homologues between these pathways and those of amniotes is an essentially unresolved problem, due largely to an incomplete understanding of how the fish forebrain is organized.

The higher-order auditory pathways of fish are poorly understood. As is the case in amphibians, the primitive condition has not been clearly established. One hypothesis that needs to be tested is that diencephalic acoustic structures in nonteleosts include at least one dorsal thalamic nucleus—the central posterior nucleus—as well as one or more subdivisions of the preglomerular complex. Whether mechanosensory lateral line efferents primitively also project to these nuclei is unknown. If, for the purposes of argument, we assume that the above hypothesis is correct, then the apparent loss of the acoustic relay through the dorsal thalamus in osteoglossomorph teleosts should be confirmed and its functional significance determined.

There is good evidence in otophysans that the central posterior nucleus receives acoustic input (Section 2.5.1). This nucleus may be homologous to the central thalamic nucleus of anurans, which likewise receives a major auditory projection from the midbrain. The anuran central thalamic nucleus projects to the striatum. The apparent projection of the central posterior nucleus to the area ventralis (possibly the subpallium) should be investigated further to determine whether this projection in fact supplies the striatum.

Auditory circuits through the preglomerular complex are particularly difficult to interpret from an evolutionary standpoint because it is not certain that this structure belongs to a single division of the diencephalon (Braford 1995). Embryological studies are needed to determine whether the preglomerular complex is derived entirely from the posterior tubercle, thus eliminating it as a potential homologue of the medial geniculate body (e.g., Murakami et al. 1986), or whether portions of it originate from the thalamus. A probable acoustic portion of the lateral preglomerular complex projects to a common telencephalic subdivision in otophysans and osteoglossomorphs—a region within the dorsomedial area. The nature of acoustic processing in that population, or in any other area of the forebrain, is unknown. Embryological and comparative studies that can evaluate whether the dorsomedial area is homologous to the pallial amygdala (Braford 1995) are needed before comparisons to amniotes can be made.

Finally, the prominence of acoustic input to the hypothalamus is an interesting finding in both anurans (Section 3.6.5) and otophysans (Section 2.5.4), although the pathways involved may not be identical. Toro-hypothalamic connections were thought to be unique to otophysans, but a similar, though significantly smaller, projection is present in an osteoglossomorph. The generality of this finding among bony fishes should be explored. In anurans, multiple pathways may be conveying acoustic input into the hypothalamus. Are these pathways specialized features related to the role vocalization plays in anuran reproduction, or are they also present in nonanurans? Furthermore, are they the precursors of midbrain-diencephalic-hypothalamic pathways in mammals that likewise originate in the inferior colliculus (LeDoux et al. 1985)?

In summary, the forebrain auditory circuits of anamniotes and amniotes may differ in two fundamental ways. First, direct auditory inputs from the diencephalon to areas that are homologous to the isocortex of mammals may either be absent or minimal in anamniotes. Intratelencephalic circuits, however, may be discovered that convey this input to isocortical homologues. Second, through a variety of pathways, at least some anamniotes may direct a larger proportion of auditory forebrain input to the hypothalamus than do amniotes. Both of these statements, however, are based on studies done in few species, and thus may ultimately prove invalid. The careful choice of species to be studied is critical to these analyses. Attention must be given to taxonomic status to maximize the possibility that the features being studied are primitive for the group. Inappropriate comparisons between structures that have evolved independently can highlight functional strategies for dealing with common selective pressures, but lead us to incorrect evolutionary conclusions.

Acknowledgments. I thank Dr. Mark Braford for helpful discussions concerning forebrain evolution and for the many ways he supported me during the writing of this chapter. Shana O'Marra kindly helped with some of the illustrations. I am also appreciative of Sandra Ronan's excellent technical assistance. This work, and many of the studies reported in it, were funded by National Science Foundation grants IBN93-0997 and IBN96-04466 and a grant to Oberlin College from the Howard Hughes Foundation.

References

Allen JM, Blaxter JHS, Denton EJ (1976) The functional anatomy and development of the swimbladder-inner ear-lateral line system in herring and sprat. J Mar Biol Assoc UK 56:471–486.

Allison JD, Wilczynski W (1991) Thalamic and midbrain auditory projections to the preoptic area and ventral hypothalamus in the green treeforg, (*Hyla cinerea*). Brain Behav Evol 38:322–331.

Altman J, Bayer SA (1980) Development of the brain stem of the rat. III. Thymidine-radiographic study of the time of origin of neurons of the vestibular and auditory neurons of the upper medulla. J Comp Neurol 194:877–904.

Ariens Kappers CU, Huber CG, Crosby EC (1967) The Comparative Anatomy of the Nervous System of Vertebrates, Including Man. New York: Hafner.

Barry MA (1987) Afferent and efferent connections of the primary octaval nuclei in the clearnose skate, *Raja eglanteria*. J Comp Neurol 266:457–477.

Bass AH (1981) Organization of the telencephalon in the channel catfish, *Ictalurus punctatus*. J Morphol 169:71–90.

Bass AH, Marchaterre MA, Baker R (1994) Vocal-acoustic pathways in a teleost fish. J Neurosci 14:4025–4039.

Bell CC (1981a) Some central connections of medullary octavolateral centers in a mormyrid fish. In: Tavolga WN, Popper AN, Fay RR (eds) Hearing and Sound Communication in Fishes. New York: Springer-Verlag, pp. 383–392.

Bell CC (1981b) Central distribution of octavolateral afferents and efferents in a teleost (Mormyridae). J Comp Neurol 195:391–414.

Bleckmann H, Niemann U, Fritzsch B (1991) Peripheral and central aspects of the acoustic and lateral line system of a bottom dwelling catfish, *Ancistrus* sp. J Comp Neurol 314:452–466.

Bodznick D, Schmidt AW (1984) Somatotopy within the medullary electrosensory nucleus of the little skate, *Raja erinacea*. J Comp Neurol 225:581–590.

Bolt JR, Lombard RE (1992) Nature and quality of the fossil evidence for otic evolution in early tetrapods. In: Webster DB, Fay RR, Popper AN (eds) The Evolutionary Biology of Hearing. New York: Springer-Verlag, pp. 377–403.

Boord RL, Montgomery JC (1989) Central mechanosensory lateral line centers among the elasmobranchs. In: Coombs S, Gorner P, Munz H (eds) The Mechanosensory Lateral Line. New York: Springer-Verlag, pp. 323–339.

Boord RL, Northcutt RG (1982) Ascending lateral line pathways to the midbrain of the clearnose skate, *Raja eglanteria*. J Comp Neurol 207:274–282.

Boord RL, Northcutt RG (1988) Medullary and mesencephalic pathways and connections of lateral line neurons of the spiny dogfish, *Squalus acanthias*. Br Behav Evol 32:76–88.

Braford MR Jr (1982) Electroreceptive and mechanoreceptive afferents of the torus semicircularis in the notopterid fish, *Xenomystus nigri*. Soc Neurosci Abstr 8:764.

Braford MR Jr (1986) African Knifefishes: The Xenomystines. In: Bullock TH, Heiligenberg W (eds) Electroreception. New York: Wiley, pp. 453–464.

Braford MR Jr (1995) Comparative aspects of forebrain organization in the ray-finned fishes: Touchstones or not? Brain Behav Evol 46:259–274.

Braford MR Jr, McCormick CA (1979) Some connections of the torus semicircularis in the bowfin, *Amia calva*: a horseradish peroxidase study. Soc Neurosci Abstr 5:193.

Braford MR Jr, McCormick CA (1993) Brain organization in teleost fishes: lessons from the electrosense. J Comp Physiol [A] 173:704–707.

Braford MR Jr, McCormick CA The lateral line system of *Typhlonectes* and other adult gymnophionan amphibians. Brain Behav Evol, in press.

Braford MR Jr, Northcutt RG (1983) Organization of the diencephalon and pretectum of the ray-finned fishes. In: Davis RE, Northcutt RG (eds) Fish Neurobiology, vol. 2. Ann Arbor: University of Michigan Press, pp. 117–163.

Braford MR Jr, Prince E, McCormick CA (1993) A presumed acoustic pathway from the ear to the telencephalon in an osteoglossomorph teleost. Soc Neurosci Abstr 19:160.

Bruce LL, Neary TJ (1995) Afferent projections to the lateral and dorsomedial hypothalamus in a lizard, *Gekko gekko*. Brain Behav Evol 46:30–42.

Bullock TE, Corwin JT (1979) Acoustic evoked activity in the brain of sharks. J Comp Physiol 129:223–234.

Bullock TE, Heiligenberg W (1986) Electroreception. New York: Wiley.

Budelli R, Macadar O (1979) Statoacoustic properties of utricular afferents. J Neurophysiol 42:1479–1493.

Carr CE, Matsubara J (1981) Central projections of the octavolateralis nerves in gymnotiform fish. Soc Neurosci Abstr 7:84.

Cochran SL, Hackett JT, Brown JL (1980) The anuran Mauthner cell and its synaptic bed. Neuroscience 5:1629–1646.

Corwin JT (1981) Peripheral auditory physiology in the lemon shark: evidence of parallel otolithic and non-otolithic sound detection. J Comp Physiol 142:379–390.

Corwin JT, Northcutt RG (1982) Auditory centers in the elasmobranch brainstem: deoxyglucose autoradiography and evoked potential recording. Br Res 236:261–273.

Crawford JD (1993) Central auditory neurophysiology of a sound producing mormyrid fish: the mesencephalon of *Pollimyrus isidori*. J Comp Physiol 172:1–14.

Dehadrai PV (1957) On the swimbladder and its relations with the internal ear in the genus *Notopterus* (Lacepede). J Zool Soc India 9:50–61.

Denton EJ, Blaxter JHS (1976) Function of the swimbladder–inner ear–lateral line system of herring in the young stages. J Mar Biol Assoc UK 56:487–502.

Dunn RF, Koester DM (1987) Primary afferent projections to the central octavus nuclei in the elasmobranch, *Rhinobatis* sp. as demonstrated by nerve degeneration. J Comp Neurol 260:564–572.

Eaton RC, Popper AN (1995) The octavolateralis system and Mauthner cell: interactions and questions. Brain Behav Evol 46:124–130.

Eaton RC, DiDomenico R, Nissanov J (1991) Role of the Mauthner cell in sensorimotor integration by the brain stem escape network. Brain Behav Evol 37:272–285.

Eaton RC, Canfield JG, Guzik AL (1995) Left-right discrimination of sound onset by the Mauthner system. Brain Behav Evol 46:165–179.

Echteler SE (1984) Connections of the auditory midbrain in a teleost fish, *Cyprinus carpio*. J Comp Neurol 230:536–551.

Echteler SE (1985) Organization of central auditory pathways in a teleost fish, *Cyprinus carpio*. J Comp Physiol [A] 156:267–280.

Fay RR (1984) The goldfish ear codes the axis of acoustic particle motion in three dimensions. Science 225:951–954.

Fay RR (1995) Physiology of primary saccular afferents of goldfish: implications for Mauthner cell response. Brain Behav Evol 46:141–150.

Fay RR, Popper AN (1975) Modes of stimulation of the teleost ear. J Exp Biol 62:379–388.

Feng AS (1986a) Afferent and efferent innervation patterns of the cochlear nucleus (dorsal medullary nucleus) of the leopard frog. Brain Res 367:183–191.

Feng AS (1986b) Afferent and efferent innervation patterns of the superior olivary nucleus of the leopard frog. Brain Res 364:167–171.

Feng AS, Lin W (1991) Differential innervation patterns of three divisions of the frog auditory midbrain (torus semicircularis). J Comp Neurol 306:613–630.

Finger TE (1980) Nonolfactory sensory pathway to the telencephalon in a teleost fish. Science 210:671–672.

Finger TE, Bullock TH (1982) Thalamic center for the lateral line system in the catfish, *Ictalurus nebulosus*. J Neurobiol 13:39–47.

Finger TE, Tong S-L (1984) Central organization of eighth nerve and mechanosensory lateral line systems in the catfish, *Ictalurus nebulosus*. J Comp Neurol 229:129–151.

Fritzsch B (1987) The inner ear of the coelacanth fish *Latimeria* has tetrapod affinities. Nature 327:153–154.

Fritzsch B (1988) Phylogenetic and ontogenetic origin of the dorsolateral auditory nucleus of anurans. In: Fritzsch B, Ryan MJ, Wilczynski W, Hetherington TE, Walkowiak W (eds) The Evolution of the Amphibian Auditory System. New York: Wiley, pp. 561–585.

Fritzsch B (1992) The water-to-land transition: evolution of the tetrapod basilar papilla, middle ear, and auditory nuclei. In: Webster DB, Fay RR, Popper AN (eds) The Evolutionary Biology of Hearing. New York: Springer-Verlag, pp. 351–375.

Fritzsch B, Wake MH (1988) The inner ear of gymnophione amphibians and its nerve supply: a comparative study of regressive events in a complex sensory system. Zoomorphology 108:210–217.

Frizsch B, Nikundiwe AM, Will U (1984) Projection patterns of lateral line afferents in anurans: a comparative HRP study. J Comp Neurol 229:451–469.

Fritzsch B, Ryan MJ, Wilczynski W, Hetherington TE, Wlakowiak W (1988) The Evolution of the Amphibian Auditory System. New York: Wiley.

Fritzsch B, Niemann U, Bleckmann H (1990) A discrete projection of the sacculus and lagena to a distinct brainstem nucleus in a catfish. Neurosci Lett 111:7–11.

Fuller PM, Ebbesson SOE (1973) Projections of the primary and secondary auditory fibers in the bullfrog (Rana catesbeiana). Soc Neurosci Abstr 3:33.

Fuzessary ZM, Feng AS (1981) Frequency representation in the dorsal medullary nucleus of the leopard frog, Rana p. pipiens. J Comp Physiol 143:339–347.

Fuzessary ZM, Feng AS (1983) Frequency selectivity in the anuran auditory medulla: excitatory and inhibitory tuning properties of single neurons in the dorsal medullary and superior olivary nuclei. J Comp Physiol 150:107–119.

Gomez Segade LA (1980) Morphology and evolution of the acoustico-lateral area in the rhombencephalon of Salamandridae I. Salamandra salamandra. Trab Inst Cajal Invest Biol Madrid 71:37–55.

Gomez Segade LA, Carrato Ibanez A (1981) Morphology and evolution of the acoustico-lateral area in the rhombencephalon of Salamandridae II. Chioglossa lusitanica. Trab Inst Cajal Invest Biol Madrid 72:111–119.

Gregory KM (1972) Central projections of the eighth nerve in frogs. Br Behav Evol 5:70–88.

Grover BR, Grusser-Cornehls U (1985) Organization of the reticulospinal projection from the isthmic region in the frog (Rana esculenta). Soc Neurosci Abstr 11:1312.

Grover BG, Sharma SC (1981) Organization of extrinsic tectal connections in the goldfish, Carassius auratus. J Comp Neurol 196:471–488.

Hall JC, Feng AS (1987) Evidence for parallel processing in the frog's auditory thalamus. J Comp Neurol 258:407–419.

Haugede-Carre F (1983) The mormyrid mesencephalon. II. The mediodorsal nucleus of the torus semicircularis: afferent and efferent connections studied with the HRP method. Br Res 268:1–14.

Highstein SM, Kitch R, Carey J, Baker R (1992) Anatomical organization of the brainstem octavolateralis area of the oyster toadfish, Opsanus tau. J Comp Neurol 319:501–518.

Kalmijn AJ (1988) Hydrodymanic and acoustic field detection. In: Atema J, Fay RR, Popper AN, Tavolga WN (eds) Sensory Biology of Aquatic Animals. New York: Springer-Verlag, pp. 83–130.

Keller CH, Maler L, Heiligenberg W (1990) Structural and functional organization of a diencephalic sensory-motor interface in the gymnotiform fish, Eigenmannia. J Comp Neurol 293:347–376.

Kevetter GA, Perachio AA (1989) Projections from the sacculus to the cochlear nuclei in the mongolian gerbil. Br Behav Evol 34:193–200.

Larsell O (1934) The differentiation of the peripheral and central acoustic apparatus in the frog. J Comp Neurol 60:473–527.

Larsell O (1967) Jansen J (ed) The Comparative Anatomy and Histology of the Cerebellum from Myxinoids Through Birds. Minneapolis: University of Minnesota Press.

Le Doux JE, Ruggerio DA, Reis DJ (1985) Projections to the subcortical forebrain from anatomically defined regions of the medial geniculate body in the rat. J Comp Neurol 242:182–213.

Lewis ER, Leverentz EL, Koyama H (1980) Mapping functionally identified auditory afferents from the peripheral origins to their central terminations. Brain Res 197:223–229.

Lombard RE (1980) The structure of the amphibian auditory periphery: a unique experiment in terrestrial hearing. In: Popper AN, Fay RR (eds) Comparative Studies of Hearing in Vertebrates. New York: Springer-Verlag, pp. 121–138.

Lombard RE, Bolt JR (1979) Evolution of the tetrapod ear: an analysis and reinterpretation. Biol J Linn Soc 11:19–76.

Lombard RE, Bolt JR (1988) Evolution of the stapes in Paleozoic tetrapods: conservative and radical hypotheses. In: Fritzsch B, Ryan MJ, Wilczynski W, Walkowiak W (eds) The Evolution of the Amphibian Auditory System. New York: Wiley, pp. 37–67.

Lowenstein O, Roberts TDM (1951) The localization and analysis of the responses to vibration from the isolated elasmobranch labyrinth: a contribution to the problem of the evolution of hearing in vertebrates. J Physiol (Lond) 114:471–489.

Luiten PGM (1981) Afferent and efferent connections of the optic tectum in the carp, *Cyprinus carpio*. Br Res 220:51–65.

Marin O, Gonzalez A, Smeets WJAJ (1997) Basal ganglia organization in amphibians: afferent connections to the striatum and the nucleus accumbens. J Comp Neurol 378:16–49.

Matesz C (1979) Central projections of the VIIIth cranial nerve in the frog. Neuroscience 4:2061–2071.

McCormick CA (1981) Central connections of the lateral line and eighth nerves in the bowfin, *Amia calva*. J Comp Neurol 197:1–15.

McCormick CA (1983a) Central projections of inner ear endorgans in the bowfin, *Amia calva*. Am Zool 23:895.

McCormick CA (1983b) Central projections of the octavolateralis nerves in the pike cichlid, *Crenicichla lepidota*. Br Res 265:177–185.

McCormick CA (1988) Evolution of auditory pathways in the Amphibia. In: Fritzsch B, Ryan MJ, Wilczynski W, Hetherington TE, Walkowiak W (eds) The Evolution of the Amphibian Auditory System. New York: Wiley, pp. 587–612.

McCormick CA (1989) Central lateral line mechanosensory pathways in bony fish. In: Coombs S, Gorner P, Munz H (eds) The Mechanosensory Lateral Line. New York: Springer-Verlag, pp. 341–364.

McCormick CA (1992) Evolution of central auditory pathways in anamniotes. In: Webster DB, Fay RR, Popper AN (eds) The Evolutionary Biology of Hearing. New York: Springer-Verlag, pp. 323–350.

McCormick CA (1997) Organization and connections of octaval and lateral line centers in the medulla of a clupeid, *Dorosoma cepedianum*. Hear Res 110:39–60.

McCormick CA (1998) Organization of the descending octaval nucleus in teleosts. Soc Neurosci Abstr 24, in press.

McCormick CA, Braford MR Jr (1988) Central connections of the octavolateralis system. In: Atema J, Fay RR, Popper AN, Tavolga WN (eds) Sensory Biology of Aquatic Animals. New York: Springer-Verlag, pp. 733–756.

McCormick CA, Braford MR Jr (1993) The primary octaval nuclei and inner ear afferent projections in the otophysan *Ictalurus punctatus*. Br Behav Evol 42:48–68.

McCormick CA, Braford MR Jr (1994) Organization of inner ear endorgan projections in the goldfish, *Carassius auratus*. Br Behav Evol 43:189–205.

McCormick CA, Hernandez DV (1996) Connections of octaval and lateral line nuclei in the medulla of the goldfish, including the cytoarchitecture of the secondary octaval population in goldfish and catfish. Br Behav Evol 47:113–137.

McCue MP, Guinan JJ Jr (1994) Acoustically responsive fibers in the vestibular nerve of the cat. J Neurosci 14:6058–6070.

Megala AL, Capranica RR (1983) A neural and behavioral study of auditory habituation in the bullfrog, *Rana catesbeiana*. J Comp Physiol 151:423–434.

Meredith GE (1985) The distinctive central utricular projections in the herring. Neurosci Lett 55:191–196.

Meredith GE, Butler AB (1983) Organization of eighth nerve afferent projections from individual endorgans of the inner ear in the teleost, *Astronotus ocellatus*. J Comp Neurol 220:44–62.

Meredith GE, Roberts BL, Maslam S (1987) Distribution of afferent fibers from end organs in the ear and lateral line in the European eel. J Comp Neurol 265:507–520.

Montgomery JC, Coombs S, Conley RA, Bodznick D (1995) Hindbrain sensory processing in lateral line, electrosensory, and auditory systems: a comparative overview of anatomical and functional similarities. Aud Neurosci 1:207–231.

Mudry KM, Capranica RR (1980) Evoked auditory activity within the telencephalon of the bullfrog, *Rana catesbeiana*. Br Res 182:303–311.

Munoz A, Munoz M, Gonzalez A, Ten Donkelaar HJ (1994) Spinothalamic projections in amphibians as revealed with anterograde tracing techniques. Neurosci Lett 171:81–84.

Munoz A, Munoz M, Gonzalez A, Ten Donkelaar HJ (1995) The anuran dorsal column nucleus: organization, immunohistochemical characterization and fiber connections in *Rana perezi* and *Xenopus laevis*. J Comp Neurol 363:197–220.

Murakami T, Fukuoka T, Ito H (1986) Telencephalic ascending acousticolateral system in a teleost (*Sebasticus marmoratus*), with special reference to the fiber connections of nucleus preglomerulosus. J Comp Neurol 247:383–397.

Neary TJ (1988) Forebrain auditory pathways in ranid frogs. In: Fritzsch B, Ryan MJ, Wilczynski W, Hetherington TE, Walkowiak W (eds) The Evolution of the Amphibian Auditory System. New York: Wiley, pp. 233–252.

Neary TJ (1995) Afferent projections to the hypothalamus in ranid frogs. Brain Behav Evol 46:1–13.

Neary TJ, Northcutt RG (1983) Nuclear organization of the bullfrog diencephalon. J Comp Neurol 213:262–278.

Neary TJ, Wilczynski W (1977) Autoradiographic demonstration of hypothalamic efferents in the bullfrog, *Rana catesbeiana*. Anat Rec 187:665.

Neary TJ, Wilczynski W (1979) Anterior and posterior thalamic afferents in the bullfrog *Rana catesbeiana*. Soc Neurosci Abstr 5:144.

Neary TJ, Wilczynski W (1986) Auditory pathways to the hypothalamus in ranid frogs. Neurosci Lett 71:142–146.

New JG, Northcutt RG (1984) Central projections of the lateral line nerves in the shovelnose sturgeon. J Comp Neurol 225:129–140.

New JG, Singh S (1994) Central topography of anterior lateral line nerve projections in the channel catfish, *Ictalurus punctatus*. Brain Behav Evol 43:34–50.

Nieuwenhuys R (1963) The comparative anatomy of the actinopterygian forebrain. J Hirnforsch 6:171–192.

Nikundiwe AM, Nieuwenhuys R (1983) The cell masses in the brainstem of the South American clawed frog *Xenopus laevis*: a topographical and topological analysis. J Comp Neurol 213:199–219.

Northcutt RG (1978) Brain organization in the cartilaginous fishes. In: Hodgson ES, Mathewson RF (eds) Sensory Biology of Sharks, Skates, and Rays. Arlington VA: Office of Naval Research, pp. 117–194.

Northcutt RG (1981) Audition and the central nervous system of vertebrates. In: Tavolga WN, Popper AN, Fay RR (eds) Hearing and Sound Communication in Fish. New York: Springer-Verlag, pp. 331–355.

Northcutt RG, Braford MR Jr (1980) New observations on the organization and evolution of the telencephalon of actinopterygian fishes. In: Ebbesson SOE (ed) Comparative Neurology of the Telencephalon. New York: Plenum, pp. 41–98.

Northcutt RG, Ronan M (1992) Afferent and efferent connections of the bullfrog medial pallium. Brain Behav Evol 40:1–16.

Oliver DL, Huerta MF (1992) Inferior and superior colliculi. In: Webster DB, Popper AN, Fay RR (eds) The Mammalian Auditory Pathway: Neuroanatomy. New York: Springer-Verlag, pp. 168–221.

Opdam P, Kemali M, Nieuwenhuys R (1976) Topological analysis of the brain stem of the frogs *Rana esculenta* and *Rana catesbeiana*. J Comp Neurol 165:307–332.

Platt C (1983) The peripheral vestibular system of fishes. In: Northcutt RG, Davis RE (eds) Fish Neurobiology, vol. 1. Ann Arbor: University of Michigan Press, pp. 89–123.

Popper AN, Edds-Walton PL (1995) Structural diversity in the inner ear of teleost fishes: implications for connections to the Mauthner cell. Brain Behav Evol 46:131–140.

Potter HD (1965) Mesencephalic auditory region of the bullfrog. J Neurophysiol 28:1132–1154.

Prasada Rao PD, Jadhao AG, Sharma SC (1987) Descending projection neurons to the spinal cord of the goldfish. J Comp Neurol 265:96–108.

Puzdrowski RL (1989) Peripheral distribution and central projections of the lateral line nerves in goldfish, *Carassius auratus*. Br Behav Evol 34:110–131.

Puzdrowski RL, Leonard RB (1991) The octavolateral region of an advanced ray, *Dasyatis sabina*. Soc Neurosci Abstr 17:319.

Puzdrowski RL, Leonard RB (1993) The octavolateral systems in the stingray, *Dasyatis sabina*. I. Primary projections of the octaval and lateral line nerves. J Comp Neurol 332:21–37.

Rock MK, Hackett JT, Les Brown D (1981) Does the Mauthner cell conform to the criteria of the command neuron concept? Brain Res 204:21–27.

Rubinson K, Skiles MP (1975) Efferent projections of the superior olivary nucleus in the frog, *Rana catesbeiana*. Brain Behav Evol 12:151–160.

Sand O (1981) The lateral line and sound reception. In: Tavolga WN, Popper AN, Fay RR (eds) Hearing and Sound Communication in Fishes. New York: Springer-Verlag, pp. 459–480.

Schellart NAM (1983) Acousticolateral and visual processing and their interaction in the torus semicircularis of the trout, *Salmo gairdneri*. Neurosci Lett 42:39–44.

Schellart NAM, Kroese ABA (1989) Interrelationship of acousticolateral and visual systems in the teleost midbrain. In: Coombs S, Gorner P, Munz H (eds) The Mechanosensory Lateral Line. New York: Springer-Verlag, pp. 421–443.

Smeets WJAJ, Nieuwenhuys R, Roberts BL (1983) The Central Nervous System of Cartilaginous Fishes. New York: Springer-Verlag.

Stipetić E (1939) Uber das Gehororgan der Mormyriden. Z Vergl Physiol 26:740–752.

Striedter GF (1990) The diencephalon of the channel catfish, *Ictalurus punctatus*: II. Retinal, tectal, cerebellar, and telencephalic connections. Br Behav Evol 36:355–377.

Striedter GF (1991) Auditory, electrosensory, and mechanosensory pathways through the diencephalon and telencephalon of the channel catfishes. J Comp Neurol 312:311–331.

Striedter GF (1992) Phylogenetic changes in the connection of the lateral preglomerular nucleus in ostariophysan teleosts: a pluralistic view of brain evolution. Br Behav Evol 39:329–357.

Striedter GF (1997) The telencephalon of tetrapods in evolution. Brain Behav Evol 49:179–213.

Torres B, Pastor AM, Cabrera B, Salas C, Delgado-Garcia JM (1992) Afferents to the oculomotor nucleus in the goldfish (*Carassius auratus*) as revealed by retrograde labeling with horseradish peroxidase. J Comp Neurol 324:449–461.

Vesselkin NP, Ermakova TV (1978) Studies on the thalamic connections in the frog *Rana temporaria* by peroxidase technique. J Evol Biochem Physiol 14:117–122.

Vesselkin NP, Ermakova TV, Kenigfest NB, Goikovic M (1980) The striatal connections in the frog *Rana temporaria*: an HRP study. J Hirnforsch 21:381–392.

Webb JF, Blum SD (1990) A swim bladder-lateral line connection in the butterflyfish genus Chaetodon (Perciformes: Chaetodontidae). Am Zool 30:98.

Wever EG (1985) The Amphibian Ear. Princeton, NJ: Princeton University Press.

White JS, Baird IL (1982) Comparative morphological features of the caecilian inner ear with comments on the evolution of amphibian auditory structures. Scanning Electron Microsc 3:1301–1312.

Wilczynski W (1978) Connections of the midbrain auditory center in the bullfrog, *Rana catesbeiana*. PhD Thesis, University of Michigan.

Wilczynski W (1981) Afferents to the midbrain auditory center in the bullfrog, *Rana catesbeiana*. J Comp Neurol 198:421–433.

Wilczynski W (1984) Central neural systems subserving a homoplasous periphery. Am Zool 24:755–763.

Wilczynski W (1988) Brainstem auditory pathways in anuran amphibians. In: Fritzsch B, Ryan MJ, Wilczynski W, Hetherington TE, Walkowiak W (eds) The Evolution of the Amphibian Auditory System. New York: Wiley, pp. 209–231.

Wilczynski W, Northcutt RG (1983) Connections of the bullfrog striatum: afferent organization. J Comp Neurol 214:321–332.

Wilczynski W, Allison JD, Marler CA (1993) Sensory pathways linking social and environmental cues to endocrine control regions of amphibian forebrains. Brain Behav Evol 42:252–264.

Will U (1986) Mauthner neurons survive metamorphosis in anurans: a comparative HRP study on the cytoarchitecture of Mauthner neurons in amphibians. J Comp Neurol 244:111–120.

Will U (1988) Organization and projections of the area octavolateralis in amphibians. In: Fritzsch B, Ryan MJ, Wilczynski W, Hetherington TE, Walkowiak W (eds) The Evolution of the Amphibian Auditory System. New York: Wiley, pp. 185–208.

Will U (1989) Central mechanosensory lateral line system in amphibians. In: Coombs S, Gorner P, Munz H (eds) The Mechanosensory Lateral Line. New York: Springer-Verlag, pp. 365–386.

Will U (1991) Amphibian Mauthner cells. Brain Behav Evol 37:317–332.

Will U, Fritzsch B (1988) The eighth nerve of amphibians: peripheral and central distribution. In: Fritzsch B, Ryan MJ, Wilczynski W, Walkowiak W (eds) The Evolution of the Amphibian Auditory System. New York: Wiley, pp. 159–183.

Will U, Luhede G, Gorner P (1985a) The area octavolateralis of *Xenopus laevis*. I. The primary afferent connections. Cell Tissue Res 239:147–161.

Will U, Luhede G, Gorner P (1985b) The area octavolateralis of *Xenopus laevis*. II. Second order projections and cytoarchitecture. Cell Tissue Res 239:163–175.

Willard FH, Martin GF (1986) The development and migration of large multipolar neurons into the cochlear nucleus of the North American opposum. J Comp Neurol 248:119–132.

Winer JA (1992) The functional architecture of the medial geniculate body and the primary auditory cortex. In: Webster DB, Popper AN, Fay RR (eds) The Mammalian Auditory Pathway: Neuroanatomy. New York: Springer-Verlag, pp. 222–409.

Zottoli SJ, van Horne C (1983) Posterior lateral line afferent and efferent pathways within the central nervous system of the goldfish with special reference to the Mauthner cell. J Comp Neurol 219:100–111.

Zottoli SJ, Bentley AP, Prendergast BJ, Rieff HI (1995) Comparative studies on the Mauthner cell of teleost fish in relation to sensory input. Brain Behav Evol 46:151–164.

6
Central Auditory Processing in Fish and Amphibians*

ALBERT S. FENG AND NICO A.M. SCHELLART

1. Introduction

Sound communication plays a vital role in the regulation of social and reproductive behaviors of fish and amphibians (see Zelick et al. Chapter 9). These two groups of animals typically communicate in two different media having different physical characteristics and constraints. Research has shown that these animals are well adapted to cope with these constraints and able to communicate effectively. This chapter summarizes the present understanding of how acoustic signals are represented in the central auditory system. The materials presented herein are built upon the comprehensive knowledge of the peripheral auditory physiology and the central auditory anatomy in these animals (see Popper and Fay, Chapter 3; Lewis and Narins, Chapter 4; and McCormick, Chapter 5).

The level of understanding of central auditory processing in these two groups of animals differs substantially. In fish, much of the work has been focused on characterizing how simple sounds are represented in the nervous system. On the other hand, research in amphibians, or in anurans specifically, has extended into how complex spectral and temporal features of natural communication signals are encoded in the brain, and how the information of sound patterns is integrated with directional information to form coherent acoustic images. In light of this, the data from fish and amphibians are presented in separate sections, representing the different coverage for the two groups of animals. At the end of the chapter, the similarities and differences in the ways acoustic signals are processed in these animals are summarized.

2. Central Auditory Processing in Fish

This section describes how neurons along the central auditory pathway encode simple acoustic stimuli, starting with a quantitative description of basic response characteristics (spontaneous activity, latency, threshold sensitivity, and firing rate), and then a detailed description of frequency coding (phase-locking, coding by spike density) and temporal coding (excitation, inhibition, adaptation, amplitude modulation). Also, the few studies (Plassmann 1985; Lu and Fay 1996) dealing with the processing of more complex stimuli (amplitude modulation, two-tone stimuli) are discussed. The mechanisms of sound localization and the mechanisms by which sound is integrated with other sensory modalities are also reviewed. Special attention is paid to the topographic representation of stimulus features (tonotopy, spatiotopy, modalities).

Auditory research in fish is complicated because the transmission medium is water, instead of air, and thus the stimulus is difficult to control. Moreover, the water flowing through the mouth over the gills makes the recordings unstable. Additionally, the firing activity of central neurons is not robust, making it difficult for the experimenter to locate and identify central auditory neurons. These are the main reasons why neurophysiological study of hearing in fish is not as advanced as in terrestrial vertebrates.

Compared to the extensive literature in fish psychophysics (see Fay 1988; Fay and Megela Simmons, Chapter 7) and in the anatomy of the peripheral (see Popper and Fay, Chapter 3) and central auditory systems (see McCormick, Chapter 5), the number of papers in central neurophysiology, especially those with a quantitative approach, is very limited. Most of the studies have focused on processing in the torus semicircularis (TS) in the midbrain. Auditory structures receiving projection from the TS have hardly been explored (Lu and Fay 1995), and the hindbrain takes an intermediate position.

Quantitative studies of auditory responses of single units in the fish medulla have been performed in the ostariophysine goldfish *Carassius auratus* (Page 1970; Enger 1973; Sawa 1976) and catfish *Ictalurus nebulosis* (Plassmann 1985); in the herring *Clupea harengus* (Enger 1967), a fish also specialized in hearing; and in the unspecialized rainbow trout *Oncorhynchus mykiss* (Wubbels et al. 1993). The hindbrain octaval column projects directly to the TS (McCormick, Chapter 5). However, the TS also receives a minor ascending projection via a small olive nucleus and a nucleus of the lateral lemniscus (McCormick 1992). Although in fish the optic tectum is reciprocally connected with the TS, auditory activity can rarely be detected in the tectum (Schellart and Kroese 1989). Therefore, auditory processing in the tectum will not be further discussed.

Our knowledge of auditory processing in the TS is mainly based on studies carried out in the goldfish (Page 1970; Page and Sutterlin 1970; Lu and Fay 1993, 1996) and the trout (Schellart 1983; Nederstigt and Schellart 1986; Schellart et al. 1987; Wubbels et al. 1995). The ostariophysine tench *Tinca tinca* (Grözinger 1967), the mormyrid *Pollimyrus isidori* (Crawford 1993), a non-ostariophysine hearing specialist (see Schellart and Popper 1992 for this distinction), and the cod *Gadus* (Horner et al. 1980) have also been investigated.

Central auditory processing has hardly been studied quantitatively at the level of single cells in nonteleost fish. However, several studies have investigated the average auditory evoked potentials and multiunit responses in these fish (Corwin 1981; Bleckmann and Bullock 1989).

2.1 Stimulation of Fish: Swim Bladder–Mediated Pressure or Direct Displacement?

The major problem involving a preparation of a partly submerged swim bladder–bearing fish in a physically confined experimental tank is quantitative control of the stimulus. In such condition, the stimulus comprises a displacement component in addition to the often dominating pressure component. This is especially the case in experiments with otophysine fish when the loudspeaker is situated in air or underwater, or when the sound projector is an oscillating membrane mounted in a wall of the tank. In these experiments, an important displacement contribution in the response cannot be excluded for certain frequencies since measurements of head vibrations have generally not been made. In some trout experiments (Table 6.1), the stimulus, generated by an oscillating membrane, has a mixed nature (Schellart and Kroese 1989; Wubbels et al. 1993). On the other hand, for experiments involving nonsubmerged trout on a vibrating table (Goossens et al. 1995; Schellart et al. 1995b), the stimulus is nearly a perfect displacement stimulus and resembles the natural stimulation of the otolith organs (Wubbels et al. 1995). The same holds for goldfish experiments with an oscillating mouth tube to which the head is attached (Fay et al. 1982). Only frequency response data, obtained with the latter two approaches can be interpreted straightforwardly.

Another problem is that weak displacement stimuli and moderate pressure stimuli, mediated by the swin bladder, can evoke lateral line responses. Since in some species projections of the auditory and lateral line system show overlap (Section 2.7.4), the contributions of these two sources of inputs are difficult to pin down. In various studies, lateral line responses cannot be excluded, since control experiments have not been done.

2.2 Basic Characteristics of Auditory Units

2.2.1 Spontaneous Activity and Latency

The spontaneous activity of auditory units within a nucleus is highly variable. In the medulla of the trout, spontaneous rates are 0 to 150 s/s (spikes/second) with a mean 27 s/s (Wubbels, et al. 1993). In other species (Page 1970; Sawa 1976; Horner et al. 1980), spontaneous rates of 80 to 200 s/s have been found. In the goldfish TS, the mean spontaneous rate is low, around

TABLE 6.1. Acoustic response features of single and multiunit recordings in the TS of nonelectroreceptive fish.

Reference	Page 1970; Page and Sutterlin 1970	Lu and Fay 1993	Grözinger 1967	Horner et al. 1980	Nederstigt and Schellart 1986; Schellart 1983; Schellart et al. 1987	Wubbels et al. 1994, 1995; Schellart, unpublished data
Animal species	Goldfish	Goldfish	Tench	Cod	Trout	Trout
Anesthesia	Tricaine	Tricaine	G 29505	None	None	None
Paralyzer	Tubocur.	Flaxedil	Flaxedil	Tubocur.	Pancuronium	Pancuronium
Electrodes	Indium and glass (h)	Indium and glass (l)	Wolfram	Glass (m)	Glass (l)	Glass (h)
Single/multi	s[1]	s	m,s	s[1]	s[1]	s
Stimulus ambiance	Water[a] p (d)	Air[b] p	Water[a] p (d)	Air[b] Mixed	Water[c] Mixed	Air[d] d
Response feature						
Spon act > 1 s/s[e]						
Occurrence	0%	37%		>50%	72%	46
Range	4[4]	13.5			10.0	8
Latency (ms)			7[5]		12[4]	10
Freq. range	90[2]–2000		200[3]–3500		40[3]–700	
Range CF	140–700	120–1020	450–800		150–350	

TABLE 6.1. *Continued.*

Reference	Page 1970; Page and Sutterlin 1970	Lu and Fay 1993	Grözinger 1967	Horner et al 1980.	Nederstigt and Schellart 1986; Schellart 1983; Schellart et al. 1987	Wubbels et al. 1994, 1995; Schellart, unpublished data
# CFs	Rarely > 1	1–3			Often [2]	
Phase-locking						
Occurrence	None	53%	Often		20%	49%
SI		0.90				1.36
PSTH						
Sustained		57%				28%
Transient		20%			Both	27%
Mixed		22%			Often	44%
Inh behavior	Perhaps	Some	Some		22%	
E^+ behavior	Rare			Some	Rare	
Habituation	Often				Often	Often
Lowest threshold						
Pressure (Pa)	0.5–26	0.0013	0.003–0.04	<<0.1	0.005–0.2	
Displacemt (nm)						<0.5 nm
Modality	Ac, V, AcV	Ac	Ac	Ac	Ac, V, AcV	Ac, V, AcV

Ac, acoustic (or acousticolateral); AcV, acoustico-visual; d, displacement; E^+, excitation after cessation of the stimulus; h, high impedance; Inh, inhibitory; l, low impedance; p, pressure; V, visual.

[a] Oscillating membrane; [b] loudspeaker; [c] vibrating submerged plate; [d] vibrating table; [e] spontaneous activity.

[1] Including recordings from the vicinity of the TS (e.g., NDT and fll); [2] artificial limit; [3] some units may be sensitive to lower frequencies; [4] estimated by applying tone bursts; [5] estimated by applying clicks; [6] estimated with a high criterion.

5 s/s, with 60% of the units showing a spontaneous rate of less than 1 s/s (Lu and Fay 1993). The latter cell type is classified as nonspontaneous. In the trout TS, the mean spontaneous activity is 7 s/s (Schellart et al. 1987), but with high impedance pipettes the average is only 4 s/s (Wubbels and Schellart, unpublished data). If the nonspontaneous units are excluded, the average rate would be 8 s/s (Table 6.1). In the central posterior nucleus (CPN) of the goldfish thalamus, the spontaneous rate, which seldom exceeds 25 s/s, shows a mean of 3 s/s (Lu and Fay 1995). In the various species, units with regular, irregular, and bursting firing patterns have been found in the medulla and TS. In general, a few central units have a regular spontaneous activity (e.g., Schellart et al. 1987; Lu and Fay 1995). In the medulla and TS of the trout, the spontaneous activity is stronger in the rostral part (Schellart et al. 1987; Wubbels et al. 1993).

The above data indicate that the spontaneous activity shows a systematic decrease along the auditory pathway from nerve fibers to the thalamic CPN. In the periphery, the spontaneous activity improves the linear performance of the system (yielding sinusoidal period histograms, irrespective of the strength of the pure tones), whereas centrally the lack of spontaneous activity is thought to improve "feature" processing (yielding all-or-none behavior).

In the medulla and TS of the trout, the shortest latency of the unit's auditory response as measured from the phase characteristics of the period histograms is 4 ms (mean ± SD of all units is 14 ± 4 ms) (Wubbels et al. 1993) and 12 ms (Schellart and Kroese 1989), respectively. For the various species investigated, the minimum latencies of TS units to clicks and tone bursts range from 4 to 12 ms (Table 6.1). However, the response latency of other units in the TS may be very long, for example, hundreds of milliseconds (Page 1970; Nederstigt and Schellart 1986). This large range observed is typical for the TS. Since the latencies of tone burst responses are dependent on the intensity and the rise time of the stimulus, a fair comparison among species and data from different laboratories is not possible. Despite these limitations, it can be concluded that, in general, the lower limit of latencies for TS units is several milliseconds longer than for neurons in the medulla (Grözinger 1967; Page 1970; Echteler 1985a).

2.2.2 Threshold Sensitivity and Firing Rate

The data from the trout medulla show that the "threshold" of the most sensitive units at 100 Hz is of the order of 5 nm or 0.3 Pa (Wubbels et al. 1993). In the medulla and the TS, the lowest pressure sensitivity for *Tinca tinca* and *Ictalurus nebulosis*, both ostariophysines, is 0.01 to 0.03 Pa (Grözinger 1967; Plassmann 1985). In the TS and CPN of goldfish, a threshold as low as 0.002 Pa has been found (Lu and Fay 1993, 1995). TS units in the trout appear to have the lowest thresholds, that is, at 30 dB below that of medullary units, thereby approximating the behavioral thresholds

(Schellart and Popper 1992). In the goldfish, the general sensitivity increases approximately 25 dB from the primary fibers to CPN units (Lu and Fay 1995). Thresholds of TS and CPN units show an inter- and intraindividual variability of 30–40 dB (Schellart and Kroese 1989; Lu and Fay 1993, 1995).

In the goldfish, Lu and Fay (1993) found that CPN units can generate a firing rate of 10 to 100 s/s and occasionally higher to stimuli around the characteristic frequency (CF) with stimulus strengths of 25 to 500 mPa. In the trout TS, for stimulation strengths of 1 to 25 Pa, the unit's firing rate is generally 5 to 30 s/s with an upper limit of 150 s/s (Nederstigt and Schellart 1986). In this species, the lowest displacement thresholds were <0.5 nm at 172 Hz (Wubbels et al. 1994).

In goldfish and trout, the dynamic range of a single TS unit is generally limited to 30 dB, but all units can respond over a range of 60 dB. In both species, a small number of units show a decrease of firing rate at the highest intensities presented. This nonmonotonic behavior is seldom observed in the earlier stages of the central auditory pathway (e.g., Page 1970).

2.3 Frequency Coding

2.3.1 Phase-Locking

Since the fish inner ear does not appear to exhibit tonotopic organization (Popper and Fay, Chapter 3), frequency information is preserved predominantly by phase-locking (i.e., synchronization of the spikes to the stimulus cycles). In this section, the strength of the phase-locking is expressed as the dimensionless index of synchronization, SI, ranging from 0 to 2 (amplitude of the first Fourier component of the period histogram divided by the mean of the histogram). Other authors use the coefficient of synchronization R (e.g., Lu and Fay 1993), which is half SI.

In the goldfish, half of the units in the medulla show phase-locking (Page 1970), whereas in the trout the proportion reaches 95% (Wubbels et al. 1993). Phase-locking has also been observed in the cod medulla (Horner et al. 1980). Figure 6.1 gives an example of a frequency tuning characteristic (FTC) with the corresponding phase characteristics.

In the TS, phase-locking was initially thought to be absent (goldfish; Page 1970) or to occur infrequently (trout; Schellart et al. 1987), but more recent studies indicate that 53% (Lu and Fay 1993) and 49% (Wubbels and Schellart, unpublished data) of TS units in the goldfish and trout, respectively, exhibit phase-locking, the difference in the type and impedance of the recording electrode probably accounts for these discrepancies. With low impedance pipettes, only 9% of the trout units showed phase-locking (Schellart and Kroese 1989). In the TS of *Gadus morhua*, with pipettes of

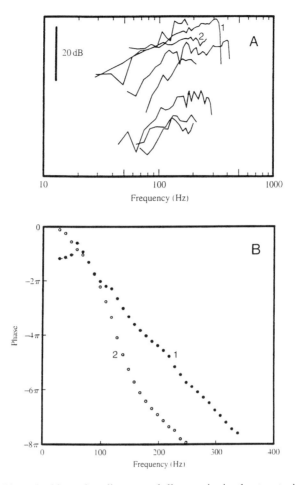

FIGURE 6.1. Phase-locking of auditory medullary units in the trout. A: Frequency tuning characteristic (FTC) obtained with a criterion of the synchronization index of 0.40. The curves have been arbitrarily shifted along the vertical axis. Numbers indicate the corresponding phase plots in B. B: Corresponding phase plots. Estimates of the latencies for unit 1 and 2 are 10 and 20 ms, respectively. (Modified from Wubbels et al. 1993, with permission from Company of Biologists Ltd.)

moderate impedance, phase-locking seems poorly developed (Horner et al. 1980). Metal electrodes are less successful for isolating phase-locked units (Lu and Fay 1993). In the TS, SI can reach a value as high as 1.96 (Schellart et al. 1987; Lu and Fay 1993; Wubbels et al. 1994). In contrast, although primary fibers in the auditory and lateral line nerves always show phase-locking, they do not reach such high SIs. Sometimes the period histogram of a TS unit shows two peaks, about 180° apart (Lu and Fay 1993; Wubbels and Schellart, unpublished data). In goldfish and trout, the distribution of the occurrence of SIs obtained with a standard stimulus is bimodal with a peak at low (i.e., weak phase-locking) and at the high SIs (Fig. 6.2). In trout, units with phase-locking show responsiveness to a broad frequency range (Schellart et al. 1987). In contrast, narrow band units don't show phase-locking. These units may obtain input from broad-band units, and by filtering complete a frequency detecting mechanism. The SI of goldfish thalamic neurons is low; it is seldom larger than 1.0 (Lu and Fay 1995).

In conclusion, the occurrence of phase-locking appears to diminish along the auditory pathway. In the TS, SI is highly variable with a small number of the units showing stronger phase-locking than in preceding structures along the auditory pathway.

2.3.2 Frequency Coding by the Mean Spike Rate

In addition to phase-locking. The mean spike rate can also be used to represent frequency information. The theoretical foundation for this encoding mechanism assumes that two neurons with different but partly overlapping frequency responsiveness are presynaptic to a third neuron with a logical AND behavior. The frequency selectivity of the third neuron depends on the degree of overlap. An alternative mechanism is excitation from one neuron and inhibition from the other. These mechanisms presumably operate for sound frequencies that are too high to be encoded by phase-locking. As the following paragraph shows, the TS contains neurons that are appropriate for realizing such a mechanism.

Various methods have been applied to quantify frequency-responsiveness. The frequency-response range is determined by the lowest and highest frequencies at which the unit gives a just-detectable change in

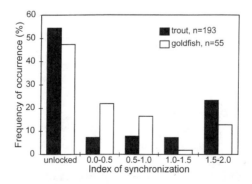

FIGURE 6.2. Distribution of indexes of synchronization (SIs) of the torus semicircularis (TS) units of the goldfish (calculated from Lu and Fay 1993) and the trout (Wubbels and Schellart, unpublished data).

the mean spike rate to a stimulus of fixed intensity (Page 1970; Schellart 1973). In addition, various laboratories used spike-rate based FTCs measured with the criterion response method. Also, response-area plots (Lu and Fay 1993, 1995, 1996) that are derived by applying tone bursts of different frequencies at a number of intensity levels have been used for this purpose, with responsiveness (spike count, z-axis) plotted as a function of frequency (x-axis) and intensity (y-axis) (Fig. 6.3). The above methods resulted in classifications of FTCs regarding the strength of frequency tuning, based on Q_{10dB} (by definition Q_{10dB} is CF/[f_1-f_2], with f_1 and f_2 representing the bandwidth of FTC at 10 dB above CF threshold), and regarding low-frequency versus high-frequency selectivity, based upon CFs (Nederstigt and Schellart 1986; Lu and Fay 1993, 1995).

For the auditory neurons in the medulla, the frequency range of response differs between species: two to four octaves in the goldfish (Page 1970; Sawa 1976), three to five octaves in *Clupea harengus* (Enger 1967), and one to three octaves in the trout (Wubbels et al. 1993). About 60% of the criterion FTCs of the goldfish medulla have a Q_{10dB} less than 0.75.

For the TS, the frequency response range is 0.5 to 4 octaves in the goldfish (Page 1970; Lu and Fay 1993) and 0.25 to 3 octaves in the trout (Schellart 1990). Combining both goldfish studies, most units (85%) have a Q_{10dB} of the criterion FTCs larger than 0.75. A small number of TS units in the goldfish (Page 1970; Lu and Fay 1993) and trout (Nederstigt and Schellart 1986), mostly broad-band units, show two CFs, or more than one response range. For some TS units, auditory responses vanishes in between two maxima, and occasionally even the spontaneous discharge is suppressed. However, since constant pressure as a function of frequency does not guarantee constant kinetics of the sound stimulus, a quantitative, detailed

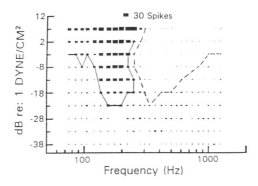

FIGURE 6.3. Response area (RA) of a spontaneous, low frequency unit of the goldfish TS. The size the filled boxes represents the number of spikes. The solid line represents a criterion frequency-threshold-curve, or FTC (mean spike rate), and the dashed line an area of inhibition. (Modified from Lu and Fay 1993.)

interpretation of the shapes of the frequency response curves, especially for nonspecialized fish (see Schellart and Kroese 1989), is not feasible. It is worth noting that none of the units in the medulla displays dual response ranges. The frequency response behavior of TS neurons considered together (Nederstigt and Schellart 1986; Lu and Fay 1993) roughly reflects the width of the behavioral audiograms (Fay 1988).

In the trout, low-frequency units (mainly lateral line input), high-frequency units (predominantly auditory input), and broad-band units could be distinguished (Nederstigt and Schellart 1986). The broad-band units appear to receive input from the ear, but for frequencies below 125 Hz some will also obtain input from the lateral line system (Schellart and Kroese 1989). TS units in the goldfish show a distinction into low-, mid-, and high-frequency units with CFs close to 155, 455, and around 855 Hz (Lu and Fay 1993). In addition, narrow and broad-band units have been found (Page 1970; Lu and Fay 1993). Frequency response ranges smaller than half an octave do not occur and mid-frequency units show less tuning. The responses are thought to be elicited exclusively by input from the otolith systems, mainly from the sacculus.

Neurons of the goldfish CPN typically show broad-band behavior with often two or three CFs, depending on the stimulus intensity, and a mean Q_{10dB} of 1.0, which is 0.5 less than the mean in the TS (Lu and Fay 1993, 1995). Since the multiple CF is more prevalent in the CPN than in the TS, the CPN (and probably the telencephalon) is likely involved in recognition of complex sounds such as the species natural calls.

A study of the TS of the mormyrid *Pollimyrus isidori* (Crawford 1993) yielded similar data as found in the goldfish, but with higher Q_{10dB}s. The frequency response data and other response features of *Pollimyrus*, and central data of *Ictalurus nebulosis* (e.g., Plassmann 1985), do not suggest that central auditory processing of electrosensory teleosts is basically different from that of nonelectrosensory teleosts.

Broad-band and narrow-band units comprise a small proportion of central auditory neurons; these neurons can potentially encode intensity, irrespective of their phase-locking behavior. A minority of TS neurons can preserve phase-locking in the auditory periphery, and the best of these neurons have a resolution up to about 0.10 ms (calculated from Schellart et al. 1987; Lu and Fay 1993). In general, spike rate is intensity, as well as frequency, dependent (e.g., Lu and Fay 1993, 1995). However, there are no indications that encoding of intensity and time is well segregated along the auditory pathway. This is in contrast to the obvious segregation observed in the electrosensory pathway of electric fish with a time resolution in TS of about 0.01 ms (Carr 1993). For both modalities, the resolution of time coding is higher in the TS than in the medulla and primary fibers. Higher auditory specialized vertebrates such as the barn owl also show a separated pathways for encoding intensity and time (Carr 1993).

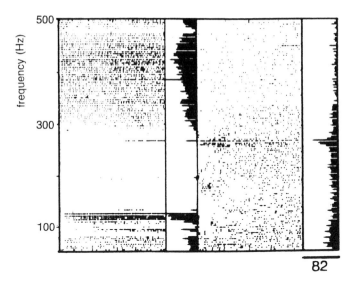

FIGURE 6.4. Frequency-time dot display of a spontaneously active, phase-locked broad-band unit of the trout TS. The figure contains the responses to nearly 140 tone bursts (left panel, on, 720 ms), each followed by a silent interval (right panel, off). Each dot represents a spike. The frequency of the tones in the bursts increases from 60 Hz (bottom) to 510 Hz (top). In the two histograms, the number of spikes elicited in each 720-ms period is given. The horizontal bar in the right lower corner gives the scaling of the histograms. Between 130 and 275 Hz, the unit exhibits strong inhibitory behavior (suppression of spontaneous spike rate); in the range 230 to 510 Hz, suppression of the spontaneous discharge during the silent intervals; and less than 130 Hz, lateral line activity (after Nederstigt and Schellart 1986).

2.4 Temporal Response Characteristics

2.4.1 Excitatory and Inhibitory Responses

The majority of central neurons show an increase in firing, that is, an excitation, during the presentation of a tone burst. In the medulla and the TS, a suppression of the spontaneous rate (when present) usually occurs during the first 0.5 to 1 s after cessation of the tone burst. Stimulus-induced inhibition is often observed in the TS (Page 1970; Schellart et al. 1987), but very rarely in the medulla (Wubbels et al. 1993). Some inhibitory units produce a spike rate higher than the spontaneous rate after the cessation of the tone burst. A small population of TS neurons show excitation to one frequency range and inhibition to another range (Figs. 6.3 and 6.4) (Nederstigt and Schellart 1986; Lu and Fay 1993). Occasionally, TS units show sustained firing even after cessation of the tone burst (Page 1970; Nederstigt and Schellart 1986). In the trout, TS units without inhibition generally have a narrow-band nature. Units with a mixed, frequency-dependent excitatory and inhibitory behavior (about 20%) are exclusively broadband units (Schellart et al. 1987). Inhibitory behavior has also been observed in the TS of *Gadus morhua* (Horner et al. 1980). Inhibition is thought to improve frequency discrimination.

Response-area plots of goldfish TS neurons occasionally show inhibitory areas, as exemplified by the unit in Fig. 6.3 (Lu and Fay 1993). Also, the response strength of a two-tone interaction experiment can be presented in a response-area plot (Lu and Fay 1996). In such experiments, the frequency of the probe signal is fixed near or at the unit's CF at a fixed intensity. The frequency of the test signal is plotted along the x-axis and its intensity along the y-axis of the response-area plot. In contrast to the single-tone response-area plots, in the two-tone response-area plots most units show inhibitory areas. These areas are adjacent to the excitatory area of the one-tone response-area plots (Lu and Fay 1996). Low-, mid-, and high-CF TS neurons exhibit these inhibitory areas often having complicated configurations, whereas in primary fibers these areas have been found only for the low-frequency type. In the TS, the most effective suppressor frequency in two-tone experiments is relatively close to the unit's CF (about 10 to 500 Hz difference), whereas for primary fibers the frequency difference is large (generally >500 Hz). In the TS, two-tone interaction (seldom excitation for a restrictive frequency band) is observed at lower intensities for the inhibitory tone, and shows stronger inhibitory effects than in primary fibers (Lu and Fay 1996).

Evaluation of the inhibitory phenomenon in goldfish and trout shows that along the auditory pathway there is substantial two-tone inhibition in the TS but not earlier. In general, two-tone interaction in the fish TS resembles that in the frog (see Section 3).

2.4.2 Transient Behavior, Adaptation, and Amplitude Modulation

In the trout, the peristimulus time histograms (PSTHs) of medullary units show that the responses to tone bursts are predominantly sustained (Wubbels et al 1993), whereas in the TS about 70% has a substantial transient component lasting 50 to 150 ms. Pure transient responses occur less frequently (Table 6.1). Similarly, in the TS of the goldfish, the transient type, with or without sustained component, is more common (Lu and Fay 1993). A minority of the units can be classified as primary-like, buildup, onset, pauser, and choppers (but not phase-locked) (Lu and Fay 1993; Wubbels and Schellart, 1997). Those types have been described earlier for the mammalian cochlear nucleus (Rhode and Greenberg 1992). The majority of the thalamic neurons shows various types of sustained PSTHs (Lu and Fay 1995). Transient and sustained behavior can be frequency dependent (Page 1970; Nederstigt and Schellart 1986). Features of the broad-band units (e.g., frequency-dependent PSTHs) indicate that their responses are produced by interaction of input from various acoustic subsystems (possibly otolith systems) with different frequency ranges.

TS units often show habituation, a decrease of the response to repeated presentation of for example tone bursts. This feature is not particularly well developed in earlier stages of auditory processing (Schellart et al. 1987).

Auditory nerve fibers generally follow the sinusoidal amplitude modulated (SAM) tone, since they have only a weak transient component and a monotonic rate-level function. Since TS units often show a nonmonotonic relation and since their PSTHs are highly variable, SAM responses are highly variable, as to be expected. In the TS of the electrosensory *Ictalurus nebulosis*, two types of units are found (Plassmann 1985). One type, the slowly adapting one, follows the SAM tone up to a SAM frequency of 60 Hz. The other, the fast adapting type, can only follow higher modulation frequencies. Its behavior can be described by differentiating the envelope with a higher sensitivity for increasing than for decreasing intensities. It is less dependent on the absolute intensity than the slowly adapting type, which is in accordance with the differentiating behavior. The transient component of the response to a SAM tone burst resembles the psychophysical sensitivity to the modulation frequency fairly well (Coombs and Fay 1985).

2.5 Processing of Directional Information

Behavioral experiments have demonstrated that ostariophysine fish and fish without hearing specializations are able to detect the direction of an underwater sound source (Schuijf and Buwalda 1980; Popper and Fay, Chapter 3). Fish cannot use interaural time, phase, and intensity differences of sound pressure to localize a sound source (Fay and Feng 1987). These cues are useless due to the small difference in times of arrival between both ears and the almost identical acoustic impedance of the fish body and the surrounding water. The temporal resolution of central neurons with the highest SIs found (see Section 2.3.2) is at least 10 times larger than the difference in times of arrival (a few microseconds for 30-cm fish). Interaural intensity differences are generally very small, at most 0.5 dB for source distances >1 m, which is well below the neurophysiological and psychophysical resolution (Fay 1988). In contrast to terrestrial vertebrates, fish make use of both the kinetic component of sound, which impinges directly onto the otolith organs, and the pressure component provided by the swim bladder, a pressure-to-displacement transducer, which stimulates the otolith organs indirectly (van Bergeijk 1967; Schuijf 1976; Schellart and de Munck 1987). Addition of direct and indirect stimulation yields different displacement orbits (ellipses). The orbit parameters depend primarily on the location of the sound source and secondly on the ear (left or right). The orbit hypothesis claims that fish use the orbit parameters for source localization (Schellart and de Munck 1987). Other theories on the mechanism of source localization are based on time analysis and separate responses to the direct and indirect stimulus (Schuijf 1976; Schuijf and Buwalda 1980; Buwalda et al. 1983; Rogers et al. 1988). Experimentally and theoretically, it seems likely that fish use the input from both ears for optimal auditory localization (Schellart and Popper 1992; Popper and Fay, Chapter 3).

The TS is the first nucleus along the auditory pathway with substantial convergence of binaural input (e.g., de Wolf et al. 1983; McCormick, Chapter 5). Therefore, studies about the processing of sound source direction have been focused on this nucleus.

In the cod (*Gadus morhua*), binaural interaction has been investigated by anodal blocking of the posterior saccular nerves during presentation of airborne sound stimuli, which reach the otolith organs mainly via the swim bladder. It was found that binaural interaction was more prominent in the TS than in the medulla (Horner et al. 1980).

In the trout, processing of directional information has been studied with a two-dimensional (2-D) (x-y) horizontal vibrating platform on which the fish was rigidly attached (Schellart et al. 1995b). By computer control, x-y cross-talk caused by small movements of the fish's head with respect to the platform was reduced to about 5% (Goossens et al. 1995). Vibrations of the skull (generally tone bursts of 172 Hz, 0.5 to 200 mms^{-2}) were measured with a miniature 3-D accelerometer.

In the trout, about 35% of the auditory units show directional selective (DS) responses in the horizontal plane (Wubbels et al. 1995). In Figure 6.5, the response to each tone burst is represented as a line of dots (spikes). The bottom histogram reveals the sustained/transient nature of the response. Since each subsequent tone burst simulates a slightly different source direction, the right-hand histogram shows the direction selectivity of the unit. The −3 dB width of the histogram is called the directional selectivity range. By definition a cosine-shaped histogram has a directional selectivity range of 90° (−3 dB points 90° apart). Such histograms have been found in hair cells (Shotwell et al. 1981) and eighth nerve fiber studies of fish (Fay 1984). However, most DS units in the TS show a range smaller than 90° (mean ± SD = 67° ± 48°, $N = 59$), which may be caused by a suppression mechanism.

SIs of DS units, which are higher than that of non-DS units, do not appear to depend on intensity and direction (Fig. 6.6). Thus, the strength of the phase-locking cannot encode sound direction. However, the mean spike rate (DS units are generally nonspontaneous) may, since this is clearly direction dependent, but intensity independent (Schellart et al. 1995a).

The present neurophysiological findings do not unequivocally support a particular model of directional hearing. However, the response of a small

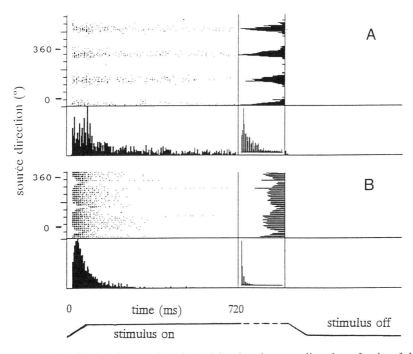

FIGURE 6.5. Spike density as a function of simulated source direction of units of the trout TS. A: Unit with a mixed transient/sustained nature and with a high directional selectivity (30 mms^2). B: Transient unit (120 mms^{-2}). For both units the stimulus was a 172-Hz tone burst (rectilinear motion). The direction of each tone burst (rotated a few degrees counter-clockwise with respect to the preceding one) is given along the vertical axis (0° is forward).

number of the DS units are affected by the direction of revolution or the shape of an elliptic displacement stimulus (Schellart et al. 1995a). These findings support the orbit hypothesis (Schellart and de Munck 1987), since these stimulus cues are the basic elements of this hypothesis.

For the central analysis of some cues of the sound signal (e.g., frequency), phase-locking is essential. The fact that 75% of the DS units are phase-locked (versus 23% of the non-DS units) suggests that this temporal information may also play a role in directional hearing in fish. That the phase-lock response of some of these units is direction dependent (Schellart et al. 1995a) provides another line of evidence supporting the role of time analysis in directional coding. Irrespective of which response features are the basic cues for source localization, it is thought that the TS plays a key role in source localization in the trout and probably in other bony fish.

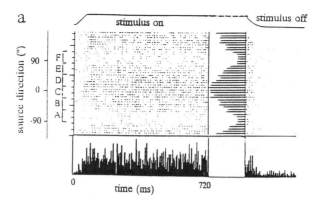

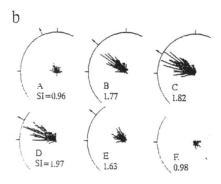

FIGURE 6.6. Response characteristics of a spontaneous, sustained DS units of the trout TS. A: The bottom histogram denotes the peristimulus time histogram (PSTH) and the right-hand histogram the number of spikes elicited during a single burst (stimulus amplitude $1.2\,\mathrm{mms}^{-2}$). B: The instant of firing of the spike response is represented in six polar period histograms, where A–F correspond to the indicted sections in A. The synchronization index (SI; range $= 0$–2) is defined as the ratio of the amplitude of the Fourier fundamental of the period histogram and the mean spike rate during stimulation. Sections A and F mainly contain spontaneous activity.

2.6 Topographic Organization of the Torus Semicircularis

2.6.1 Tonotopy

In fish, mapping of auditory frequency has not been adequately studied. In the carp *Cyprinus carpio*, high-frequency units are found more rostrally and low-frequency units more caudally (Echteler 1985b). In the trout TS, it has been found that high-frequency units are mainly located rostrally, whereas broad-band units occur preferentially in the middle and caudal parts of the TS (Schellart et al. 1987). A well-developed tonotopic organization, which is typically found in the inferior colliculus of higher vertebrates, is probably not the general organizing principle for fish species. In part this is because the overall range of frequency sensitivity is often very narrow and the population of well-tuned units with one CF is small.

Auditory fibers and caudal brain stem auditory units of fish often show less frequency selectivity compared to TS units (Fay 1978, 1990; Lu and Fay 1993). At central levels, frequency selectivity can be realized by neuronal interaction or temporal filtering. By definition, broad-band units (low Q_{10dB}, often two or three CFs) are not the elements to construct a frequency map. However, phase-locked broad-band units may provide input to narrow-band units and broad-band units not displaying time-locking may form part of a frequency detecting mechanism (see Section 2.3) or may play a role in "recognizing" complex sounds.

2.6.2 Spatiotopy

Mapping of auditory space has been investigated for the horizontal plane in the trout TS (Wubbels et al. 1995). The DS units found in a single-electrode track of some hundreds of microns have the same preferred direction. Figure 6.7A presents the preferred directions of 32 clusters of DS units and the locations of their recording sites (projected on the horizontal plane). The medial part of the TS appears to code the rostrocaudal direction (0°). In the rostrolateral part, the 45° direction predominates, and in the caudolateral part, −45° (Wubbels et al. 1995).

In the ostariophysine goldfish, there is evidence that the pressure and displacement components of the sound are not processed homogeneously in the TS (Fay et al. 1982). It is not clear whether this segregation is (1) related to some segregation of the projections of the various otolith systems, (2) playing a role in frequency analysis, or (3) playing a key role in directional hearing. This finding does not imply that a similar inhomogeneity should be expected for nonspecialized fish.

For airborne sound, mapping of auditory space has been demonstrated in the midbrain of birds and mammals (Irvine 1992). Although fish, like other aquatic vertebrates, are subject to different biophysical constraints, auditory space also appears to be mapped in the midbrain (Wubbels et al. 1995).

2.6.3 Multimodality and Topography of Modalities

The significance of multisensory integration is improvement of signal detection, localization or orientation, and recognition. Improvement of deductibility, leading to easier perception of an object, may be important under near-threshold conditions. In the midbrain, interaction between the visual and acousticolateral systems has been reported for the ventral part of the goldfish TS (Page and Sutterlin 1970) and the trout TS (Schellart 1983). In addition to an algebraic-like interaction (possibly due to dendritic summation), some bimodal units show a more complex type of interaction

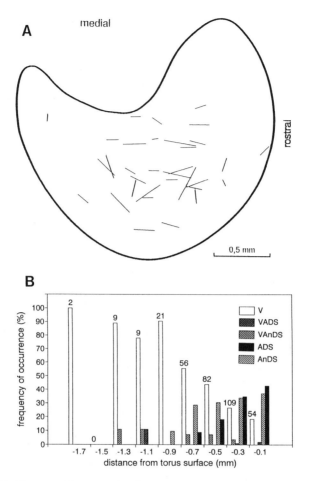

FIGURE 6.7. Topographic organization. A: Map of preferred directions of directional selective (DS) units in the trout TS (projection on the plane parallel to the torus-midbrain ventricle interface). The direction of preference (rectilinear stimulus) is denoted by a line segment, the length of which is determined by the directional selectivity ranges of the DS units found in a single electrode track (Wubbels, et al. 1995, supplemented with new data, with permission from Elsevier Science Ireland Ltd., Clare Ireland.) B: distribution of recording depths of visual units (V), non-DS bimodal units (VAnDS), DS bimodal units (VADS), non-DS units (AnDS), and DS units (ADS). The columns represent the normalized probability of occurrence. Numbers on top of the bars represent numbers of units per bin.

(possibly based on presynaptic facilitation). This type of interaction is of instantaneous nature and can be interpreted as fast, visually induced auditory arousal (Schellart 1990). Since the TS projects to the reticular formation (Schellart 1990), this fast arousal behavior may play a role in sensory motor control. The functional significance of the correlated occurrence of visual sensitivity and the narrow-band auditory character is not clear (Page and Sutterlin 1970; Schellart et al. 1987).

In goldfish, recordings have been made from acoustico-visual (AcV) units in the ventral portion of the TS (Page and Sutterlin 1970). In the trout, AcV units, less numerous than the unimodal types, occur in the deep as well as superficial portions of the TS, with the highest occurrence being in the rostral part (Schellart et al. 1987). In goldfish and trout, about 80% of the units located in the ventral TS are purely visual; only a few unimodal acoustic units can be found in this locus. The dorsal portion of the TS appears to have a high concentration of acoustic neurons (Page 1970; Wubbels et al. 1995). This is consistent with the result of a horseradish peroxidase study in the trout showing that the ventral TS (and the most rostral part) receives tectal input and that the rostral and ventral regions of the TS are reciprocally interconnected (Schellart, unpublished data). Non-DS units of the trout are located in the upper 800 μm of the TS and DS units mainly in the upper 400 μm (Wubbels et al. 1995) (Fig. 7B); the upper 400 μm comprises two layers of neurons with small somata (Cuadrado 1987).

Besides the general rostrocaudal and dorsoventral segregations, there is also a finer structure-function relationship. Neighboring units often have the same modality, resulting in separate clusters or columns of visual, acoustic (or acousticolateral) units, and, less frequently, AcV units (Schellart et al. 1987; Wubbels et al. 1995). The clusters are oblong and oriented perpendicular to the surface of the TS with a rostrocaudal width of about 0.1 mm and a vertical length of (at least) 0.2 and 0.4 mm for acoustic and visual clusters, respectively. DS and non-Ds units occur in separate clusters.

Central processing from the hindbrain to at least the TS does not appear to be different in a significant way among the small number of species investigated, hearing specialists as well as nonspecialists. A remaining question is the extent to which minor differences in topographic organization, (e.g., tonotopy and sound localization) may exist. These differences may be related to lifestyle (e.g., feeding behavior and sound communication), and environment (e.g., fresh versus marine waters, ambient noise levels) (Schellart 1992).

The trout TS shows some retinotopic organization (Schellart and Rikkert 1989), but apparently this map is not in register with the auditory space map. In contrast, visuo-acoustic spatiotopy in the midbrain has been established for various higher vertebrates with the two maps in close register with one another (Kuwada and Yin 1987).

2.7 Are Acoustic and Lateral-Line Processing in the Medulla and Torus of Fish Integrated?

The central auditory and lateral-line pathways in fishes are in close proximity, so that some of their structures are anatomically intermingled (Schellart and Kroese 1989; McCormick, Chapter 5). Accordingly, auditory sensitive and lateral-line sensitive neurons have been encountered in the same part or layer of a nucleus, and thus the discovery of multimodal neurons in a single nucleus. A typical example of such structures is the TS, where auditory, vestibular, visual, lateral-line, and, in some species, electroreceptive information is processed.

The processing of mechanosensory lateral-line information and auditory information appears to be separated at the level of the medulla (Bleckmann and Bullock 1989; Wubbels et al. 1993), although in some rare instances auditory-visual and lateral-line–auditory units have been encountered (Caird 1978, Wubbels et al. 1993). The separation is in line with the observation that there is little or no overlap of the primary projections of both systems (see Schellart and Kroese 1989; Popper and Fay, Chapter 3).

In the TS, the situation is less clear. In the carp (*Cyprinus carpio*), the medial part of the TS processes acoustical but not lateral-line information (Echteler 1985a). However, it is questionable whether the airborne stimulus was strong enough to stimulate the lateral-line organs (see Schellart and Kroese 1989). In goldfish, acoustic responses have been recorded from the medial TS as well as from the lateral TS (about 20% of the total width) (Page 1970). Although the physical stimuli are not carefully delineated in this study, control experiments were made to exclude acoustic responses that were contaminated by lateral-line responses. In the catfish (*Ictalurus nebulosis*), there is strong evidence that the two modalities and the electrosensory modality are processed in rather distinct areas of the TS (see Schellart and Kroese 1989). The data from the trout TS, however, show that the medial TS processes acoustic as well as lateral-line information (Schellart and Kroese 1989).

Results of neuroanatomical studies (for a review see Schellart and Kroese 1989) reveal that the extent of overlap of projections from both systems to the TS is strongly species dependent. In the trout, there is considerable overlap, but the modality of the most lateral part is an open question. In *Sebastiscus* there is a good segregation and in the *Cyprinus carpio* there is little overlap.

In conclusion, depending on the species, the medial part of the TS is acoustic or acousticolateral. Whether in some species the lateral part processes acoustic information in addition to lateral-line information is not clear.

2.8 Neurophysiology of the Efferent System

The efferent system comprises a paired, octavolateral efferent nucleus (OEN), which is located in the hindbrain lateral to the medial longitudinal fasciculus (McCormick, Chapter 5). Its functional significance is unclear. In general, efferent fibers that innervate hair cells modulate the afferent response in various end organs (Highstein 1991; Roberts and Meredith 1992). The number of efferent fibers is much lower than the number of afferent fibers. Since probably nearly all otolithic hair cells obtain efferent input, each efferent neuron must innervate hundreds of hair cells (Popper and Saidel 1990). However, the exact role of the efferent system in inner ear functioning remains to be determined. An interesting question is the functional difference between the OEN neurons, that innervate the inner ear bilaterally (Schellart et al. 1992) versus those that primarily innervate the ipsilateral inner ear. Other remaining questions are (1) Does the OEN have a topographic organization with respect to the end organ (auditory and vestibular)? (2) Under what conditions (intensity, frequency, etc., of the acoustic stimulation, other modality) does the efferent system operate? (3) Are there major differences between hearing specialists and nonspecialists?

2.9 Summary of Information Processing Along the Fish Auditory Pathway

In fish, various response parameters change along the central auditory pathway. The following are some of the most notable changes, assuming that the findings in the goldfish thalamus hold qualitatively for other teleosts:

- The occurrence and strength of spontaneous activity decreases systematically.
- The occurrence of phase-locking decreases.
- The SI averaged over all units decreases, but the highest SIs are found in some units in the TS.
- The auditory sensitivity increases.
- The variability of frequency behavior increases up to the TS, but high-frequency selectivity is lost in the CPN.
- There is an increase in transient responses to stimulus onset (excitation) and offset (inhibition).
- The occurrence of frequency-dependent inhibition is typical for TS neurons.
- Up to the TS there is an increase of the occurrence of bimodality, especially with the visual modality.

In many respects, the TS has well-developed topographic organizations. The most obvious ones are

- the existence of a spatiotopic map (shown for the rainbow trout) and probably a global tonotopic map;
- the occurrence of directionally selective units in the superficial layers and nondirectionally selective auditory units in deep layers.

3. Central Auditory Processing in Amphibians

This section describes our current understanding of the central neural mechanisms that underlie sound pattern recognition and sound localization in amphibians. The emphasis is on anurans, particularly those frog species that communicate by airborne sounds (for studies on underwater sound processing, see review by Elepfandt 1996). Selected issues of major interest to auditory neurobiologists are highlighted instead of presenting an exhaustive review of all existing data.

3.1 Characteristics of Frog Calls and Behaviorally Relevant Features

Acoustic signals are emitted in various behavioral contexts and the anuran natural sounds can be classified into different functional categories, for example, advertisement, territorial, release and distress calls, etc. (see Zelick et al. Chapter 9). The advertisement (or mating) calls are typically produced by males and have stereotyped power spectrums and temporal patterns. Females of numerous different species have been shown to have the ability to discriminate calls of conspecific males from those of sympatric males. In a mixed chorus, a female frog can single out a male among the many calling males participating in the chorus, and approach this male, bypassing other males en route. This behavioral response demonstrates the effectiveness by which frogs discriminate and localize sounds. Their ability to perform auditory scene analysis is even more impressive when one considers that the interaural distance between the frog ears is very small and thus the physical binaural cues for sound localization for these animals are minute.

Figure 6.8 illustrates the diversity of the vocal signals of North American ranid frogs. The mating call of Northern leopard frogs (*Rana pipiens pipiens*) comprises a train of sound pulses of 15- to 20-ms duration with a pulse repetition rate of about 20 pulses per sec (pps), having spectral energy concentrated at 375 Hz, 1100 Hz, and 1500 to 1700 Hz. In contrast, the release call of this same species consists of a train of broad-band sound pulses of shorter duration (<8 ms) with a faster repetition rate (~40 pps). The mating call of *R. p. pipiens* differs from that of bullfrogs (*R. catesbeiana*) or of plains leopard frogs (*R. blairi*), the two frog species that are often found to share their breeding habitats. The bullfrog mating call consists of single notes that are long (>500 ms), whereas that of plains leopard frogs consists of several notes having a duration of about 80 ms (this note duration is intermediate between that of Northern leopard frogs and bullfrogs). Similarly, the repetition rate as well as the rise and fall times of the note for plains leopard frogs are intermediate between the same parameters found in the

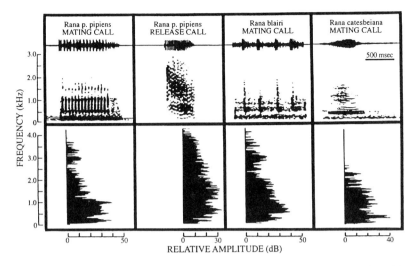

FIGURE 6.8. Diversity of vocal signals of ranid frogs in North America. (From Hall and Feng 1988, with permission of Elsevier Science–NL, Amsterdam, the Netherlands.)

Northern leopard frogs and bullfrogs. Finally, the power spectrum of the mating calls of these three ranid frogs are distinctly different.

The salient features of frog's mating calls have been well characterized for many ranid and hylid frogs (see Zelick et al. Chapter 9 for details). The power spectrum (Capranica 1965; Gerhardt 1974), or one (or more) of the temporal characteristics of the call such as the pulse repetition rate (Blair 1964; Loftus-Hills and Littlejohn 1971; Gerhardt 1978; Diekamp and Gerhardt 1995), the calling rate (Schneider 1982), the pulse number (Fouquette 1975), the pulse duration (Narins and Capranica 1978), or the rise time (Gerhardt and Doherty 1988; Diekamp and Gerhardt 1995), is important in call discrimination. Other studies, however, underscore the importance of both the spectral as well as the temporal features (Walkowiak and Brzoska 1982; Gerhardt and Doherty 1988). There is evidence that frogs can employ a variety of acoustic cues available to them, but these cues interact and become significant under different conditions (Gerhardt and Doherty 1988; Diekamp and Gerhardt 1995).

3.2 Neural Representations of Behaviorally Important Sound Features

3.2.1 Representations in the Frog Auditory Periphery

Since this information is treated in detail by Lewis and Narins in Chapter 4, only a cursory review will be included here in order to appreciate the manner by which signal representations are transformed from the peripheral to the central auditory system.

Auditory nerve fibers in frogs possess V-shaped frequency-threshold curves (Frishkopf et al. 1968; Feng et al. 1975). For each fiber, a distinct characteristic frequency (CF) can be identified to which the fiber is most sensitive. Those fibers that are tuned to low frequencies display pronounced two-tone suppression (i.e., response to fiber's CF at 10 dB above threshold can be suppressed completely by the simultaneous presence of a second tone of higher frequencies). Fibers tuned to high and intermediate frequencies generally do not exhibit, or display much weaker, two-tone suppression (see Chapter 4). In response to a tone burst, a fiber typically shows a sustained firing throughout the duration of the stimulus; there is a pronounced and consistent phasic component that is followed by a lower level of sustained firing due to adaptation. Frog's auditory fibers show varying degrees of adaptation (Megela and Capranica 1981; Megela 1984).

Auditory nerve fibers function basically as envelope detectors (Rose and Capranica 1985; Feng et al. 1991). In response to an increase in signal duration, auditory nerve fibers typically show a corresponding increase in their firing duration and thus an increase in the mean spike count (Fig. 6.9A).

In response to SAM stimuli, auditory nerve fibers give rise to time-locked discharges to the individual modulation cycle. The time-locking, as measured quantitatively by the synchronization coefficient (range = 0 to 1, with a value of 1 representing perfect firing synchrony), is usually robust at low modulation frequencies and the synchronization coefficient deteriorates at higher AM frequencies. Thus, the synchronization coefficient is a low-pass function of the modulation frequency (Hillery and Narins 1984; Rose and Capranica 1985; Feng et al. 1991). In other words, auditory nerve fibers give low-pass sync-based modulation transfer functions (Fig. 6.9B).

In response to a different type of AM stimuli, the pulsed-amplitude–modulated (PAM) stimuli (characterized by a train of sound pulses), the upper limit of time-locking is determined not by the AM rate itself, but by the silent gap between tone pulses (Feng and Lin 1994a). The upper limit is higher when tone pulses comprising the PAM stimuli have shorter durations (Fig. 6.9C).

The spike count also depends on whether the AM signal is PAM or SAM (Feng et al. 1991). When presented with PAM tones, auditory nerve fibers show an increase in the mean spike count with the modulation frequency, thereby giving rise to high-pass modulation transfer functions (Fig. 6.9F). The high-pass response function is attributed to the fact that both the average stimulus energy and the duty cycle increase with the modulation frequency. When presented with SAM tones, however, these same fibers give rise to all-pass count-based modulation transfer functions because the average energy for such signals is independent of the modulation frequency (Fig. 6.9E).

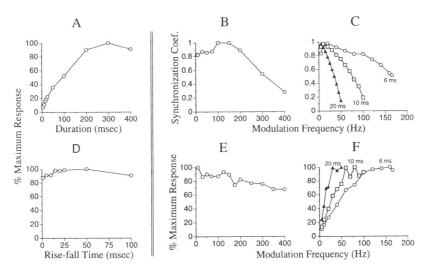

FIGURE 6.9. Representative response characteristics of eighth nerve fibers to changes in the temporal parameters of sound stimuli: duration (A), rise-fall time (D), and AM frequency of sinusoidal amplitude modulated (SAM) stimuli (B and E) and of PAM stimuli (C and F). For panels C and F, response functions deriving from pulsed-amplitude–modulated (PAM) stimuli having different pulse durations are plotted separately and labeled accordingly. (Modified from Feng et al. 1991; Fend and Lin 1994a.)

Finally, in response to tone bursts of a constant duration having a wide range of rise-fall times, auditory nerve fibers show little change in the mean firing rate or mean spike count (Fig. 9D) (Feng et al. 1991). Taken together, these results are consistent with data from other vertebrate species showing that auditory nerve fibers function primarily as envelope detectors.

3.2.2 Representations in the Central Auditory System

3.2.2.1 Dorsal Nucleus in the Medulla

All auditory nerve fibers terminate in the ipsilateral dorsal (or dorsolateral) nucleus in the medulla oblongata (see McCormick, Chapter 5). Fibers innervating the high-frequency–sensitive organ (i.e., the basilar papilla) terminate in the dorsomedial region of the nucleus, whereas fibers innervating the low-frequency–sensitive organ (i.e., the amphibian papilla) terminate in the ventrolateral region of the nucleus (Fuzessery and Feng 1981; Lewis et al. 1980). The dorsal nucleus is thus tonotopically organized (see McCormick, Chapter 5). Further, neurons in the dorsal nucleus preserve faithfully the frequency tuning characteristics of auditory nerve fibers, that is, they have V-shaped frequency-threshold curves with low-

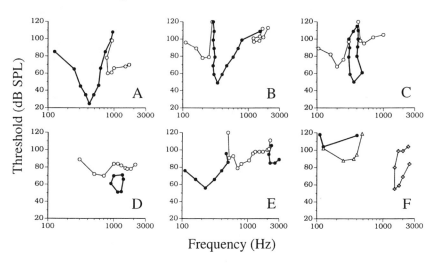

Frequency (Hz)

FIGURE 6.10. Representative frequency-tuning characteristics of neurons in the dorsal nucleus (A), superior olivary nucleus (B), TS (C, D, E), and posterior thalamic nucleus (F). The excitatory frequency tuning curves are shown by thick lines and filled circles, and the inhibitory tuning curves by thin lines and open circles. A neuron's inhibitory tuning curve was determined using a two-tone stimulation paradigm with one tone fixed at the neuron's CF and the second tone varied in its frequency and amplitude. TS neuron in E was excitable by a low-frequency tone, or by a high-frequency tone. Thalamic neuron in F showed poor excitation by a single tone (filled circles) but was easily excitable by two tones presented simultaneously; the curve with open plus symbols was obtained when one tone was fixed at 200 Hz and 90 dB SPL (below threshold of excitation to this tone), and the curve with open triangle symbol was obtained when one tone was fixed at 1700 Hz and 90 dB SPL. (Modified from Fuzessery 1983.)

frequency–sensitive neurons displaying two-tone suppression (Fig. 6.10A) (Fuzessery and Feng 1983a).

Unlike auditory nerve fibers, however, the temporal discharge patterns of neurons in the dorsal nucleus are diverse (Hall and Feng 1990; Feng and Lin 1994b). Although one-half of neurons in this nucleus give primary-like firing patterns in response to tone bursts at the unit's CF, that is, a pronounced onset response followed by varying levels of steady-state discharge (Fig. 6.11A–C), the remaining neurons produce onset response (Fig. 6.11D), phasic burst pattern (Fig. 6.11F–I), or onset response followed by a pause and subsequent steady-state discharges (i.e., pauser pattern; Fig. 6.11E), or a chopper firing pattern (Fig. 6.11F, G). As described later, neurons in the dorsal nucleus, having different firing patterns, possess different temporal processing capacities.

Single neurons in the frog dorsal nucleus display phase-locked discharges to tones at frequencies of up to 800 Hz (Feng and Lin 1994b). This limit is

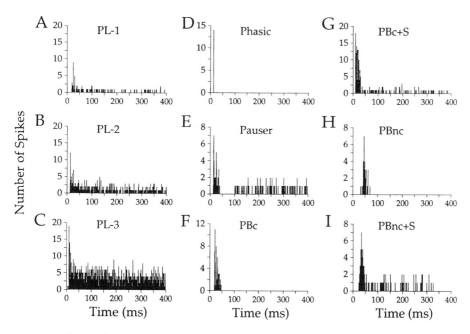

FIGURE 6.11. Diversity of temporal discharge patterns in the central auditory system of Northern leopard frogs. These patterns are taken from neurons in the superior olivary nucleus, but most of these patterns are observable from other auditory centers as well. PL-1, PL-2, and PL-3, the three classes of primary-like units having different adaptation rates, with PL-1 representing units that have the most rapid adaptation and PL-3 the slowest adaptation; PB, phasic-burst units; PBc and PBnc, phasic-burst units showing chopping and nonchopping discharge patterns; S, the additional presence of sustained firing that follows the early onset response. (Reprinted from Condon et al. 1995, with permission of Elsevier Science–NL, Amsterdam, the Netherlands.)

as high as for the auditory nerve fibers (Hillery and Narins 1984). The capacity of phase-locking to tones is correlated with the functional classification and the steady-state firing rate of the neuron in the dorsal nucleus (Feng and Lin 1994b). For example, the upper cutoff frequency of phase-locking to tones is highest for primary-like neurons that display high steady-state firing rates. In contrast, for the majority of neurons in the dorsal nucleus, and especially the phasic and chopper neurons, the phase-locking capacity to tones is poor, having upper cutoff frequencies of <300 Hz. This limit is considerably lower than that of eighth nerve fibers (see Lewis and Narins, Chapter 4) and furthermore it covers only a fraction of the total spectrum of frog's natural calls. Therefore, it is highly unlikely that frogs can make use of information deriving from interaural comparison of ongoing time (i.e., sound carrier) for localizing sounds in the auditory space. Sound localization must rely instead on interaural comparison of the difference in the timing of the AM envelope of complex natural sounds, or in the sound intensity (after a substantial amplification of the difference in intensity via the pressure gradient–receiving mechanism of the peripheral auditory system; see below). The pulsatile nature of frog vocal signals is thus not only useful for sound pattern recognition but also for sound localization.

The frequency range of firing synchrony to the envelope of complex SAM stimuli is also different for the different functional cell classes, as evidenced by the variation in the units' sync-based modulation transfer functions (Feng and Lin 1994b). Although neurons in the dorsal nucleus typically display low-pass, sync-based, modulation transfer functions, the mean upper cutoff frequency of phase-locking to AM stimuli differs among cell types, and ranges between 155 and 251 Hz. Primary-like neurons have lower, whereas phasic, phasic-burst, pauser, and chopper neurons have higher, upper cutoff frequencies. Taken together, it appears that although the non–primary-like neurons in the dorsal nucleus show inferior phase-locking to tones when compared to the primary-like neurons, they display superior phase-locking to AM stimuli and thus may play a more important role in the coding of sound pattern and direction (see above).

The emergence of new temporal discharge patterns at the dorsal nucleus also produces an increase in AM selectivity when a different response metric, the firing rate or spike count, is evaluated (Hall and Feng 1991; Feng and Lin 1994b). The primary-like and pauser neurons in the dorsal nucleus as well as primary auditory nerve fibers typically show an increase in the mean spike count with an increase in the modulation frequency of pulsed-AM stimuli and hence high-pass modulation transfer functions (Fig. 6.12A). The phasic and phasic-burst neurons, on the other hand, exclusively display band-pass modulation transfer functions showing optimal spike counts to a narrow range of AM frequencies (Fig. 6.12B). Thus, the transition from the peripheral to the central auditory system is characterized by an increase in AM selectivity. As described below, the number of primary-like neurons decreases systematically as one ascends beyond the dorsal nucleus, and concomitantly the number of neurons displaying high-pass, count-based, modulation transfer functions, a primary-like characteristic, also decreases systematically.

Taken together, these results show that signal representation in the time domain undergoes significant transformation at the first central auditory station. At present, it is unclear how neurons in the dorsal nucleus construct the different functional cell classes that possess different temporal response properties. A recent anatomical study (Feng and Lin 1996) indicates that the neuronal architecture of the frog dorsal nucleus is complex, approximating that of the cochlear nucleus of birds and mammals. This study points to the possibility that the different functional cell classes in the dorsal nucleus may have a morphological origin (e.g., synaptic connectivity) as well as a biophysical origin.

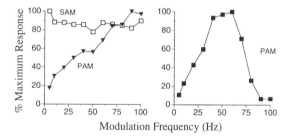

Modulation Frequency (Hz)

FIGURE 6.12. Representative AM response characteristics of primary-like neurons. (A: high-pass in response to PAM stimuli and all-pass to SAM stimuli) and phasic neurons (B: band-pass to PAM stimuli) in the dorsal and superior olivary nuclei. PAM stimuli are 400 ms in duration, and consist of a gated tone burst (at the unit's CF) of 400-ms duration that is modulated with 10-ms pulses at different modulation rates. SAM stimuli differ from PAM in that the modulating stimulus is a sine wave at different frequencies. The abscissa represents the modulation frequency of PAM and SAM stimuli. The ordinate represents the relative mean spike counts at different modulation frequencies (reference is the maximal response to the stimuli in the set). (Modified from Hall and Feng 1991; Condon et al. 1991.)

3.2.2.2 Superior Olivary Nucleus

Neurons in the dorsal nucleus project bilaterally to the superior olivary nucleus, as well as to the superficial reticular nucleus and to the torus semicircularis (TS) (Feng 1986a,b). Projection to the superior olivary nucleus is tonotopically organized; low to high frequencies are represented primarily dorsoventrally in the superior olivary nucleus. Although the majority of neurons (81%) in the superior olivary nucleus exhibit V-shaped frequency-threshold curves, approximately 19% of neurons have closed or W-shaped frequency-threshold curves (Feng and Capranica 1978; Fuzessery and Feng 1983a). Furthermore, all neurons in the superior olivary nucleus, not just those tuned to low frequencies, show two-tone suppression with inhibitory tuning curves along one, or both, flanks of the excitatory frequency-threshold curve (Fig. 6.10B). Thus, there appears to be a fair amount of convergence in the ascending auditory projection in the caudal brain stem leading to the diversity in the unit's frequency selectivity.

Signal representation in the time domain at the superior olivary nucleus changes only slightly from the dorsal nucleus (Condon et al. 1991, 1995). Neurons in the superior olivary nucleus exhibit primary-like, phasic, phasic-burst, and pauser temporal discharge patterns (Condon et al. 1995). The primary-like neurons in the superior olivary nucleus, like their counterparts in the dorsal nucleus, typically give higher spike counts in response to tone bursts of longer durations, or to PAM stimuli at high modulation frequencies, and are insensitive to stimulus rise-fall time (Condon et al. 1991). In contrast, phasic neurons in the superior olivary nucleus consistently give a single spike to each tone burst irrespective of the stimulus duration. Also, they show a greater selectivity to the modulation rate of PAM stimuli. Many phasic neurons in the superior olivary nucleus give band-pass, count-based, modulation transfer functions and display a preference for stimuli with short rise-fall times. One notable difference with the phasic neurons in the dorsal nucleus is the range of tone frequency or AM frequency over which phase-locking can be observed. The capacity of phase-locking degrades significantly in the superior olivary nucleus. The upper cutoff frequencies are noticeably lower for superior olivary neurons when compared to neurons in the dorsal nucleus.

3.2.2.3 Torus Semicircularis (TS)

As described in Chapter 5, the TS receives bilateral projections from both the dorsal nucleus and the superior olivary nucleus (Rubinson and Skiles 1975; Pettigrew 1981; Wilczynski 1981; Feng 1986b; Feng and Lin 1991). The TS has five subdivisions and three of which (the principal, laminar, and magnocellular nuclei) participate in auditory information processing. In the principal nucleus of the TS, low frequencies are primarily represented in the central region and higher frequencies at outer regions of the nucleus (Fuzessery 1983; Mohneke 1983; Eggermont and Epping 1986; Feng 1986a,b; Schneider et al. 1986; Walkowiak and Luksch 1994).

The frequency-threshold curves of TS neurons are considerably more complex than at lower centers, and many more TS neurons exhibit W-shaped frequency-threshold curves than at the superior olivary nucleus (Fig. 6.10E; Bibikov 1974a,b; Walkowiak 1980; Fuzessery and Feng 1982; Hermes et al. 1982). At the same time, as in the superior olivary nucleus, two-tone suppression is pronounced and can be observed in all TS neurons (Fig. 6.10C–E). The shape of the unit's frequency-threshold curve and the occurrence of two-tone-suppression appear to be attributed to γ-aminobutyric acid (GABA)ergic inhibition (Hall 1994). Evidence for this finding is that an application of a minute amount of bicuculline, a GABA-a antagonist, decreases the unit's frequency selectivity in the TS.

Different from the dorsal and the superior olivary nuclei, 14% of TS neurons respond poorly to single tones irrespective of the tone frequency and level (Fuzessery and Feng 1982). These same neurons, however, respond vigorously to specific combinations of tones and hence represent a neural analogue of the logical AND operation (see Fig. 6.11F as an example for thalamic neurons). The TS appears to be the first nucleus along the frog ascending auditory pathway where some of its neurons require the simultaneous presence of multiple tones for excitation.

Temporal coding in the TS has been investigated extensively. The TS plays an important role in the processing of time-varying signals. Gooler and Feng (1992) showed that it is here that neurons responding selectively to a narrow range of stimulus duration are first observed (Fig. 6.13A).

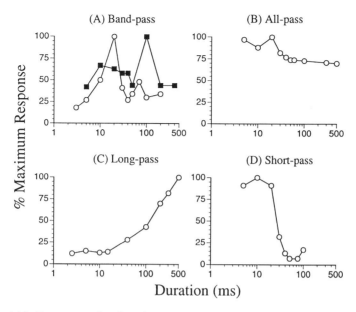

FIGURE 6.13. Representative duration response functions for neurons in the frog TS: band-pass (A), all-pass (B), long-pass (C), and short-pass (D). The abscissa represents the duration of a tone burst (in ms). The ordinate represent the relative mean spike counts at different stimulus durations (reference is the maximal response to the stimuli in the set). (Modified from Gooler and Feng 1992.)

Different neurons respond optimally to different durations. When the duration response functions of brain stem neurons are evaluated as a whole, it is apparent that computation of stimulus duration is achieved through a systematic change in the response function (Table 6.2). Whereas all eighth nerve fibers display long-pass duration response functions (Fig. 6.13C), neurons in the dorsal nucleus show more diverse response functions, with primary-like neurons giving rise to long-pass functions and phasic neurons to all-pass functions (Fig. 6.13B). At the superior olivary nucleus, the dichotomy at the dorsal nucleus is not completely preserved. A small fraction of superior olivary neurons are found to respond preferentially to short-duration stimuli, displaying short-pass duration response functions (Fig. 13D). Finally, at the TS and the thalamus (see next section), neurons displaying tuned response to duration are observed. While the underlying mechanism for duration selectivity in frogs is not presently understood, the similarity in its encoding scheme for sound duration with the auditory system of bats (Casseday et al. 1994) indicates the possibility that it too may be mediated by a delayed-concidence detection scheme as found in bats. However, this hypothesis requires direct experimental validation in the frog auditory system.

TABLE 6.2. Distributions of the different types of duration response functions for neurons in the eighth nerve, dorsal nucleus (DN), superior olivary nucleus (SON), torus semicircularis (TS), and central thalamic nucleus (Thal-C) of the Northern leopard frog.

	8th N	DN	SON	TS	Thal-C
Short-pass	0	0	0	8% (4)	61% (19)
Long-pass	100% (30)	78% (31)	76% (45)	75% (38)	13% (4)
Band-limited	0	0	0	12% (6)	19% (6)
All-pass	0	22% (9)	24% (14)	6% (3)	0
Others	0	0	0	0	6% (2)
Total # of cells	(30)	(40)	(59)	(51)	(31)

Numbers in parentheses indicate the number of cells belonging to the different functional cell types. Percents may not sum to 100% because of rounding.

Additionally, the synchrony code for AM representation at the periphery and the lower brain stem is further transformed into the rate code in the TS. TS neurons seldom can fire in a time-locked fashion to an AM frequency of >50 Hz (Bibikov and Gorodetskaya 1980; Walkowiak 1980; Rose and Capranica 1985; Epping and Eggermont 1986a,b; Gooler and Feng 1992; see review by Feng et al. 1990). Thus, encoding of AM signals depends mostly on the rate code. When the unit's spike count (or firing rate) is used as a response metric, TS neurons exhibit high selectivity to AM rates as do superior olivary neurons (Table 6.3). In response to PAM stimuli, essentially all eighth nerve fibers display high-pass modulation transfer functions due to the progressive increase in signal duty cycle with increasing AM rate, as described earlier. At the dorsal and superior olivary nuclei and the TS, the majority of primary-like and phasic-burst neurons display high-pass modulation transfer functions (66%, 58%, and 64%, respectively), but some display all-pass and others display band-pass modulation transfer functions (Fig. 6.14A,B). One-quarter to one-fifth of neurons (phasic and primary-like neurons) in these neural centers, however, give band-pass modulation transfer functions; these neurons respond selectively to a narrow range of AM rate. The remaining small population of neurons give low-pass (Fig. 6.14C) and band-suppression (Fig. 6.14D) modulation transfer functions. At the next level (i.e., the auditory thalamus), about half of the neurons display band-pass responses to PAM stimuli, indicating the specialization of this center for AM coding (see below).

3.2.2.4 Thalamus

In the auditory thalamus, auditory processing in the frequency and time domains is mediated by two separate structures. The auditory thalamus comprises two distinct nuclei: the central and the posterior nuclei. The central thalamic nucleus receives its auditory input primarily from the magnocellular and principal nuclei of the TS, whereas projections to the

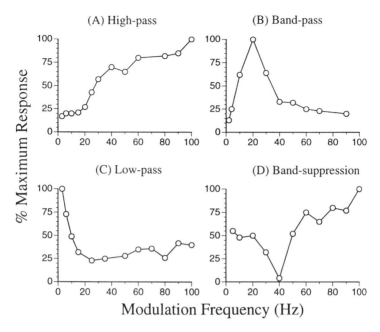

FIGURE 6.14. Diversity in count-based modulation transfer functions among TS neurons in response to PAM stimuli: high-pass (A), band-pass (B), low-pass (C), and band-suppression (D). The abscissa represents the modulation frequency and the ordinate represents the relative mean spike counts at different modulation frequencies (reference is the maximal response to the stimuli in the set). See the caption of Figure 6.12 for description of stimulus parameters. (Modified from Gooler and Feng 1992.)

TABLE 6.3. Distributions of the different types of modulation transfer functions (in response to PAM stimuli) for neurons in the doral nucleus (DN), superior olivary nucleus (SON), torus semicircularis (TS) and central thalamic nucleus (Thal-C) of the Northern leopard frog.

	8th N	DN	SON	TS	Thal-C
Low-pass	0	0	3% (3)	5% (2)	23% (7)
High-pass	98% (52)	66% (41)	58% (59)	63% (24)	26% (8)
Band-pass	0	27% (17)	24% (25)	21% (8)	45% (14)
Band-suppression	0	0	6% (6)	5% (2)	6% (2)
All-pass	0	6% (4)	6% (6)	5% (2)	0
Others	2% (1)	0	3% (3)	0	0
Total # of cells	(53)	(62)	(102)	(38)	(31)

Numbers in parentheses indicate the number of cells belonging to the different functional cell types. Percents may not sum to 100% because of rounding.

posterior thalamic nucleus derive mainly from the laminar nucleus of the TS (Hall and Feng 1987). Neurons in the posterior thalamic nucleus appear to be specialized for processing the spectral features of the species vocalization, the advertisement call in particular (Hall and Feng 1987; Feng et al. 1990). An evoked potential recorded from the posterior thalamic nucleus, in response to a specific tone combination, is much larger than the sum of evoked potentials to individual tones (Hall and Feng 1987). Further, one-third of the recorded single neurons in the posterior thalamic nucleus display AND response properties (Fig. 6.10F). These neurons respond strongly to signals that contain both a low- and a high-frequency spectral component (namely, the presumed salient feature of the species advertisement call) but not to either of these components alone (Fuzessery and Feng 1983b). When compared to response characteristics of brain stem auditory neurons, it seems clear that the frequency selectivity of auditory neurons undergoes dramatic transformation from the periphery to the auditory thalamus, that is, from the simple V-shaped frequency-threshold curve at the auditory nerve, to the more complex AND responses seen at higher levels, giving rise to response preference to the spectral features of natural calls.

In contrast to the posterior thalamic nucleus, the central thalamic nucleus appears to be involved primarily with processing of temporal sound features. Neurons in the central thalamic nucleus are broadly tuned and show little selectivity to the spectrum of the acoustic signals (Hall and Feng 1987). However, there is evidence that neurons in this nucleus are selective to specific temporal features of the call, such as the duration (see Table 6.2), or the repetition rate (see Table 6.3), of sound pulses (Hall and Feng 1986). Furthermore, for neurons exhibiting response selectively to signal duration (or pulse repetition rate), the duration (or pulse repetition rate) to which these cells respond optimally differs among neurons (Feng et al. 1990). At present, it is unclear whether or not neurons that are tuned to the different temporal features are clustered in different regions of the nucleus, and whether or not the sound features are represented topographically within the nucleus. This is an area of research that warrants further attention.

Neurons in the central thalamic nucleus also exhibit response selectivity to natural frog calls, and further there is evidence that this selectivity may be attributed to the units' selectivity to one or more temporal sound features (Feng et al. 1990). For some neurons, the call to which a unit responds maximally is one of the species calls. However, for other neurons it is the calls of sympatric species that elicit the optimal responses. Single-unit studies in the TS of the *Hyla versicolor* (Diekamp and Gerhardt 1995) yield a similar result, indicating that the frog auditory system is not entirely dedicated to processing conspecific acoustic signals. In other words, conspecific and sympatric acoustic signals likely excite different populations of auditory neurons in the frog central auditory system whereby call discrimination is likely accomplished by a population code.

Taken together, data from systematic studies in the derived frogs reveal that analysis of complex sound features is carried out in parallel by temporal and spectral filter-neurons in two separate populations of the thalamus, and that these filter-neurons are constructed in a hierarchical manner along the central auditory pathways. Thus the parallel processing scheme evinced in the frog central auditory system appears similar to those observed in the central auditory system of echolocating bats (Suga 1990) and in the central visual system in mammals (Livingstone and Hubel 1988).

Acoustic representation in the forebrain of anurans, or other groups of amphibians, is poorly understood. This is a major gap in our understanding. Due to the limited amount of factual information available, this topic is not covered in this chapter.

3.3 Mechanisms of Sound Localization

3.3.1 Acoustically Guided Orienting Response

As described earlier, for a gravid female to single out a calling male among the many males participating in a chorus, she must be able to localize sound accurately. Behavioral tests in the field reveal that gravid females indeed can accurately locate a calling male (or a loudspeaker used to broadcast the species mating call) in both the horizontal as well as in the vertical plane (Feng et al. 1976; Rheinlaender et al. 1979; Gerhardt and Rheinlaender 1982; Passmore et al. 1984; Klump and Gerhardt 1989; Jørgensen and Gerhardt 1991). The phonotactic response comprises a series of jumps that are directed toward the sound source. The response follows a zigzag pattern, with the frog making fine adjustment in the jumping direction such that the slight error introduced by each hop (i.e., the deviation from the line that connects the frog and the sound source) is corrected during the subsequent hop. Prior to each hop, a female often scans her head laterally, and after listening to one or more croaks the head and body are aligned toward the sound source. The head scanning improves the accuracy of the orienting response; the mean jump angle is $11.8°$ when preceded by head scanning, and $17.6°$ in the absence of head scanning (Rheinlaender et al. 1979). Sound localization in the vertical plane is less precise when compared to the acuity in the horizontal plane (Passmore et al. 1984). To locate an elevated sound source, the frog usually also tilts its head upward before and after the lateral head scan (Gerhardt and Rheinlaender 1982).

3.3.2 Processing of Sound Direction

The ability to locate sound depends on both ears being intact (Feng et al. 1976). Blocking the sound input to one ear undermines the frog's sound localization ability. An interpretation of this finding is that frogs need two ears for the creation of a receiver that is directionally sensitive, that is, a combination pressure/pressure-gradient receiver (Feng and Shofner 1981). The physical basis of this receiver has been described in several review articles (Palmer and Pinder 1984; Fay and Feng 1987; Eggermont 1988). Briefly, because the two ears in frogs are communicated through the mouth and the middle-ear cavities, they are coupled acoustically, forming a push-pull like receiver, that is, a pressure-gradient receiver. The actual behavior of this receiver approximates neither that of a pure pressure receiver not that of a pure pressure-gradient receiver, but rather it behaves as a combination pressure/pressure-gradient receiver that is directionally sensitive (Beranek 1954).

There is evidence indicating that nontympanic pathways are also important for the directionality of the frog's ear (Vlaming et al. 1984; Wilczynski et al. 1987; Narins et al. 1988; Ehret et al. 1990; Jørgensen 1991; Jørgensen and Gerhardt 1991; Jøgensen et al. 1991; Keilwerth and Ehret 1991; Christensen-Dalsgaard and Narins 1993). Lung and other pathways are believed to play a role in giving rise to the directionality of the frog ear (see Lewis and Narins, Chapter 4).

The directionality of frog's ear is such that eighth nerve fibers arising from the ear produce differential responses to tone bursts originating from different sound azimuths (Feng 1980; Feng and Shofner 1981; Wang et al. 1996; Jøgensen and Christensen-Dalsgaard 1997). At low frequencies (65–450 Hz), the fibers display a figure-8 directivity pattern, with sound coming from the two sides having equal effectiveness in exciting the fibers but sound from the frontal field being highly ineffective (Fig. 6.15A). At higher frequencies, the directivity pattern is ovoidal, with sound from the ipsilateral side having strong excitatory effect, but the excitation is progressively weaker as the sound is rotated toward the contralateral side (Fig. 6.15B). This asymmetrical response, for sound frequencies encompassing most of the frog's audible range, is equivalent to a change in sound pressure level of 3 to 8 dB, which is far greater than the actual physical differences in the sound levels at the two ears (Fay and Feng 1987). Thus the pressure/pressure-gradient mechanism essentially produces an "amplification" of the interaural sound levels (see below). Furthermore, since a 1-dB change in sound level elicits an average shift of 0.1 to 0.6 ms in firing latency in eighth nerve fibers (Feng 1982), a difference in perceived sound level of 3 to 8 dB translates into a 0.3- to 4.8-ms difference in excitation times between the two eighth nerves.

Interaural level and time differences of such magnitudes are readily discriminated by frog central auditory neurons (Kaulen et al. 1972; Feng and Capranica 1976, 1978; Bibikov 1977; Melssen and Epping 1990, 1992; Gooler et al. 1996). For example, many neurons in the dorsal and superior olivary nuclei and the TS are sensitive to small differences in the interaural level of binaural stimuli presented dichotically, for example, 0 to 5 dB (Feng and Capranica 1976, 1978; Melssen and Epping 1990; Gooler et al. 1996). Some neurons in these auditory centers are also sensitive to small interaural time differences in the range of 0.1 to 5.0 ms (Feng and Capranica 1976,

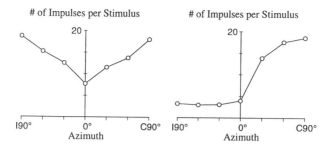

FIGURE 6.15. Directional response characteristic of low-frequency (A) and high-frequency (B) sensitive eighth nerve fibers in response to tone bursts (at unit's CF) presented from a free-field loudspeaker. (From Feng 1980, with permission, Copyright 1980 Acoustical Society of America.)

1978; Melssen and Epping 1992), a range of difference in excitation times that exists between the two eighth nerves (see above).

Binaural processing of interaural level and time differences gives rise to sensitivity to a change in the sound direction in free-field. Many single neurons in the frog TS show notable changes in the firing rate (or spike count) when presented with tones originating from different directions (Feng 1981; Carlisle and Pettigrew 1984). The majority of directionally sensitive neurons in the TS are maximally excited by sounds originating from the contralateral field (or by binaural stimuli that favor the contralateral ear), and responses are progressively weakened when the sound source is moved toward the ipsilateral field (Fig. 6.16A). Other TS neurons, however, show maximal (or in other cases minimal) response from the frontal field, with sound originating from the lateral fields having less effectiveness in exciting these neurons (Fig. 6.16B).

To date, there is no conclusive evidence for the presence of a space map in the TS. Earlier studies have suggested, on the basis of evoked-potential data, that there may be a dorsocaudal representation of auditory space from the frontal to the posterior field (Pettigrew et al. 1978, 1981; Pettigrew and Carlisle 1984). However, these data are compromised by the inclusion of data from recording sites that are outside of the boundary of the TS. Thus, whether or not a space map is present in the frog TS remains to be determined.

In the absence of a space map, what computational scheme can frogs utilize to extract directional information? Two schemes, not mutually exclusive, have been hypothesized previously (see review in Fay and Feng 1987). One involves the rate code. Namely, by virtue of the fact that most TS neurons are maximally excited when sound is presented from the contralateral sound field and that the firing rate diminishes systematically when the sound source is moved toward the ipsilateral side (Feng 1981), the absolute firing rate of central auditory neurons can provide information of sound direction. An interesting aspect of this encoding scheme is that because the

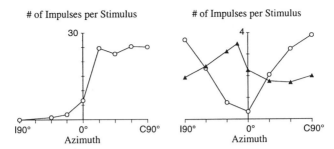

FIGURE 6.16. Two dominant types of directional response characteristics for neurons in the frog TS in response to tone bursts (at unit's CF) presented from a free-field loudspeaker (the scale of one of the response functions on the right panel has been modified for the purpose of conceptual presentation). (Modified from Feng 1981.)

change in firing rate is typically steeper for a change in sound direction in a narrow section of the frontal field (Fig. 6.16A), the directional coding is likely most accurate for this sector of auditory space. Such azimuth-dependent localization acuity has been shown previously for different vertebrate species (Knudsen et al. 1979; Brown et al. 1982). However, at present there is no direct evidence for direction dependency in localization acuity for the anuran.

Alternatively, the frog may utilize a population code for determining the sound direction. That TS neurons display a graded change in the firing rate with sound direction also implies that sound direction can be represented by the relative levels of excitation on the two sides of the brain (van Bergeijk 1962). For example, when sound is presented from the right side of the animal, the number of neurons excited on the left TS as well as the level of excitation of these neurons are greater than the number of neurons excited on the right TS and their levels of excitation. Moreover, this difference in numbers and in the level of excitation is a function of the angular deviation from the midline. When the number and the level of excitation are about equal on both sides, this can be interpreted as sound originating in the frontal field. It is likely that both the rate code and the population code are utilized in the encoding of sound direction.

3.4 Auditory Scene Analysis

As described earlier, coherent perception of sound pattern and location, or analysis of auditory scene, is vital for frogs that typically communicate in complex acoustic environments. At present, it is unclear how the directional information is integrated with the information about sound patterns to produce a unitary percept of where and what the sound is. The underlying mechanisms for auditory scene analysis is poorly understood. This represents an area of research requiring further investigations.

Several recent studies have investigated how sound direction influences central auditory processing (Gooler et al. 1993; Xu et al. 1994, 1996). These investigators have shown that sound direction can influence the frequency as well as temporal selectivities of many TS neurons. Specifically, when the direction of a free-field loudspeaker is rotated from the contralateral to the ipsilateral field, the frequency selectivity as well as the temporal selectivity (i.e., the range of AM frequency to which the unit responds) of many TS neurons increase progressively. If we assume that extraction of auditory information in frogs is carried out actively, the head scanning that improves the localization acuity can presumably also sharpen the sound analysis in both the frequency and the time domains. Improvement in frequency and time domain analysis would enhance the frog's ability to recognize sound patterns in a complex auditory environment.

There is evidence that the direction-dependent sharpening of frequency selectivity is attributed to binaural inhibition (Gooler et al. 1996). These investigators found that when the ipsilateral ear (i.e., the ear that exerts inhibition to most TS neurons) is attenuated monaurally, the free-field frequency selectivity of TS neurons becomes direction independent. In other words, no sharpening can be observed when binaural inhibition is suppressed. Additionally, preliminary results from Xu and Feng (unpublished observation) further indicate that GABA is one of the neurotransmitters involved in binaural inhibition. Iontophoretic application of bicuculline, a GABA-a antagonist, into the TS abolishes the direction-dependent sharpening of frequency selectivity of TS neurons.

It therefore appears that binaural inhibition has a dual role in auditory processing. Previously, binaural inhibition has been shown to be a key mechanism for creating spatial receptive field (Fujita and Konishi 1991; Takahashi and Keller 1992). The new results described above indicate that it additionally plays a role in the analysis of sound spectrum. Thus, the neural mechanism that underlies coding of sound direction is shared with that responsible for coding of sound pattern. Whether or not this sharing is a vehicle by which the nervous system constructs a unitary percept of individual auditory objects remains to be determined. The actual mechanisms that underlie coherent perception is completely obscure at this time. Clearly, an understanding of these mechanisms represents one of the biggest challenges for auditory scientists in the coming decades.

3.5 Summary of Information Processing Along the Amphibian Auditory Pathway

In amphibians, various response properties change progressively along the central auditory pathway. Here is a list of some of the most notable changes:

- a decrease in phase-locking to tones and to the modulation envelope of AM stimuli;
- an increase in the variability in frequency-threshold curves and in two-tone suppression;
- a systematic change in the transient response to tone bursts, that is, an increase in the occurrence of phasic response up to the level of the TS, which is followed by the predominant presence of tonic response in the auditory thalamus;
- the existence of tonotopic map at the lower brain stem (its presence in the auditory thalamus, however, is unclear);
- a decrease in the correlation between a unit's temporal discharge pattern and a unit's response selectivities to temporal attributes;
- the TS represents an important processing center, and TS neurons exhibit selectivities to temporal and/or spectral sound features;
- an increases in functional segregation such that temporal and spectral processing of auditory features are mediated by two separate auditory nuclei at the thalamus.

4. Evaluation of Recent Knowledge and Remaining Questions

In fishes, studies in recent years have significantly advanced our understanding of central auditory processing. However, a number of fundamental questions remain unresolved. Two questions are most urgent and present the greatest challenge. One of them, specific for swim bladder—bearing fish, is whether or not the pressure information and direct particle-displacement information are processed along separate pathways? Processing by separate channels can provide the basis for directional hearing. Furthermore, is such a separation evidence for the functionally different types of hearing organs (e.g., otophysans, fish with other hearing adaptations and nonspecialized fish)? The second question relates to the extent to which the information from distinct areas of the otolithic maculae is integrated (summation, inhibition, facilitation, etc.), monaurally and binaurally at various stations along the central auditory pathway. These questions are difficult to resolve, since pure pressure stimuli and pure monaural stimuli are difficult to present without irreversible elimination of a part of the system (see Section 2.1).

In fishes and amphibians, the torus semicircularis is an important center for processing of acoustic information as well as other sensory modalities. Response characteristics of neurons in the TS resemble those of the inferior colliculus of mammals. In amphibians, the thalamic auditory nuclei receiving projections from the TS seem to be even more specialized than the TS; here the analysis of complex sound features is mediated by two separate populations of neurons, with one population devoted to temporal and the other to spectral processing. In fishes, however, it is unclear whether or not different stimulus features are processed along separate pathways at any level. Also, the specific functional roles of the medulla and of the diencephalon remain to be determined for fishes.

For both groups of animals, the role the telencephalon plays in auditory processing is completely unknown; this is a significant gap in our understanding. It would be important to determine whether the auditory part of the telencephalon, insofar as its function is concerned, is equivalent to a primitive auditory cortex, which in turn provides input to associative centers as observed in the mammalian cerebrum.

For both groups of animals, temporal and spectral filter-neurons are constructed in a hierarchical manner along the central auditory pathways. Neurons in the lower brain stem respond to sound in a similar manner when compared to the eighth nerve fibers, but those in the upper brain stem have greater response selectivities, for example, frequency-dependent excitatory/inhibitory behavior; W-shaped frequency-threshold curves or highly nonlinear two-tones interaction (such as logical AND response); and stronger selectivities to AM rate and stimulus duration. This general organization scheme is similar to that of higher vertebrates. For fishes, the similarity is interesting in light of the fact that their inner ear and the accessory hearing structures are so different from those of other vertebrates.

In amphibians, as in higher vertebrates, tonotopy can be observed along the central auditory pathway. The tonotopy is a result of systematic central projections of afferents arising from lower centers that are themselves tonotopically organized. However, for fishes, the presence of tonotopy has yet to be demonstrated at any level of the central auditory pathway. The fish's otolith system itself does not exhibit a tonotopic organization.

In fishes and amphibians, TS neurons show response preference to sound direction. However, spatiotopy, which has been shown for fish, birds, and mammals, remains to be shown in the amphibians. With respect to selectivity for sound direction, the fish TS is functionally stratified and shows column-like organizations. The topographic organization of other auditory nuclei, however, remains to be investigated.

For fishes, it is unclear how the frequency selectivity arises along the central auditory pathway to enable analytic and synthetic listening, as evidenced by the goldfish (Popper and Fay 1993), and presumably by all species with vocalizations. Furthermore, it is not known whether or not there are basic differences between species with poor and excellent development of vocalizations.

In fishes and amphibians, the orderly mapping of computed stimulus features has not been studied systematically. It is unclear whether such a map exists in the central auditory system of these two groups of animals, as has been observed in other vertebrates. Also the role of the efferent system in hearing is completely unknown. For fishes and some amphibians, the understanding of the relationship between the descending system and the Mauthner system is another area requiring future studies.

To date, our understanding of the pharmacological basis of information processing in the central auditory system of fishes and amphibians, and of the physiological and biophysical properties of the different cell types, is still at its infancy. These lines of investigation are important for the overall understanding of the underlying mechanisms for information processing. Also, at present it is unclear how the auditory system integrates the temporal and spectral features to form a unitary percept of complex sounds. This process may take place in the forebrain or in sensorimotor processing areas. Similarly, the mechanisms that underlie coherent perception of individual auditory objects in complex scenes are poorly understood.

Acknowledgments. Research in Albert Feng's laboratory is supported by grants from the Research Board and the Beckman Institute of the University of Illinois, and the National Institute on Deafness and Other Communication Disorders (R01 DC 00663), National Institutes of Health. Research in Nico Schellart's laboratory is supported by a grant (810-401-17) from the Dutch Foundation of Life Sciences (NWO).

References

Beranek LL (1954) Acoustics. New York: McGraw-Hill.

Bibikov NG (1974a) Encoding of the stimulus envelope in peripheral and central regions of the auditory system of the frog. Acustica 31:310–314.

Bibikov NG (1974b) Impulse activity of torus semicircularis neurons of the frog *Rana temporaria.* J Evol Biochem Physiol 10:36–41.

Bibikov NG (1977) Dependence of the binaural neurons reaction in the frog torus semicircularis on the interaural phase difference. Sechenov Physiol J USSR 63:365–373.

Bibikov NG, Gorodetskaya ON (1980) Single unit responses in the auditory center of the frog mesencephalon to amplitude-modulated tones. Neirofiziologiia 12:264–271.

Blair WF (1964) Isolating mechanisms and interspecies interactions in anuran amphibians. Q Rev Biol 20:334–344.

Bleckmann H, Bullock TH (1989) Central nervous physiology of the lateral line, with special reference to cartilaginous fishes. In: Coombs S, Görner P, Münz H (eds) The Mechanosensory Lateral Line: Neurobiology and Evolution. New York: Springer-Verlag, pp. 387–408.

Brown CH, Schessler T, Moody D, Stebbins W (1982) Vertical and horizontal sound localization. J Acoust Soc Am 72:1804–1811.

Buwalda RJA, Schuijf A, Hawkins AD (1983) Discrimination by the cod of sounds from opposing directions. J Comp Physiol [A] 150:175–184.

Caird DM (1978) A simple cerebellar system: the lateral line lobe of the goldfish. J Comp Physiol [A] 127:61–74.

Capranica RR (1965) The Evoked Vocal Response of the Bullfrog: A Study of Communication in Anurans. Research monographs #33. Cambridge: MIT Press.

Carlisle S, Pettigrew AG (1984) Auditory responses in the torus semicircularis of the cane toad, *Bufo marinus*. II. Single unit studies. Proc R Soc Lond [B] 222:243–257.

Carr CE (1993) Processing of temporal information in the brain. Annu Rev Neurosci 16:223–243.

Casseday JH, Ehrlich D, Covey E (1994) Neural tuning for sound direction: role of inhibitory mechanisms in the inferior colliculus. Science 264:847–850.

Christensen-Dalsgaard J, Narins PM (1993) Sound and vibration sensitivity of VIIIth nerve fibers in the frogs *Leptodactylus albilabris* and *Rana pipiens pipiens*. J Comp Physiol [A] 172:653–662.

Condon CJ, Chang SH, Feng AS (1991) Processing of behaviorally relevant temporal parameters of acoustic stimuli by single neurons in the superior olivary nucleus of the leopard frog. J Comp Physiol [A] 168:709–725.

Condon CJ, Chang SH, Feng AS (1995) Classification of the temporal discharge patterns of single auditory neurons in the frog superior olivary nucleus. Hear Res 83:190–202.

Coombs S, Fay RR (1985) Adaptation effects on amplitude modulation detection: behavioral and neurophysiological assessment in the goldfish auditory system. Hear Res 19:57–71.

Corwin JT (1981) Audition in elasmobranchs. In: Tavolga WN, Popper AN, Fay RR (eds) Hearing and Sound Communication in Fishes. New York: Springer-Verlag, pp. 81–105.

Crawford JD (1993) Central auditory neurophysiology of a sound-producing fish: the mesencephalon of *Pollimyrus isidori* (Mormyridae). J Comp Physiol [A] 172:139–152.

Cuadrado MI (1987) The cytoarchitecture of the torus semicircularis in the teleost, *Barbus meridionalis*. J Morphol 191:233–245.

de Wolf FA, Schellart NAM, Hoogland PV (1983) Octavolateral projections to the torus semicircularis in the trout, *Salmo gairdneri*. Neurosci Lett 38:209–213.

Diekamp B, Gerhardt HC (1995) Selective phonotaxis to advertisement calls in the gray treefrog *Hyla versicolor*: behavioral experiments and neurophysiological correlates. J Comp Physiol [A] 177:173–190.

Echteler SM (1985a) Organization of central auditory pathways in a teleost fish, *Cyprinus carpio*. J Comp Physiol [A] 156:267–280.

Echteler SM (1985b) Tonotopic organization in the midbrain of a teleost fish. Brain Res 338:387–391.

Eggermont JJ (1988) Mechanisms of sound localization in anurans. In: Fritsch B, Ryan MJ, Wilczynski W, Hetherington TE, Walkowiak W (eds) The Evolution of the Amphibian Auditory System. New York: Wiley, pp. 307–336.

Eggermont JJ, Epping WJM (1986) Sensitivity of neurons in the auditory midbrain of the grassfrog to temporal characteristics of sound. III. Stimulation with natural and synthetic mating calls. Hear Res 24:255–268.

Ehret G, Tautz J, Schmitz B, Narins PM (1990) Hearing through the lungs: lung-eardrum transmission of sound in the frog *Eleutherodactylus coqui*. Naturwissenschaften 77:192–194.

Elepfandt A (1996) Underwater acoustics and hearing in the clawed frog, *Xenopus*. In: Tinsley RC, Kobel HR (eds) The Biology of *Xenopus*. Oxford: Clarendon Press, pp. 177–193.

Enger PS (1967) Hearing in herring. Comp Biochem Physiol 22:527–538.

Enger PS (1973) Masking of auditory responses in the medulla oblongata of goldfish. J Exp Biol 59:415–424.

Epping WJM, Eggermont JJ (1986a) Sensitivity of neurons in the auditory midbrain of the grassfrog to temporal characteristics of sound. I. Stimulation with acoustic clicks. Hear Res 24:37–54.

Epping WJM, Eggermont JJ (1986b) Sensitivity of neurons in the auditory midbrain of the grassfrog to temporal characteristics of sound. II. Stimulation with amplitude modulated sound. Hear Res 24:55–72.

Fay RR (1978) Coding of information in single auditory-nerve fibers of the goldfish. J Acoust Soc Am 63:136–146.

Fay RR (1984) The goldfish ear codes the axis of acoustic particle motion in three dimension. Science 225:951–954.

Fay RR (1988) Hearing in Vertebrates: a Psychophysics Data Book. Winnetka, IL: Hill-Fay Associates.

Fay RR (1990) Suppression and excitation in auditory nerve fibers of the goldfish, *Carassius auratus*. Hear Res 48:93–110.

Fay RR, Feng AS (1987) Mechanisms for directional hearing among non mammalian vertebrates. In: Yost WA, Gourevitch G (eds) Directional Hearing. New York: Springer-Verlag, pp. 179–213.

Fay RR, Hillery CM, Bolan K (1982) Presentation of sound pressure and particle motion in the midbrain of goldfish. Comp Biochem Physiol 71:181–191.

Feng AS (1980) Directional characteristics of the acoustic receiver of the leopard frog (*Rana pipiens*): a study of eighth nerve auditory responses. J Acoust Soc Am 68:1107–1114.

Feng AS (1981) Directional response characteristics of single neurons in the tours semicircularis of the leopard frog (*Rana pipiens*). J Comp Physiol [A] 144:419–428.

Feng AS (1982) Quantitative analysis of intensity-rate and intensity-latency functions in peripheral auditory nerve fibers of northern leopard frogs (*Rana p. pipiens*). Hear Res 6:242–246.

Feng AS (1986a) Afferent and efferent innervation patterns of the cochlear nucleus (dorsal medullary nucleus) of the leopard frog. Brain Res 367:183–191.

Feng AS (1986b) Afferent and efferent innervation patterns of the superior olivary nucleus of the leopard frog. Brain Res 364:167–171.

Feng AS, Capranica RR (1976) Sound localization in anurans. I. Evidence of binaural interaction in dorsal medullary nucleus of bullfrogs (*Rana catesbeiana*). J Neurophysiol 39:871–881.

Feng AS, Capranica RR (1978) Sound localization in anurans II. Binaural interaction in superior olivary nucleus of the green treefrog (*Hyla cinerea*). J Neurophysiol 41:43–54.

Feng AS, Gerhardt HC, Capranica RR (1976) Sound localization behavior of the green treefrog (*Hyla cinerea*) and the barking treefrog (*H. gratiosa*). J Comp Physiol [A] 107:241–252.

Feng AS, Hall JC, Gooler DM (1990) Neural basis of sound pattern recognittion in anurans. Prog Neurobiol 34:313–329.

Feng AS, Hall JC, Siddique S (1991) Coding of temporal parameters of complex sounds by frog auditory nerve fibers. J Neurophysiol 65:424–445.

Feng AS, Lin WY (1991) Differential innervation patterns of three divisions of frog auditory midbrain (torus semicricularis). J Comp Neurol 306:613–630.

Feng AS, Lin WY (1994a) Detection of gaps in sinusoids by frog auditory nerve fibers: importance in AM coding. J Comp Physiol [A] 175:531–546.

Feng AS, Lin WY (1994b) Phase-locked response characteristics of single neurons in the frog "cochlear nucleus" to steady-state and sinusoidally-amplitude-modulated tones. J Neurophysiol 72:2209–2221.

Feng AS, Lin WY (1996) The neuronal architecture of the dorsal nucleus (cochlear nucleus) of the frog *Rana pipiens pipiens*. J Comp Neurol 366:320–334.

Feng AS, Narins PM, Capranica RR (1975) Three populations of primary auditory fibers in the bullfrog (*Rana catesbeiana*): their peripheral origins and frequency sensitivities. J Comp Physiol [A] 100:221–229.

Feng AS, Shofner WP (1981) Peripheral basis of sound localization in anurans. Acoustic properties of the frog's ear. Hear Res 5:201–216.

Fouquette MJ Jr (1975) Speciation in chorus frogs. I. Reproductive character displacement in the *Pseudacris nigrita* complex. Syst Zool 24:16–23.

Frishkopf LA, Capranica RR, Goldstein MH (1968) Neural coding in the frog's auditory system—a teleological approach. Proc IEEE 56:969–980.

Fujita I, Konishi M (1991) The role of GABAergic inhibition in processing of interaural time difference in the owl's auditory system. J Neurosci 11:722–739.

Fuzessery ZN (1983) Neural correlates of mating call selectivity in the leopard frog, *Rana p. pipiens*. Ph.D. Thesis, University of Illinois, Urbana-Champaign.

Fuzessery ZN, Feng AS (1981) Frequency representation in the dorsal medullary nucleus of the leopard frog, *Rana p. pipiens*. J Comp Physiol [A] 143:339–349.

Fuzessery ZN, Feng AS (1982) Frequency selectivity in the anuran auditory midbrain: single unit responses to single and multiple tones. J Comp Physiol [A] 146:471–484.

Fuzessery ZN, Feng AS (1983a) Frequency selectivity in the anuran medulla: excitatory and inhibitory tuning properties of single neurons in the dorsal medullary and superior olivary nuclei. J Comp Physiol [A] 150:107–119.

Fuzessery ZN, Feng AS (1983b) Mating call selectivity in the thalamus and midbrain of the leopard frog (*Rana p. pipiens*): single and multi-unit analyses. J Comp Physiol [A] 150:333–344.

Gerhardt HC (1974) The significance of some spectral features in mating call recognition in the green treefrog (*Hyla cinerea*). J Exp Biol 61:229–241.

Gerhardt HC (1978) Temperature coupling in the vocal communication system of the gray treefrog *Hyla versicolor*. Science 199:992–994.

Gerhardt HC, Doherty JA (1988) Acoustic communication in the gray treefrog, *Hyla versicolor*: evolutionary and neurobiological implications. J Comp Physiol [A] 162:261–278.

Gerhardt HC, Rheinlaender J (1982) Localization of an elevated sound source by the green treefrog. Science 217:663–664.

Gooler DM, Feng AS (1992) Temporal coding in the frog auditory midbrain: the influence of duration and rise-fall time on the processing of complex amplitude-modulated stimuli. J Neurophysiol 67:1–22.

Gooler DM, Condon CJ, Xu JH, Feng AS (1993) Sound direction influences the frequency-tuning characteristics of neurons in the frog inferior colliculus. J Neurophysiol 69:1018–1030.

Gooler DM, Xu JH, Feng AS (1996) Binaural inhibition is important in shaping the free-field frequency selectivity of single neurons in the frog inferior colliculus. J Neurophysiol 76:2580–2594.

Goossens JHLM, Wubbels RJ, Schellart NAM (1995) Computer-controlled vibrations of a biological subject in the nm range. Med Biol Eng Comput 33:732–735.

Grözinger B (1967) Elektro-physiologische Untersuchugen an der Hirbahn der Schleie (*Tinca tinca* [L]). Z Vergl Physiol 57:44–76.

Hall JC (1994) Central processing of communication sounds in the anuran auditory system. Am Zool 34:670–684.

Hall JC, Feng As (1986) Neural analysis of temporally patterned sounds in the frog's thalamus: processing of pulse duration and pulse repetition rate. Neurosci Lett 63:215–220.

Hall JC, Feng AS (1987) Evidence for parallel processing in the frog's auditory thalamus. J Comp Neurol 258:407–419.

Hall JC, Feng AS (1988) Influence of envelope rise time on neural responses in the auditory system of anurans. Hear Res 36:261–276.

Hall JC, Feng AS (1990) Classification of the temporal discharge patterns of single neurons in the dorsal medullary nucleus of the northern leopard frog. J Neurophysiol 64:1460–1473.

Hall JC, Feng As (1991) Temporal processing in the dorsal medullary nucleus of the northern leopard frog (*Rana pipiens pipiens*). J Neurophysiol 66:955–973.

Hermes DJ, Eggermont JJ, Aertsen AMHJ, Johannesma PIM (1982) Spectrotemporal characteristics of single units in the auditory midbrain of the lightly anesthetized grass frog (*Rana temporaria* L.) investigated with tonal stimuli. Hear Res 6:103–126.

Highstein SM (1991) The central nervous system efferent control of organs of balance and equilibrium. Neurosci Res 12:13–30.

Hillery C, Narins PM (1984) Neurophysiological evidence for a traveling wave in the amphibian ear. Science 225:1037–1039.

Horner K, Sand O, Enger PS (1980) Binaural interaction in the cod. J Exp Biol 85:323–331.

Irvine DRF (1992) Physiology of the auditory brain stem. In: Popper AN, Fay RR (eds) The Mammalian Auditory Pathway: Neurophysiology. New York: Springer-Verlag, pp. 153–231.

Jørgensen MB (1991) Comparative studies of the biophysics of directional hearing in anurans. J Comp Physiol [A] 169:591–598.

Jørgensen MB, Chistensen-Dalsgaard J (1997) Directionality of auditory nerve fiber responses to pure tone stimuli in the grassfrog, *Rana temporaria* I. Spike rate responses. J Comp Physiol [A] 180:493–502.

Jørgensen MB, Gerhardt HC (1991) Directional hearing in the gray tree frog *Hyla versicolor*: eardrum vibrations and phonotaxis. J Comp Physiol [A] 169:177–183.

Jørgensen MB, Schmitz B, Christensen-Dalsgaard (1991) Biophysics of directional hearing in the frog *Eleutherodactylus coqui*. J Comp Physiol [A] 168:223–232.

Kaulen R, Lifschitz W, Palazzi C, Adrian H (1972) Binaural interaction in the inferior colliculus of the frog. Exp Neurol 37:469–480.

Keilwerth E, Ehret G (1991) The lung of treefrogs as "outer ears" serving in sound localization? Verh Dtsch Zool Ges 84:347.

Klump GM, Gerhardt HC (1989) Sound localization in the barking treefrog. Naturwissenschaften 76:35–37.

Knudsen EI, Blasdel GG, Konishi M (1979) Sound localization by the barn owl (*Tyto alba*) measured with the search coil technique. J Comp Physiol [A] 133:1–11.

Kuwada S, Yin TCT (1987) Physiological studies of directional hearing. In: Yost WA, Gourevitch G (eds) Directional Hearing. New York: Springer-Verlag, pp. 146–176.

Lewis ER, Leverenz EL, Koyama H (1980) Mapping functionally identified auditory afferents from the peripheral origins to the central terminations. Brain Res 197:223–229.

Livingstone MS, Hubel DH (1988) Segregation of form, color, movement, and depth: anatomy, physiology, and perception. Science 240:740–749.

Loftus-Hills JJ, Littlejohn MJ (1971) Pulse repetition rate as the basis for mating call discrimination by two sympatric species of *Hyla*. Copeia 1:154–156.

Lu Z, Fay RR (1993) Acoustic response properties of single units in the torus semicircularis of the goldfish, *Carassius auratus*. J Comp Physiol [A] 173:33–48.

Lu Z, Fay RR (1995) Acoustic response properties of single units in the central posterior nucleus of the thalamus of the goldfish, *Carassius auratus*. J Comp Physiol [A] 176:747–760.

Lu Z, Fay RR (1996) Two-tone interaction in auditory nerve fibers and midbrain neurons of the goldfish, *Carassius auratus*. Audit Neurosci 2:257–273.

McCormick CA (1992) Evolution of central auditory pathways in anamniotes. In: Webster DB, Fay RR, Popper AN (eds) The Evolutionary Biology of Hearing. New York: Springer-Verlag, pp. 323–350.

Megela AL (1984) Diversity of adaptational patterns in responses of eighth nerve fibers in the bullfrog, *Rana catesbeiana*. J Acoust Soc Am 75:1155–1162.

Megela AL, Capranica RR (1981) Response patterns to tone bursts in peripheral auditory system of anurans. J Neurophysiol 46:465–478.

Melssen WJ, Epping WJM (1990) A combined sensitivity for frequency and interaural intensity difference in neurons in the auditory midbrain of the grassfrog. Hear Res 44:35–50.

Melssen WJ, Epping WJM (1992) Selectivity for temporal characteristics of sound and interaural time difference of auditory midbrain neurons in the grassfrog. Hear Res 60:178–198.

Mohneke R (1983) Tonotopic organization of the auditory midbrain nuclei of the midwife toad (*Alytes obstetricans*). Hear Res 9:91–102.

Narins PM, Capranica RR (1978) Communicative significance of the two-note call of the treefrog, *Eleutherodactylus coqui*. J Comp Physiol [A] 127:1–9.

Narins PM, Ehret G, Tautz J (1988) Accessory pathway for sound transfer in a neotropical frog. Proc Natl Acad Sci USA 85:1508–1512.

Nederstigt LJA, Schellart NAM (1986) Acoustical activity in the torus semicircularis of the trout *Salmo gairdneri*. Pflugers Arch 406:151–157.

Page CH (1970) Electrophysiological study of auditory responses in the goldfish brain. J Neurophysiol 33:116–128.

Page CH, Sutterlin AM (1970) Visual and auditory responses in the goldfish tegmentum. J Neurophysiol 33:129–136.

Palmer AR, Pinder AC (1984) The directionality of the frog ear described by a mechanical model. J Theor Biol 110:205–215.

Passmore NI, Capranica RR, Telford SR, Bishop PJ (1984) Phonotaxis in the painted reed frog (*Hyperolius marmoratus*). The localization of elevated sound source. J Comp Physiol [A] 154:189–197.

Pettigrew AG (1981) Brainstem afferents to the torus semicircularis of the Queensland cane toad (*Bufo marinus*). J Comp Neurol 202:59–68.

Pettigrew AG, Carlisle S (1984) Auditory response in the torus semicircularis of the cane toad (*Bufo marinus*). I. Field potential studies. Proc R Soc Lond [B] 222:231–242.

Pettigrew AG, Chung S-H, Anson M (1978) Neurophysiological basis of directional hearing in amphibia. Nature 272:138–142.

Pettigrew AG, Anson M, Chung S-H (1981) Hearing in the frog: a neurophysiological study of the auditory response in the midbrain. Proc R Soc Lond [B] 212:433–457.

Plassmann W (1985) Coding of amplitude-modulated tones in the central auditory system of catfish. Hear Res 17:209–217.

Popper AN, Fay RR (1993) Sound detection and processing by fish: a critical review and major research questions. Brain Behav Evol 41:14–38.

Popper AN, Saidel WM (1990) Variations in receptor cell innervation in the saccule of a teleost fish ear. Hear Res 46:211–227.

Rheinlaender J, Gerhardt HC, Yager DD, Capranica RR (1979) Accuracy of phonotaxis by the green treefrog (*Hyla cinerea*). J Comp Physiol [A] 133:247–255.

Rhode W, Greenberg S (1992) Physiology of the cochlear nuclei. In: Popper AN, Fay RR (eds) The Mammalian Auditory Pathway: Neurophysiology. New York: Springer-Verlag, pp. 94–152.

Roberts BL, Meredith GE (1992) The efferent innervation of the ear: variations on an enigma. In: Webster DB, Fay RR, Popper AN (eds) The Evolutionary Biology of Hearing. New York: Springer-Verlag, pp. 185–210.

Rogers PH, Popper AN, Hastings MA, Saidel WM (1988) Processing of acoustic signals in the auditory system of bony fish. J Acoust Soc Am 83:338–349.

Rose GJ, Capranica RR (1985) Sensitivity to amplitude modulated sounds in the anuran auditory nervous system. J Neurophysiol 53:446–465.

Rubinson K, Skiles MP (1975) Efferent projections of the superior olivary nucleus in the frog, *Rana catesbeiana*. Brain Behav Evol 122:151–160.

Sawa M (1976) Auditory responses from single neurons of the medulla oblongata in the goldfish. Bull Jpn Soc Sci Fish 45:141–152.

Schellart NAM (1983) Acoustical and visual processing and their interaction in the torus semicircularis of the trout, *Salmo gairdneri*. Neurosci Lett 42:39–44.

Schellart NAM (1990) The integration of visual and acoustic processing in the teleost subtectum. In: Guthrie DM (ed) Higher Order Sensory Processing. Manchester, England: Manchester University Press, pp. 49–74.

Schellart NAM (1992) Interrelations between the auditory, the visual and the lateral line systems; a mini-review of modeling sensory capabilities. Neth J Zool 42:459–477.

Schellart NAM, Kroese ABA (1989) Interrelationship of acousticolateral and visual systems in the teleost midbrain. In: Coombs S, Görner P, Münz H (eds) The Mechanosensory Lateral Line: Neurobiology and Evolution. New York: Springer-Verlag, pp. 421–443.

Schellart NAM, Munck JC de (1987) A model for directional and distance hearing in swimbladder-bearing fish based on the displacement orbits of hair cells. J Acoust Soc Am 82:822–829.

Schellart NAM, Popper AN (1992) Functional aspects of the evolution of the auditory system of actinopterygian fish. In: Webster DB, Fay RR, Popper AN (eds) The Evolutionary Biology of Hearing. New York: Springer-Verlag, pp. 295–322.

Schellart NAM, Rikkert WEI (1989) Response features of visual units in the lower midbrain of the rainbow trout. J Exp Biol 144:357–375.

Schellart NAM, Kamermans M, Nederstigt LJA (1987) An electrophysiological study of the topographical organization of the multisensory torus semicircularis of the rainbow trout. Comp Biochem Physiol 88A:461–469.

Schellart NAM, Prins M, Kroese ABA (1992) The pattern of trunk lateral line afferents and efferent in the rainbow trout (*Salmo gairdneri*). Brain Behav Evol 39:371–380.

Schellart NAM, Wubbels RJ, Goossens JHLM (1995a) Coding of sound source direction in the trout lower midbrain. Eur J Neurosci Suppl 8:75 (32.20).

Schellart NAM, Wubbels RJ, Schreurs W, Faber A Goossens JHLM (1995b) Two-dimensional vibrating platform in nm range. Med Biol Eng Comput 33:217–220.

Schneider H (1982) Phonotaxis bei Weibchen des Kanarischen Laubfrosches, *Hyla meridionalis*. Zool Anz Jena 208:161–174.

Schneider H, Schneichel W, Walkowiak W (1986) Frequency representation in the midbrain of anuran amphibians. Verh Dtsch Zool Ges 79:294–295.

Schuijf A (1976) The phase model of directional hearing in fish. In: Schuijf A, Hawkins AD (eds) Sound Reception in Fish. Amsterdam: Elsevier, pp. 63–86.

Schuijf A, Buwalda RJA (1980) Underwater localization: a major problem in fish acoustics. In: Popper AN, Fay RR (eds) Comparative Studies of Hearing in Vertebrates. New York: Springer-Verlag, pp. 43–77.

Shotwell SL, Jacobs R, Hudspeth AJ (1981) Directional sensitivity of individual vertebrate hair cells to controlled deflection of their hair bundles. NY Acad Sci 374:1–10.

Suga N (1990) Biosonar and neural computation in bats. Sci Am 262:60–68.

Takahashi TT, Keller CH (1992) Commissural connections mediate inhibition for the computation of interaural level difference in the barn owl. J Comp Physiol [A] 170:161–169.

van Bergeijk WA (1962) Variation on a theme of Bekesy: a model of binaural interaction. J Acoust Soc Am 34:1431–1437.

van Bergeijk WA (1967) The evolution of vertebrate hearing. In: Neff WD (ed) Contributions to Sensory Physiology. New York: Academic Press, pp. 1–49.

Vlaming MSMG, Aertsen AMHJ, Epping WJM (1984) Directional hearing in the grassfrog (*Rana temparasia* L.): I. Mechanical vibrations of tympanic membrane. Hear Res 14:191–201.

Walkowiak W (1980) The coding of auditory signals in the torus semicircularis of the fire-bellied toad and the grassfrog: responses to simple stimuli and to conspecific calls. J Comp Physiol [A] 138:131–148.

Walkowiak W, Brzoska J (1982) Significance of spectral and temporal call parameters in the auditory communication of male grassfrogs. Behav Ecol Sociobiol 11:247–252.

Walkowiak W, Luksch H (1994) Sensory motor interfacing in acoustic behavior of anurans. Am Zool 34:685–695.

Wang J, Ludwig TA, Narins PM (1996) Spatial and spectral dependence of the auditory periphery in the Northern leopard frog. J Comp Physiol [A] 178:159–172.

Wilczynski W (1981) Afferents to the midbrain auditory center in the bullfrog, *Rana catesbeiana*. J Comp Neurol 198:421–433.

Wilczynski W, Resler C, Capranica RR (1987) Tympanic and extra-tympanic sound transmission in the leopard frog. J Comp Physiol [A] 161:659–669.

Wubbels RJ, Kroese ABA, Schellart NAM (1993) Response properties of lateral line and auditory units in the medulla oblongata of the rainbow trout (*Oncorhynchus mykiss*) J Exp Biol 179:77–92.

Wubbels RJ, Schellart NAM (1997) Neural coding of sound direction in the auditory midbrain of the rainbow trout. J Neurophysiol 77:3060–3074.

Wubbels RJ, Schellart NAM, Goossens JHLM (1994) Response characteristics of direction selective acoustic units in the trout lower midbrain. In: Elsner N, Breer H (eds) Proceedings of the 22nd Göttingen Neurobiology Conference, vol. 2. Stuttgart: Georg Thieme Verlag, p. 357.

Wubbels RJ, Schellart NAM, Goossens JHLM (1995) Mapping of sound direction in the trout lower midbrain. Neurosci Lett 199:179–182.

Xu JH, Gooler DM, Feng AS (1994) Single neurons in the frog inferior colliculus exhibit direction-dependent frequency selectivity to iso-intensity tone bursts. J Acoust Soc Am 95:2160–2170.

Xu JH, Gooler DM, Feng AS (1996) Effects of sound direction on the processing of amplitude-modulated signals in the frog inferior colliculus. J Comp Physiol [A] 178:435–445.

7
The Sense of Hearing in Fishes and Amphibians

RICHARD R. FAY AND ANDREA MEGELA SIMMONS

1. Introduction

For humans, the act of hearing results in a set of experiences that can lead to knowledge, but may or may not lead to overt behaviors. Ordinary experience suggests that most humans share these experiences and acquired knowledge, and thus share a sense of hearing. However, hearing in other species can be inferred only from behaviors that may or may not reveal experience and knowledge. If we are careful not to anthropomorphize, as many of us have been taught, our view of hearing in nonhuman animals tends to be tied to the behaviors most easily observed and understood, such as predator avoidance, prey identification, courtship, and vocal social interaction. Since experience and knowledge are impossible to observe directly, we may tend to deny their existence in other species, particularly those with which we do not readily identify, and those that are most distantly related to us. This makes it difficult for us to evaluate and understand the sense of hearing in other species in terms other than naturally occurring, sound-related behaviors. We may be led to believe, for example, that hearing in a given species or class can be fully explained as an adaptation for initiating and directing behaviors that occur in close temporal association with those sound sources that seem to require a prompt response, that is, those thought to be of "biological significance." In this view, we are probably fated to regard the sense of hearing in these species as simplified or impoverished compared with our own. Whether this view is accurate or not, it arises from our general ignorance of the experiences and knowledge that nonhuman animals may have with respect to audible sounds, their sources, and the "auditory scene."

The literature on fishes and amphibians presents two very different approaches to understanding the sense of hearing in taxa that are very distantly related to us. The majority of experiments on hearing in fish use behavioral control and psychophysical methods to measure detection and discrimination thresholds. These paradigms of experimental psychology have been borrowed from the quantitative and systematic study of human hearing. While these methods permit quantitative comparisons among species, including humans, the requirements that the learned behavior observed be simple, well defined, and highly controlled practically forces the view of hearing revealed as simple and impoverished compared with our everyday auditory experience. On the other hand, these experiments on fishes have so far failed to demonstrate fundamental differences between fishes and humans in the functions of hearing. Several examples of this parallelism in hearing are presented in this chapter. A significant challenge remains to develop experimental methods that can be applied to fishes, humans, and other vertebrates in a quantitative analysis of the more complex aspects of the sense of hearing.

The quantitative study of hearing in anuran amphibians (frogs and toads) has been hampered by the difficulty in training these animals in associative learning procedures (e.g., McGill 1960; Capranica 1965; Simmons and Moss 1995; but see Elepfandt 1985). The study of hearing in anurans has thus been dominated by an ethological approach in which species-specific behaviors are observed in response to complex communication sounds, or to simpler sounds that elicit stereotyped behaviors in more or less natural settings. Two different ethological techniques have been used to measure sound perception abilities in anurans. The evoked calling technique is based on observations that male anurans vocalize in response to playbacks of signals resembling conspecific advertisement or aggressive calls (e.g., Capranica 1965). The selective phonotaxis method is based on observations that female anurans orient to and approach the source of a conspecific advertisement call (e.g., Gerhardt 1978). Both evoked calling and selective phonotaxis are robust responses and can be studied both in the field and in the laboratory. Data obtained using these measures have proven valuable in identifying the acuity of perception for both spectral and temporal features of conspecific vocal signals. Since communication signals are stereotyped within a species and tend to be more variable across species, the view that emerges from these studies could suggest that a species' sense of hearing may have evolved to process biologically relevant vocal sounds (matched filter hypothesis; Capranica and Moffat 1983). In this view, species differences and uniqueness are not only expected, but may be forced by the ethological paradigm. In some cases, an experimenter is not permitted to observe those aspects of hearing that are shared among species and classes because the sounds used and the behaviors measured are, by definition, highly derived or species-specific rather than primitive or shared.

Ethological methods are of limited usefulness in measuring perceptual responses to signals that may be audible but do not resemble a species-specific communication signal. For example, female frogs tested in a two-choice design do not typically approach a sound source that resembles a communication signal of a heterospecific male frog; however, they sometimes approach the source of a heterospecific call in a one-stimulus (no choice) design (Gerhardt and Doherty 1988). This means that a lack of a phonotactic response or the lack of a calling response by a male frog does not necessarily imply a failure to perceive a sound, and exclusive reliance on these methods may underestimate perceptual abilities of anurans. There are some data on auditory perception in anurans obtained using a neutral psychophysical response (Megela-Simmons et al. 1985). However, there are many fewer psychophysical data available for frogs than for fish, and we are left with a sharp contrast between views of hearing in these different animals that arises from the assumptions inherent in the experimental paradigms that have been applied. A significant challenge remains to be able to compare quantitatively the sense of hearing in different taxa using some common assumptions and methods.

This chapter presents a large set of experimental results on the hearing capacities of fishes and amphibians. For the reasons outlined above, it is often difficult to draw close parallels between these groups in terms of the questions asked and the answers obtained. For example, it remains unclear whether and to what extent hearing plays fundamentally different adaptive roles among fish species, or the extent to which a general amphibian sense of hearing exists that may be shared with fishes and other vertebrates. Given these kinds of questions, we shall attempt to draw parallels and highlight species differences when the data permit.

2. Sound Detection Pathways

2.1 Fishes—Sound Pressure and Acoustic Particle Motion

Sound is detected by one or more of the otolith organs (saccule, lagena, and utricle) found in all fishes (see Popper and Fay, Chapter 1). These organs contain a patch of hair cell receptors overlaid by a solid otolith having a density of about 3 g/cc. As sound passes through a fish and brings its tissues into motion, the otoliths are thought to move at a different phase and amplitude due to their greater mass, the stiffness of their attachment to the hair cells and support structures, and their inertia. In this way, a relative displacement of the otolith occurs that is in proportion to acoustic particle motion (displacement, velocity, or acceleration), vector quantities having magnitude and direction. All otolith organs in all species tend to respond to sound-induced motions of the fish's body.

Otolith organs can respond with great sensitivity to accelerations due to gravity (0 Hz), and to sinusoidal motions up to several hundred Hertz. Neurophysiological thresholds for whole-body, sinusoidal motions can be as low as 0.1 nm at 100 Hz for *Carassius auratus* (goldfish) and *Opsanus tau* (toadfish) (Fay 1984; Fay and Edds-Walton 1997).

In many fish species, the otoliths may also receive a displacement input from the swim bladder or other gas-filled chamber near the ears. Since motions of the swim-bladder wall are created by changes in the bladder's volume as sound pressure fluctuates, this input to the ears is proportional to sound pressure, a scalar quantity having magnitude but not direction. Thus, many fishes may respond to both acoustic pressure and particle motion. Species having a particularly efficient mechanical coupling between the gas bladder and the otolith organs (i.e., the hearing specialists) tend to have very high sensitivity to sound pressure and may hear in a frequency range up to 3 to 5 kHz.

In experimental investigations of the sensitivity and frequency range of hearing, detection thresholds are defined as the lowest stimulus levels that result in statistically reliable conditioned or other behavioral responses. Meaningful absolute thresholds are difficult to obtain and interpret for those fish species that detect both sound pressure and particle motion since it may not be clear whether the threshold should be defined as a minimum detectable sound pressure or particle motion amplitude. This question is made particularly problematic because the ratio between sound pressure (p) and particle velocity (v) declines as the distance between the source and receiving fish becomes less than a wavelength or so (the wavelength of 100 Hz is 15 m). Consult Kalmijn (1988) and Rogers and Cox (1988) for excellent quantitative treatments of these near-field effects.

At a given frequency, if the sound pressure sensitivity of a species is relatively high, as it is for the hearing specialists, the sound pressure level at threshold may be an appropriate measure of sensitivity. However, if the ears respond to particle motion at threshold, as is the case for hearing generalists, the sound pressure at threshold is not an appropriate description of the species' sensitivity, and could reflect only differences in p/v that existed in the acoustic test field. In a typical test tank (even meters in dimension), p/v values may vary with frequency, the distance between the listener and the source (or nearby reflector such as the water surface), and the presence of standing waves, where p/v values may vary between zero and infinity. Furthermore, if an animal were primarily sensitive to particle motions, its thresholds would depend on the direction(s) of particle motion with respect to that most effective in stimulating the otolith organs. An interpretable particle motion threshold requires the measurement of the direction(s) as well as the magnitude of the stimulus motion.

In general, most detection thresholds for fishes reported in the literature have been specified in terms of sound pressure levels, and not particle motion amplitudes (Fay 1988), in spite of the fact that it is unknown for many species whether the sound pressure threshold is appropriate. Calibrated, off-the-shelf transducers for underwater particle motion are not available.

2.2 Anurans—Tympanic and Extratympanic Pathways

In anurans, sound pressure can be detected by either a tympanic pathway or an extratympanic pathway to the three inner ear organs (amphibian papilla, basilar papilla, saccule; see Lewis and Narins, Chapter 4). The tympanic system is driven by direct acoustic stimulation of the external tympanum, which is flush with the side of the animal's head. A cartilaginous columella couples the tympanum to the otic capsule (Wever 1985). The sensitivity of the tympanic pathway to free-field sound stimulation extends over the frequency range of about 200 to about 6000 Hz, depending on the species and on the size of the animal (Hetherington 1992). In addition, the lateral body wall overlying the lung vibrates in response to free-field sound. These vibrations are transmitted to the internal surface of the tympanic membrane through the mouth cavity and eustachian tubes (Narins et al. 1988). The sensitivity of this pathway is limited to frequencies below about 1 kHz in most species, and the peak amplitude of response of the lateral body wall, as measured by laser vibrometry, typically occurs at a frequency lower than that of the tympanum (Narins et al. 1988; Hetherington 1992).

The extratympanic pathway involves use of the shoulder and lateral head surface to transmit sound energy to the inner ear via the opercularis system (Wilczynski et al. 1987). This pathway is most efficient in transmission of low frequency sounds and vibrations. The peak sensitivity of the extratympanic pathway varies with body size, and thus shows a different frequency response for small and large animals (Hetherington 1992).

Larval anurans (tadpoles) undergo a period of metamorphoses during which the auditory system acquires the ability to detect airborne sounds. Physiological recordings from the auditory midbrain (Boatright-Horowitz and Simmons 1997) reveal that tadpoles are sensitive to free-field sounds. The peripheral transduction pathway mediating this auditory sensitivity changes as the animal metamorphoses from a tadpole into a frog. In early stages of larval development, environmental sound induces pressure changes in the inner ear directly through the oval window (Hetherington 1987), via a fenestral pathway (Boatright-Horowitz and Simmons 1997). The extratympanic opercularis system forms in later larval stages, and, in bullfrogs, the tympanic system does not function until after metamorphosis, when the external tympanum first appears on the side of the head. The progressive development of these pathways is reflected in changes in auditory sensitivity across the larval period, including a striking transient loss of neural responsiveness as the forming opercularis system blocks the oval window. Over the time course of metamorphic development, neural responses from the tadpole's midbrain show auditory sensitivity similar to that of hearing specialist fish. After metamorphosis, neural responsiveness undergoes further refinement, as the tympanic pathway matures and the froglet grows in body size (Boatright-Horowitz and Simmons 1995). The role of the bronchial columella, which connects the lung sacs to the round window in early stage tadpoles (Witschi 1949), is unknown.

3. Audiograms

3.1 Fishes

The only behavioral thresholds reported for fishes in the literature that could be interpreted are sound pressure thresholds for hearing specialists (Fig. 7.1), and particle motion thresholds for several hearing generalist species that were specifically demonstrated to be detecting particle motion in the experiment conducted (Fig. 7.2). Even in both these cases, however, there has been wide variation among thresholds determined for the same species by different investigators in different laboratories (see Fay 1988).

As shown in Figure 7.1, hearing specialists detect sound pressure with the lowest thresholds in the range between 50 to 75 dB with respect to 1 μPa, in a frequency range between 200 and 2000 Hz. Pressure sensitivity generally rolls off at frequencies below 200 to 300 Hz, and at frequencies above 400 to 1000 Hz, depending on species. Underwater sound pressure thresholds may be compared with thresholds for terrestrial animals by expressing them in terms of intensity in watts cm^{-2}, taking the acoustic impedance of the medium into consideration (Fay 1988). By this measure, the most sensitive hearing specialists among fishes have approximately the same sensitivity as

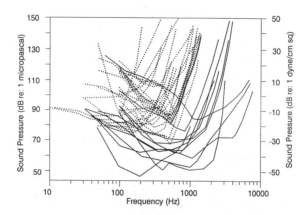

FIGURE 7.1. Behavioral sound pressure audiograms for over 50 species of fish taken from an extensive review (Fay 1988). Dark lines are hearing specialists and dotted lines are species not known to have special connections between a gas bladder and the ears (hearing generalists). Thresholds are given in dB re: 1 μPa (left ordinate) and in dB re: 1 dyne cm^{-2} (right ordinate). See Fay (1988) for complete references and identification of species.

the most sensitive mammals (e.g., cat) and birds (e.g., *Tyto alba*, barn owl). There are reports from field behavioral studies that clupeids (Dunning et al. 1992; Nestler et al. 1992) and *Gadus morhua* (Astrup and Mohl 1993) act to avoid intense ultrasound (160 to 200 dB re: 1 μPa, between 110 and 140 kHz) under certain circumstances. Mann and colleagues (1997) have reported conditioned responses from a clupeid to tones at frequencies over 100 kHz. The mechanisms for these responses to ultrasound are unknown.

Figure 7.2 shows the audiograms for generalized species that respond only to particle motion at threshold. These audiograms look somewhat

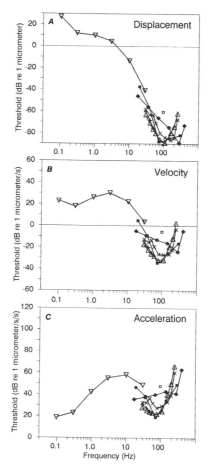

FIGURE 7.2. Behavioral particle motion audiograms for *Gadus morhua* (cod) (Buerkle 1969 [◆], Chapman and Hawkins 1973 [+], Offutt 1973 [■], Sand and Karlsen 1986 [∇]); *Pleuronectes platessa* (plaice) (Chapman and Sand 1974 [△]); *Limanda limanda* (dab) (Chapman and Sand 1974, [×]); *Astronotus ocellatus* (oscar) (Lu et al. 1996, [single □]). Panels A, B and C plot thresholds in terms of displacement, velocity, and acceleration, respectively.

different depending on whether the thresholds are defined in terms of particle displacement (panel A), or acceleration (panel C). At 100 Hz, displacements of 0.04 to 1.0 nm (root mean square) can apparently be detected, depending on species. These values are quite small, corresponding to the acoustic particle motion in the far field at a sound pressure level below 80 dB re: 1 μPa, to a fraction of the diameter of a hydrogen atom, and to the amplitude of motion of the mammalian basilar membrane at the threshold of hearing (Allen 1997). Displacement sensitivity at 100 Hz of 0.1 nm has also been measured for primary saccular afferents of *C. auratus* (Fay 1984) and *Opsanus tau* (Fay and Edds-Walton 1997). The lowest displacement thresholds occur between 50 and 300 Hz. In terms of displacement, sensitivity rolls off steeply at frequencies below 100 Hz. However, in terms of acceleration, sensitivity remains good down to frequencies as low as 0.1 Hz (Sand and Karlsen 1986), Kalmijn (1988) has argued that it may be more useful to describe these functions in terms of acceleration level at threshold. Behavioral studies on infrasound detection and its mechanisms have also been reported by Karlsen (1992a,b) for *Pleuronectes platessa* (plaice) and *Perca fluviatilis* (perch), and by Knudsen and colleagues (1992, 1994) for *Salmo salar* (salmon).

3.2 Anurans

Audiograms have been measured for only a few anuran species, and few trends can be described from these data. Strother (1962) and Weiss and Strother (1965) measured an unconditioned galvanic skin response to sound in two species, *Rana catesbeiana* (bullfrog), and *Hyla cinerea* (green treefrog). The electrodermal response typically shows relatively high and variable thresholds across frequency. For *R. catesbeiana*, the best sensitivity of this response occurs around 100 Hz, with thresholds of 44 to 64 dB SPL re: 20 μPa (mean of 56 dB; Fig. 7.3A). In four other ranid species—the grass frog (*R. temporaria*), pond frog (*R. lessonae*), water frog (*R. ridibunda*), and the hybrid water frog (*R. esculenta*)—sensitivity of the electrodermal response varies with frequency according to a low-pass filter shape, with most sensitive frequencies around 1000 to 1500 Hz (Brzoska et al. 1977; Brzoska 1980). The lowest thresholds that could be obtained were at around 40 dB SPL for *R. temporaria*, increasing to about 80 dB SPL for the least sensitive species, *R. ridibunda* (Fig. 7.3A).

Audiograms measured using the reflex modification technique in *R. catesbeiana* show a U shape rather than the low-pass or nonselective shape of the electrodermal response functions, and indicate greater sensitivity (Megela-Simmons et al. 1985; Simmons and Moss 1995; Fig. 7.3A). Maximal sensitivity for *R. catesbeiana* occurs in the frequency range of 600 to 1000 Hz, which matches the peak frequency sensitivity of the tympanum of

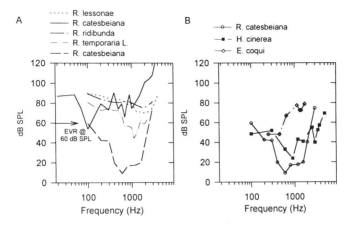

FIGURE 7.3. A: Audiograms for five species of Rana. Data for *R. lessonae, R. ridibunda, R. Temporaria* L., and *R. catesbeiana* were collected using the electro-dermal response, and are replotted from Strother (1962), Brozska et al. (1977), and Brzoska (1980). Data for *R. catesbeiana* were collected using the reflex modification technique and are replotted from Megela-Simmons et al. (1985). The threshold for elicitation of the evoked vocal response (EVR) is 60 dB SPL (Megela-Simmons 1984) and is indicated on the figure by a horizontal arrow. Thresholds measured using the reflex modification technique are lower than those obtained using the electrodermal response. B: Audiograms from three species of frogs. Data for *R. catesbeiana* and *Hyla cinerea* were collected using the reflex modification technique and are replotted from Megela-Simmons et al. (1985). Data for *E. coqui* are based on the suppression of the evoked vocal response to tonal stimuli and are replotted from Zelick and Narins (1982). *H. cinerea* is the only species that shows maximal sensitivity at a frequency (900 Hz) representing a spectral peak in its advertisement call.

this species (Hetherington 1992). It is also within the range of best frequency response of several species of hearing-specialist fishes (Fig. 7.1). Reflex modification data show that *R. catesbeiana* is not maximally sensitive to low frequency (around 200 Hz) energy present in the species adver-tisement call, as predicted from a matched filter hypothesis (Capranica and Moffat 1983). Thresholds measured to pure tones using reflex modi-fication are lower than those measured from studies of the evoked vocal response to complex advertisement calls, which indicate a threshold for a calling response between 50 and 60 dB SPL (Megela-Simmons 1984). Thresholds estimated using evoked calling more likely represent masked rather than absolute thresholds because they are measured at natural calling sites with background noise present; moreover, they do not necessarily reflect only perceptual thresholds but also a threshold for a motor response.

The hearing sensitivity of *H. cinerea* measured by reflex modification extends to higher frequencies than that of *R. catesbeiana* (Megela-Simmons et al. 1985; Fig. 7.3B). This is consistent with the differences in the relative sizes of these two species (adult snout-vent length for *H. cinerea* is about 5 cm compared with 10 to 18 cm in *R. catesbeiana*), and from differences in the range of frequencies present in their vocal signals (*H. cinerea* advertisement calls contain energy up to about 4000 Hz, while *R. catesbeiana* advertisement calls contain energy up to about 2000 Hz). The frequency range of maximal sensitivity for *H. cinerea* encompasses the frequency range for one of the spectral peaks (about 900 Hz) in its advertisement call. Moreover, the audiogram shows a second dip in threshold around 3000 Hz, near the dominant high-frequency peak in the call. These data suggest that the hearing sensitivity of *H. cinerea* is consistent with a matched filter model (Capranica and Moffat 1983). The shape and sensitivity of the audiogram also reflects the summation of the combined peak frequency responses of the tympanic and extratympanic surfaces of this species (Hetherington 1992).

An audiogram for *Eleutherodactylus coqui* (Puerto Rican coqui frog) was estimated using a behavioral technique measuring the suppression of the evoked calling response to stimuli timed to overlap with the male's own response (Zelick and Narins 1982). Behavioral auditory threshold functions from these data show the frequency and level of a tone required to evoke a particular criterion of call suppression. At tone frequencies of 1100 to 1200 Hz, corresponding to the frequency of the "co" note of the male's advertisement call, more intense tones (levels of 60 to 85 dB SPL) are required to suppress calling than at lower frequencies of 200 to 300 Hz (Zelick and Narins 1982; Fig. 7.3B). This does not necessarily mean that the frogs are less sensitive to tones of 1100 Hz than to tones of other frequencies, but rather could reflect the different communicative significance of these different sound frequencies to the animal. The high thresholds for call suppression might also reflect interference from other sound sources present in the natural environment.

In general, audiograms from anurans show that the hearing of these animals is limited to a low-frequency range below about 6000 Hz, with sensitivity somewhat less than that measured in other vertebrate species (Fay 1988). The poor high-frequency sensitivity of anurans reflects the poor high-frequency response of their middle ear and the mechanics of the peripheral auditory organs (see Lewis and Narins, Chapter 4). Because behavioral audiograms are based on free-field presentation of sounds that stimulate both the tympanic and extratympanic pathways, behavioral thresholds might be expected to differ from those based on eighth nerve thresholds to sounds, which are measured using closed-field presentations that stimulate the tympanic pathway only. It is also unclear how factors such as the animal's posture and the anesthesia used in physiological experiments might differentially affect the operation of the tympanum-columella, lung-eardrum, and extratympanic pathways.

4. Effects of Sound Duration on Sound Detection

4.1 Fishes

The threshold for detecting a brief sound generally declines as sound duration increases in most vertebrate species studied (Fay 1992a), including *C. auratus* (Fay and Coombs 1983), and cod (*Gadus morhua*; Hawkins 1981). There are two ways that the effect shown in Figure 7.4 can be understood. First, to the extent that sensory systems respond in proportion to stimulus energy (the product of duration and intensity), this effect follows. Second, stimuli of longer duration offer the opportunity for a greater number of "looks" or independent decisions about their presence (Viemeister and Wakefield 1991). To a first approximation, threshold functions of duration for *C. auratus* (Fay and Coombs 1983) are power functions with an exponent (slope) of about −1.0, meaning that a 10-fold increase in duration produces a 10-dB reduction in threshold. At long durations, these functions are limited by the time constant of a hypothetical neural integrator that combines neural activity, or the decisions based on this activity, over time. For *C. auratus*, durations greater than about 400 ms do not lead to lower thresholds.

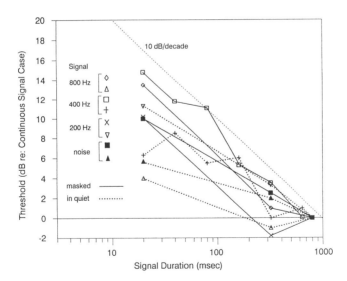

FIGURE 7.4. Behavioral thresholds as a function of signal duration in *C. auratus* (Fay and Coombs 1983). Data were obtained for tones of different frequency (open symbols) and noise (filled symbols) presented in quiet (dotted lines) or against a broad-band noise background (solid lines). Functions obtained against a background of noise tend to lie parallel to the 10-dB/decade line (equal energy). Functions obtained in nominal quiet tend to have a lower slope. Asymptotic thresholds are reached for signals with durations greater than 300 to 600 ms.

There are some conflicting data in the literature regarding the effects of sound duration on threshold. Popper (1972) obtained flat threshold functions of duration for *C. auratus* tested in quiet conditions, indicating a lack of temporal summation. Fay and Coombs (1983) demonstrated that the threshold function of duration has a lower slope for sound detection in quiet conditions than in a background of masking noise (illustrated in Fig. 7.4). These differing results remain unexplained.

4.2 Anurans

Two experiments examining sensitivity to stimulus duration in frogs have yielded different curves for duration sensitivity. Narins and Capranica (1978) measured the percent of "co" responses emitted by a vocalizing male *E. coqui* in response to playbacks of synthetic "co" notes of different durations. Animals vocalized more in response to tone durations of 100 ms, the duration of the natural "co" note, than to either shorter or longer durations (Fig. 7.5A). Although threshold declined with stimulus intensity

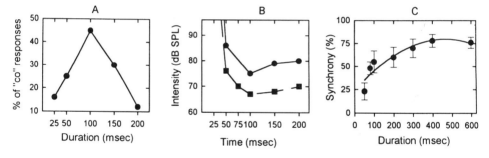

FIGURE 7.5. Effects of stimulus duration on evoked calling responses in two species of frogs. A: Percent of "co" responses by vocalizing male *E. coqui* given to playbacks of "co" stimuli of different durations. Frequency of the "co" note is about 1100 Hz. Animals vocalize more in response to stimuli of 100-ms duration, which is the duration of the natural "co" note. Data are based on responses from seven males and are replotted from Narins and Capranica (1978). B: Stimulus intensity required to evoke a criterion level of evoked vocal response by male *E. coqui* to "co" notes of different durations. Effective intensity declines with increasing stimulus duration up to a duration of 100 ms, then increases slightly at longer durations. Data from two animals are plotted separately (Narins and Capranica 1978). C: Percent of evoked calling responses by male Central American treefrogs (*Hyla ebraccata*) that are synchronized to one-note advertisement calls (about 3000 Hz) of different durations. Each data point is the mean (± standard deviation) from 11 frogs. The natural advertisement and aggressive calls of this species have a duration of about 150 to 200 ms; advertisement and aggressive calls of sympatric species have durations ranging from about 100 to 400 ms. Data are replotted from Schwartz and Wells (1984).

as predicted by an energy detector model, sensitivity was greatest to sounds of 100-ms duration, and decreased slightly at longer durations (Fig. 7.5B). Schwartz and Wells (1984) found that the ability of male Central American treefrogs (*Hyla ebraccata*) to synchronize their vocal responses to playbacks varies with stimulus duration (Fig. 7.5C). If it is assumed that increased synchrony of response indicates greater sensitivity, then the data from *H. ebraccata* are consistent with those from fishes in showing that sensitivity increases with stimulus duration over the range of about 50 to 400 ms, then levels off. These data are more consistent with duration sensitivity shown by fishes (Fig. 7.4) and other vertebrates (Fay 1988). For *H. ebraccata*, these data also reflect the animals' sensitivity to both conspecific and heterospecific vocalizations present in their natural environments.

5. Effects of Interfering Sounds and Masking

5.1 Fishes

The detectability of a given sound (signal) can be determined by the presence of another, interfering sound (masker), in all vertebrate species studied (Fay 1992a). In fish as in other vertebrates, the masker is most effective when it is simultaneously present with the signal (simultaneous masking), but can also exert interfering effects if it ends before signal onset (forward masking), or begins after signal offset (backward masking) (Popper and Clarke 1979). In human listeners and in *C. auratus*, simultaneous masking is more effective when the signal and masker have simultaneous onsets compared to the case in which the signal comes on 30 ms or so after the masker onset (Coombs and Fay 1989). In the everyday world, most sounds to be detected (signals) are usually masked by other environmental sounds (noise). Once the principles of masking are known, predictions about the detectability of a given sound can be made based on measurements of the signal levels and the levels of background noise.

In general, masking can be understood by assuming that signal detection is based on a decision on whether an hypothetical detection channel (e.g., an "auditory filter," possibly but not necessarily an auditory nerve fiber) is activated by noise alone, or by a signal plus noise (the output of the channel is usually greater when the signal is present with the noise). Fletcher (1940) presented a scheme and assumptions for understanding a simple case of masking in which the signal is a pure tone, and the masker is a flat-spectrum noise of wide bandwidth. First, he assumed that the signal tone is detected at threshold by monitoring the output of the auditory filter centered on the signal frequency. Thus, only the masking noise falling within this filter's passband is effective in interfering with tone detection. Second, he assumed that at threshold, the power of the tone signal is equal to the power of the noise that the filter passes. Thus, a narrow filter would admit a sinusoid at the filter's center frequency, but would tend to reject, or "filter out," noise components falling outside the filter's band. In this case, the signal-to-noise ratio (S/N) at the output of the filter would be high and the signal would be relatively detectable. A wide, nonselective filter would admit the signal as well, but would also admit more noise components (a greater noise power), and the S/N would be less favorable for detection.

Under these assumptions, the width of the filter can be estimated by determining the S/N for a tone at masked threshold. The width of the filter can be estimated as a bandwidth $(B) = 10^{((S/N)/10)}$, where S/N is the level of the tone signal at threshold (in dB) minus the spectrum level (dB/Hz) of the noise. For example, in *C. auratus* it has been found that a 500-Hz tone must be 20 dB higher in level than the spectrum level of a masking noise for the tone to be detected (Fay 1974a). Thus, the bandwidth of the hypothetical detection filter is 100 Hz (i.e., $10^{(20/10)}$). The interpretation is that only a 100-Hz-wide band of noise, surrounding the tone in frequency, is effective in masking because only this band is admitted by the filter that is optimally tuned to admit the tone signal.

Figure 7.6 shows S/N values as a function of tone signal frequency for *C. auratus* and *Gadus morhua*. In general, the data show that as signal frequency increases, the bandwidths of the auditory filters increase, from about 20 Hz for a 100-Hz signal to 250 Hz for a 1200-Hz signal. The S/N values at threshold tend to be constant as a function of noise level. These data are qualitatively and quantitatively similar to those of other vertebrates tested, including humans (Fay 1988). This has been interpreted to suggest that sound detection is mediated by an auditory filter bank in all vertebrates.

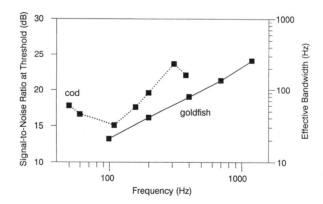

FIGURE 7.6. Behavioral critical masking ratios (S/N ratios) for *Gadus morhua* (cod) (Chapman and Hawkins 1973) and *C. auratus* (goldfish) (Fay 1974a) as a function of signal frequency. Signal-to-noise ratio is the signal level at threshold minus the spectrum level (level per Hz) of the masking noise. The right ordinate indicates equivalent bandwidths of auditory filters based on the assumptions of Fletcher (1940).

Other experiments have been designed to measure the shapes of the auditory filters using masking methods. In one experiment on *G. morhua* (Hawkins and Chapman 1975), narrow-band (10-Hz-wide) noise maskers at various center frequencies were used. In the presence of each masker, the thresholds for tones of various frequencies were determined. Figure 7.7 shows these signal thresholds plotted as a function of masker frequency. For each masker, signal thresholds trace an inverted V function, with the most masking occurring when the signal and masker frequency were equal. These and similar results on other species (e.g., Buerkle 1969; Tavolga 1974; Fay et al. 1978; Hawkins and Johnstone 1978; Coombs and Popper 1981; McCormick and Popper 1984) indicate that fishes detect sounds using frequency-selective filters. In practical terms, this means that a sound of a given frequency will be potentially masked only by noise components that are similar to the signal frequency. Noise components outside the bandwidth of auditory filters will have little interfering effect on signal detection.

5.2 *Anurans*

Frogs, as well as fish, detect sounds using frequency-selective filters; however, the few data available on frequency selectivity suggest that the width of these filters are wider in frogs than in most other vertebrates. Moss and

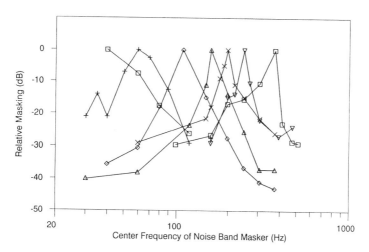

FIGURE 7.7. Behaviorally determined auditory filter shapes for *Gadus morhua* (cod) (Hawkins and Chapman 1975). For each curve, a tone signal was fixed in frequency (corresponding to the peak of each function), and a 10-Hz-wide band of noise served as a masker. Signal threshold was measured as a function of the center frequency of the masker noise band. Symbols serve to help separate the functions visually.

Simmons (1986) and Simmons (1988a) estimated S/N values for detection of tones in noise for *H. cinerea* and *R. catesbeiana* using the reflex modification technique (Fig. 7.8). For both species, the overall function describing the relationship between filter bandwidth and stimulus frequency differs in shape from the simple increase with frequency typically observed in fishes (Fig. 7.6) and in most other vertebrates (Fay 1988). For *H. cinerea*, the S/N function is W-shaped, with greatest frequency resolution at two frequencies, 900 Hz and 3000 Hz. Data for *R. catesbeiana* show an increase in S/N ratio with tone frequency from about 100 to 600 Hz, then again from 1200 to 3000 Hz, with a dip at about 1000 Hz. The most sensitive S/N values for both anuran species (20–22 dB for *H. cinerea*, 17 dB for *R. catesbeiana*) lie in the frequency region around 1000 Hz. These values are consistent with the widths of frequency selective filters for fishes (Fig. 7.6), and for other vertebrate species (Fay 1988) at similar center frequencies. The estimates for *H. cinerea* in this frequency range are also consistent with estimates

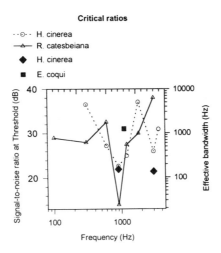

FIGURE 7.8. Signal-to-noise ratios at threshold (y-axis, left) and equivalent bandwidths (y-axis, right) for detection of tones in noise by three species of frogs. Data for *R. catesbeiana* are replotted from Simmons (1988a) and are based on thresholds using the reflex modification technique. Data for *H. cinerea* (circles) are replotted from Moss and Simmons (1986) and are also based on thresholds using the reflex modification technique. Two data points (stars) for *H. cinerea* are replotted from Ehret and Gerhardt (1980) and are based on thresholds estimated from the selective phonotaxis response of the female. Lowest S/N ratios for *H. cinerea* occur at frequencies (900 and 3000 Hz) important for species-specific communication. Although *R. catesbeiana* also has a peak in selectivity at 1000 Hz this frequency does not represent a spectral peak in the advertisement call. The data point from *E. coqui* is based on suppression of the evoked calling response and is replotted from Narins (1982).

obtained using a selective phonotaxis technique (Ehret and Gerhardt 1980). The S/N ratios at frequencies lower than about 1000 Hz are considerably larger in both *H. cinerea* and *R. catesbeiana* than in *C. auratus* (Fig. 7.6). The increased selectivity at or around 900 Hz for both anuran species might represent some specialized neural processing in the auditory system in this frequency range. This might reflect, for example, a specialization of the receptor surface of the amphibian papilla in this frequency range or the influence of mechanical tuning of the tectorial membrane (Moss and Simmons 1986). An alternative explanation for these effects is that decision efficiency (Patterson and Moore 1986) may vary in species-specific ways among anurans. The increased selectivity of *H. cinerea* at 900 and 3000 Hz, the two dominant spectral peaks in the species advertisement call, might represent an adaptation for filtering biologically relevant signals from noise in a manner consistent with a matched filter hypothesis (Capranica and Moffat 1983).

Narins (1982) estimated S/N values using the evoked calling response for signals presented against masking noise. The noise level at which calling might be suppressed or is no longer synchronous with the playback can be used as an estimate of the efficacy of masking. For *E. coqui*, the S/N value (tone level required for call suppression minus background noise level) is between 31 and 40 dB (bandwidth of 1250–10,000 Hz) at a center frequency of 1000 Hz, depending on the absolute level of the tone. When the bandwidth of the masking noise was varied, an estimate of a critical bandwidth of 27 dB (501 Hz) was obtained (Narins 1982). Overall, the estimates of frequency selectivity from these experiments indicate somewhat poorer selectivity at about 1000 Hz for *E. coqui* than for *H. cinerea* or *R. catesbeiana*, and poorer selectivity than observed for other vertebrates at a similar center frequency (Fay 1988; see Fig. 7.6). Whether this is a true species difference or related to the technique used to measure the response is unclear. The high level of background noise in the natural habitat present during evoked calling experiments may have contributed to masking and therefore may have overestimated the S/N values.

Narins (1983) and Schwartz and Wells (1983) attempted to estimate auditory filter shapes in frogs by measuring vocal responses to signals presented against narrow-band masking noise. As expected, maskers centered around the signal frequency produced a greater effect on the evoked vocal response than maskers of similar bandwidth centered around either higher or lower frequencies. The filter functions derived from these experiments are asymmetrical in shape, probably due to acoustic interference from both conspecific and heterospecific males on the evoked vocal response of the target male (Narins 1983; Schwartz and Wells 1983). Noise bandwidth was not varied in these experiments, and so precise estimates of the width of the detection filters could not be obtained.

6. Effects of Signal and Masker Source Separation on Signal Detection

6.1 Fishes

Signal detectability can depend on the spatial locations of the signal and noise sources. In humans and other mammals (Fay 1988), a noise can lose some of its effectiveness as a masker if its source is spatially separated (in azimuth) from the signal source. In terrestrial animals, this sort of masking release occurs as a result of central binaural processing; neural activity resulting from a particular interaural time or level difference can be separated from activity resulting from another interaural time delay using a cross-correlation network (Jeffress 1948; Carr and Konishi 1990). This phenomenon has been termed the "cocktail party effect" because people unable to access this processing due to a hearing loss in one ear tend to avoid cocktail parties and similar situations where one must "hear out" a conversation in the presence of many competing voices.

Fishes also show a release from masking when the noise source is spatially separated from the signal source. Figure 7.9 summarizes this sort of data for *G. morhua* and *Melanogrammus aeglefinus* (haddock). As the

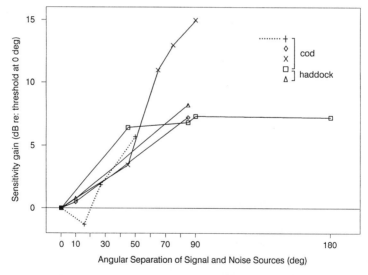

FIGURE 7.9. Behaviorally determined masked thresholds for tone signals in *Gadus morhua* (cod) and *Melanogrammus aeglefinus* (haddock) as a function of the angular separation between the signal source and the noise source (Chapman 1973, Chapman and Johnstone 1974, Hawkins and Sand 1977, Buwalda 1981) in azimuth (solid lines) and elevation (dotted line). The ordinate indicates a release from masking with reference to the masked threshold at a 0° separation between the signal and masker sources.

angular separation between signal and noise sources widens, the signal threshold declines (sensitivity is gained), indicating that the effective noise power has also declined (by up to 15 dB). This effect probably depends on the directional response characteristics of hair cells responding directly to acoustic particle motion (e.g., Fay 1984). In addition to frequency filtering, then, fishes also may also use space-domain filtering to enhance signal detectability in a noise background.

6.2 Anurans

The directionality of the anuran auditory system also aids in the perceptual segregation of concurrent sounds; however, in the animals that have been studied to date, the effects are not large. Schwartz and Gerhardt (1989) examined the role of spatial separation of sound sources on the phonotactic response of *H. cinerea* to synthetic advertisement calls presented with concurrent noise. When the noise sources are separated from the signal sources by either 45° or 90°, there was a release from masking of 3 dB. Female gray treefrogs (*H. versicolor*) discriminate sound sources separated by 120° but not those separated by 45°; discrimination is abolished when the intensity of the separated sources is decreased by 3 dB (Schwartz and Gerhardt 1995). In the same species, neurons in the auditory midbrain show a release from masking of about 9 dB at similar angular separations. In other phonotactic experiments with smaller frogs, angular separations of signal from noise sources do not increase detectability (Schwartz 1993). The 3-dB release from masking shown behaviorally in two anuran species is less than that seen for fish at similar angular separations (Fig. 7.9); whether this represents a true species difference or a difference in experimental technique is not clear.

7. Level Discrimination

7.1 Fishes

Level discrimination is measured by asking listeners to discriminate between sounds differing only in level or intensity. The smallest difference in level required for reliable discrimination is the level discrimination threshold (LDT). The ability to detect a change in the overall level of a sound is a simple yet important hearing function for all animals. Not only does this ability have obvious survival value (e.g., its role in the perception of source distance and changes in distance), but it is also important for the identification and localization of sources through their characteristic spectral shapes (e.g., perceiving the relative amplitudes of multiple frequency components; Green 1988). Level discrimination also plays an important role in the detection of sound in noisy backgrounds since signal detection in noise can be a decision about an increment in level at one or several auditory filters (Fay and Coombs 1983).

Figure 7.10 summarizes LDTs for *C. auratus* (Fay 1985, 1989b). For most of these experiments, animals were exposed to tone bursts of constant level and were conditioned to respond when the level changed. For burst durations between 10 and 200–300 ms, LDTs for both tones (solid lines) and noise (dotted lines) decline nearly linearly with log duration (about −3 dB per decade of duration), reaching an asymptote of 2 to 3 dB at durations greater than about 200 ms. Thresholds for detecting increments in continuous sounds are lower than for pulsed sounds. Increment thresholds for continuous tones are very small (about 0.1 dB; see also Fay 1980), and independent of duration.

For long-duration pulsed tones, LDTs are independent of frequency and remain constant with increasing overall level, consistent with Weber's law (Fay 1989a—data not shown in Fig. 7.10). Weber's law states that the just-detectable change in stimulus magnitude is proportional to the overall stimulus magnitude. When LDTs are expressed in dB, Weber's law predicts a constant LDT. In general, LDT values and the effects of overall level, duration, and frequency are similar in *C. auratus* to those of other vertebrates studied (reviewed in Fay 1988, 1992a).

There are some conflicting data in the literature on the size of the LDT and the effect of frequency on the LDT. Chapman and Johnstone (1974) found that the LDT for *G. morhua* and *M. aeglefinus* increased from 1.3 dB at 50 Hz to 9.5 dB at 380 Hz (a frequency approaching the upper limit of the

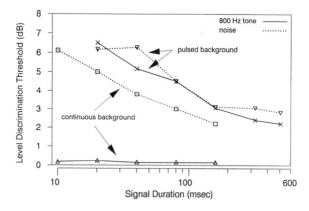

FIGURE 7.10. Behavioral sound level difference thresholds for *C. auratus* (goldfish) as a function of signal duration (Fay 1989a). Thresholds were determined for tone (solid lines) and noise (dotted lines) signals, either as repeated pulses, or as ramped, amplitude increments in a continuous background. Except for increments in a continuous tone, thresholds decline with sound pulse duration at about 3 dB/decade up to a duration for 200 to 300 ms.

hearing range). Jacobs and Tavolga (1967) measured LDTs for *C. auratus* that were 4 to 5 dB at frequencies in the middle of the hearing range and rose to over 10 dB at 1500 Hz. The reasons for the differences between these results and those summarized above are not clear.

7.2 Anurans

The LDT for tone bursts at one frequency was measured by Zelick and Narins (1983) using the suppression of the evoked vocal response as a measure of discriminability. Calling male *E. coqui* were presented with 1100 Hz tone bursts that varied in amplitude over a 1- to 1.5-second interval in a 2.5-second-long playback. Males tended to call only within the time interval of the reduced stimulus amplitude. Only 16% of animals shifted their responses to occur within this interval when the amplitude was reduced by 1 to 3 dB, but 59% shifted their calling in response to decreases in amplitude of 4 to 6 dB. Zelick and Narins (1983) suggested that the intensity difference limen in *E. coqui* lies between 4 and 6 dB at 1100 Hz. LDTs at other frequencies could not be tested using this paradigm. These estimates are within the range of those measured for *C. auratus* (Fig. 7.10). The influence of stimulus duration on the LDT in *E. coqui* is not known.

Several studies of level discrimination in anurans examined the animal's ability to detect changes in the level of a complex sound source resembling a conspecific vocal signal. Experiments by both Capranica (1965) and Megela-Simmons (1984) showed that vocalizing male *R. catesbeiana* could detect differences of 10 dB in the amplitude of conspecific advertisement calls; smaller differences in level were not tested. Gerhardt and Klump (1988) reported that female *H. cinerea* could discriminate an advertisement call whose sound pressure level differed by 6 dB from background chorus noise. These data imply an intensity difference limen of 6 dB. Other studies of the selective phonotactic response (Fellers 1979; Arak 1983; Dyson and Passmore 1988a) showed that female frogs can discriminate between complex sound sources differing in intensity by 2 to 6 dB. This is consistent with the estimates of 4 to 6 dB measured by Zelick and Narins (1983) for pure tone level discrimination.

8. Frequency Discrimination

8.1 Fishes

Frequency discrimination is measured by asking listeners to discriminate between pure tones differing only in frequency. Frequency discrimination has been of great theoretical interest in studies of hearing in fishes since the time of von Frisch (1936). Otolith organs lack an analogue of the basilar membrane, and it appears unlikely that a traveling wave or other macromechanical frequency selective process takes place along the sensory epithelium. Thus, a macromechanical place principle of frequency analysis (von Bekesy 1960) seems unlikely and any frequency analytic capacities would have to be explained by other mechanisms, such as hair cell tuning (Crawford and Fettiplace 1981), hair cell micromechanics (Fay 1997), or time-domain processing (Fay 1978). In this way, fishes may provide a valuable preparation for investigating frequency analysis in the absence of a macromechanical place principle.

The smallest difference in frequency (df) required for reliable discrimination is the frequency discrimination threshold (FDT). To reduce the possibility that listeners base their judgments on the psychological dimension of loudness, the levels of the tones to be discriminated are adjusted to be the same level above absolute threshold, and in some cases the levels are varied randomly from trial to trial in an attempt to make a potential loudness cue irrelevant.

Figure 7.11 shows FDTs determined as a function of frequency for several species. In general, FDTs increase with frequency (f), maintaining a df/f (Weber ratio) of about 0.04 for *C. auratus*. These data suggest that the

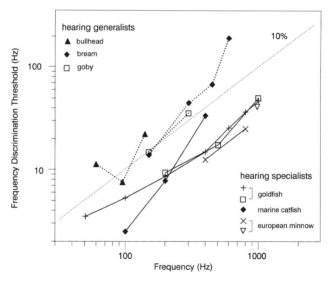

FIGURE 7.11. Behavioral frequency difference thresholds plotted as a function of frequency for hearing specialists (solid lines) and hearing generalists (dotted lines). *Phoxinus laevis* (European minnow); Wohlfahrt 1939 [▽] and Dijkgraaf and Verheijen 1950 [×]. *Carassius auratus* (goldfish); Jacobs and Tavolga 1968 [□] and Fay 1970a [+]. *Arius felis* (marine catfish); Tavolga 1982 [◆]. *Sargus annularis* (bream) [◆] and *Gobius niger* (goby) [□]; Dijkgraaf 1952. *Cottus scorpius* (bullhead) [▲]; Pettersen 1980. In general, the functions tend to parallel the equal-percentage line in accord with Weber's law.

hearing specialists shown are more sensitive to frequency changes than are the hearing generalists. FDTs tend to remain constant with changes in overall level (Fay 1989b). These features of the FDT in fishes are essentially similar to those of all vertebrates tested (Fay 1992a). In particular, the increase in df with f (the slopes of the functions in Fig. 7.11) are similar in all vertebrates, and the values of df for *C. auratus* are within the upper range of values determined for mammals and birds (reviewed in Fay 1974b, 1988, 1992a).

The question of the mechanisms underlying frequency discrimination in fishes has not been completely resolved. The early (e.g., Fay 1970a) suggestion that tone frequency is represented in patterns of interspike-intervals in auditory (saccular) nerve fibers has received support by measurements showing that the temporal error with which primary saccular afferents phase-lock to tones is approximately equal to the behaviorally measured df throughout the frequency range of hearing (Fay 1978). On the other hand, primary afferents are frequency-selective to some degree (Fay and Ream 1986; Fay 1996), and this selectivity is enhanced at or below the auditory midbrain (Lu and Fay 1993), apparently through lateral inhibition (Lu and Fay 1996). Thus, in fishes as in all other vertebrates investigated, an across-cell representation of frequency exists in addition to a temporal representation at frequencies below about 3000 Hz, and it cannot be ruled out that a tonotopic representation plays some role in behavioral frequency analysis by fishes.

8.2 *Anurans*

There are no complete psychophysical studies examining frequency discrimination of tonal stimuli across the entire audible spectrum in anurans. Estimates of the ability of anurans to discriminate frequencies can be obtained from studies using either selective phonotaxis or evoked calling responses to tonal stimuli or complex stimuli in which one component frequency is being varied. The absence of a discriminative response to two stimuli differing slightly in frequency can be used to infer the size of the frequency discrimination limen. However, these estimates are probably overestimates because other features of the signal besides frequency (e.g., intensity, temporal relationship between the stimuli) can affect responses in these experiments (Gerhardt 1987; Dyson and Passmore 1988a,b). Moreover, when one frequency in a multicomponent stimulus is changed, the perceived timbre of the sound might also change. This can complicate straightforward interpretations of preferences as being due to discrimination of frequency alone rather than to a difference in a more complex acoustic attribute such as timbre.

Estimates of the frequency discrimination limen for anurans are plotted in Figure 7.12. Frequency differences are presented in terms of percent change from the center frequency, using the most sensitive estimate for a particular species. Data collected from selective phonotaxis experiments represent the smallest frequency difference between a standard stimulus (center frequency) and an alternative stimulus of different frequency that the female discriminates by an approach response. Data collected from evoked calling experiments represent the smallest frequency difference producing a change in the number of evoked calls or the synchrony of the calling response. Because of the constraints of the design of these studies, only one or two center frequencies were tested for a particular species. Moreover, only a small number of frequency differences were tested, and the size of the difference was motivated more by analysis of the spectral content of calls of conspecific and heterospecific species than by an attempt to find the smallest perceptible frequency difference (Schwartz and Wells 1984).

Figure 7.12 also includes estimates obtained from studies examining the ability of the African clawed frog (*Xenopus laevis*) to discriminate two frequencies of water wave stimulation using associative learning techniques (Elepfandt 1985, 1986; Elepfandt et al. 1985). Approach toward one source elicited a positive food reward, while approach toward the second source produced a noxious stimulus. Animals were able to discriminate stimuli in

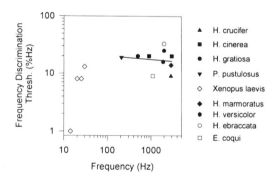

FIGURE 7.12. Relative frequency difference limens for nine species of frogs based on thresholds estimated from the selective phonotactic response of the female or the evoked vocal response of the male. Frequency differences are represented as percent detectable change in frequency from a standard. Each data point represents the most sensitive estimate for that particular species. The line through the data points for frequencies greater than 100 Hz is the best-fitting regression. Data are calculated and plotted from Narins and Capranica 1978 (Puerto Rican coqui frog, *Eleutherodactylus coqui*); Gerhardt 1981a (the barking treefrog, *Hyla gratiosa*); Ryan 1983 (the tungara frog, *Physalaemus pustulosus*); Doherty and Gerhardt 1984 (the spring peeper, *H. crucifer*); Schwartz and Wells 1984 (Central American treefrog, *Hyla ebraccata*); Elepfandt et al. 1985 (the African clawed frog, *Xenopus laevis*); Gerhardt 1987 (the green treefrog, *H. cinerea*); Dyson and Passmore 1988b (the painted reed frog, *Hyperolius marmoratus*); Gerhardt and Doherty 1988 (the gray treefrog, *H. versicolor*).

the frequency range of 5 to 30 Hz, achieving success rates as high as 80%. Relative discrimination limens measured using this technique range from 1% at 14 Hz to 15% at 30 Hz. The function relating the discrimination limen to frequency in this species does not follow Weber's law (Elepfandt et al. 1985).

Together, these various studies estimate discrimination limens for anurans within the range of 1% to 39% for frequencies from 14 to 4000 Hz (Fig. 7.12), if the estimates for discrimination of water wave frequencies (Elepfandt et al. 1985) are included. If these estimates are removed, then the range of difference limens over frequencies from about 200 to 3000 Hz falls between 9% and 33%. The best-fitting regression line through these data indicate that the relative frequency discrimination limen (% Hz) stays constant over center frequency. This implies that Weber's law holds for frequency discrimination across different anuran species; however, the data do not imply that Weber's law would hold for discrimination across different frequencies in the same species. Overall, these estimates are above the means, but within the range of values measured psychophysically for fishes (Fig. 7.11). However, frequency discrimination threshold estimates are higher in anurans than those reported for birds and mammals at comparable frequencies (Fay 1988). Because most of the data for anurans were collected using a communication response, the estimates for these animals are conservative; difference limens measured psychophysically are likely to be smaller. On the other hand, these differences in the ability to discriminate frequencies may reflect differences in the inner ear organs between anurans on the one hand and birds and mammals on the other hand. The inner ear organs in frogs, similar to the fish otolith organs, contain no basilar membrane. Although the anuran inner ear contains frequency selective filters, the width of these filters, measured either behaviorally (Fig. 7.8) or physiologically (Freedman et al. 1988), are typically broader in anurans than in other vertebrates, even within the same frequency range. As is also the case for fishes, eighth nerve fibers innervating the inner ear organs may encode frequency by a temporal coding mechanism based on phase-locked discharges to the stimulus envelope rather than by a place principle of frequency analysis (Schwartz and Simmons 1990).

9. Temporal Pattern Discrimination

9.1 Gap Detection

When organisms are motivated to detect brief, broad-band sounds, it would be advantageous to have a temporal integrator with high temporal resolution (short time constant) (Green 1973). Minimum integration times have been estimated for *C. auratus* as the shortest detectable silent "gap" in an ongoing noise. The gap threshold is about 35 ms (Fay 1985), compared with thresholds of 2 to 4 ms for mammals and birds. This large threshold probably reflects the relatively narrow, low frequency hearing range for *C. auratus* compared with mammals and birds, since the statistical variability of noise amplitude grows considerably as noise bandwidth narrows. If the detection of gaps is limited by the variability with which the gap is encoded as an envelope fluctuation, species with poor high-frequency hearing, such as fishes, would be expected to have relatively large gap detection thresholds.

Psychophysical studies specifically examining gap detection have not been reported in anurans. Data examining sensitivity to call repetition rate suggest that female frogs prefer high rather than low calling rates, even if these high calling rates are beyond those typically emitted by conspecific males. Preferred calling rates have silent intervals or gaps between individual calls of 40 to 240 ms, depending on the species (Klump and Gerhardt 1987; Gerhardt 1991). These estimates are consistent with the notion that species with poor high-frequency hearing would have relatively large gap detection thresholds. Preferences were not tested against continuous calls or noise, however, and the upper limit of calling rates that can be discriminated from continuous noise has not been tested in any species. Because perception of call rate by anurans is not as constrained by species-specific properties of signals as are other acoustic parameters such as duration or frequency, experiments explicitly testing gap detection abilities in anurans might provide an estimate of more generalizable aspects of auditory functioning in these animals.

9.2 Time Interval or Repetition Rate Discrimination

These experiments ask listeners to discriminate between periodic sounds differing in the time interval (period) between identical acoustic events. For *C. auratus* (Fay 1982), the just-discriminable change in the period of repeated clicks is a power function of the standard repetition period at which the discrimination was measured (5 to 50 ms), with an exponent of about 2. Fay and Passow (1982) showed that the period discrimination threshold can be systematically raised by adding random temporal variability to the repetition period. At high levels of temporal jitter, the period discrimination threshold asymptotes at the standard deviation of the imposed jitter. This result suggests that *C. auratus* estimates the repetition period directly rather than counting events occurring within a given integration time. These data also indicate that the internal noise normally limiting this time interval measurement grows approximately as the square of the repetition period. Time interval or repetition rate discrimination per se has not been studied in anurans.

9.3 Repetition Noise (Echo) Discrimination

In many acoustic environments, sound reaches the ears by at least two pathways: directly from the source, and by one or more reflections, such as from the air/water surface. For ongoing ambient noise, which may include information of immediate interest (Rogers et al. 1989), the direct and reflected sounds overlap in time and combine to produce a signal that contains information about the echo delay and thus about the structure of the reverberant environment. For humans, a sound and its echo produce a pitch that can be matched by a pure tone equal in frequency to the inverse of the echo delay (Yost et al. 1978). In architectural acoustics, such sounds are said to have "coloration," and are clues to the location of reflecting surfaces and the size of the room.

An experimental analysis of these perceptual effects uses flat-spectrum, broad-band noise as a source, and the identical waveform added with a delay (and possibly attenuation) as the echo. One of the most interesting experiments using this repetition noise (RN) stimulus asks listeners to discriminate between RNs having different echo delays. Delay discrimination thresholds increase linearly with the standard delay at which the discrimination is measured (0.5 to 50 ms), with a Weber fraction of 0.02 to 0.03 (Yost et al. 1978).

In a set of experiments on *C. auratus* using repetition noise (Fay et al. 1983), it was found that delay discrimination thresholds also increase linearly with a Weber fraction of 0.05 to 0.06. Thus, this species is nearly as sensitive as human listeners in discriminating RN delay. Neurophysiological studies on saccular afferents using the same stimuli revealed that echo delay is robustly encoded in the distribution of time intervals between evoked spikes. This suggests that the sound quality of RN in *C. auratus* (possibly analogous to pitch perception in human listeners) arises from the analysis of the stimulus waveform in the time domain.

10. Sound Source Localization

10.1 Fishes

Much has been written and speculated about sound source localization in fishes, but very few behavioral experiments have been done that measure localization acuity. Briefly, most authors believe that localization is not possible by fishes unless the otolith organs receive input directly from acoustic particle motion. The reason for this is that particle motion is a vector quantity that is coded by directionally sensitive hair cells (Fay 1984; Fay and Edds-Walton 1997). Pressure-mediated input to the ears, on the other hand, comes equally to the two ears and always from the direction of the swim bladder or other gas chamber. Particle motion processing is most likely at low frequencies (<300 Hz), at high levels (above 80 dB re: 1 μPa sound pressure), and close to the sound source (within a wavelength or so).

There are two kinds of experiments that could quantitatively measure localization acuity: one observes animals moving to choose (e.g., approach or orient toward) a particular source in a forced-choice experiment; the other measures the smallest angle between two sources that still permits the animal to discriminate between them (the minimum audible angle, MAA). There are few quantitative studies using the first kind of experiment (e.g., Schuijf and Siemelink 1974; Schuijf 1975). However, there are some quantitative data for *G. morhua* in the second kind of experiment, shown in Figure 7.13. These say that *G. morhua* can discriminate between sources separated by 10° to 20° in azimuth or elevation if the signal level is sufficiently high. Demonstrations have been made that *G. morhua* also can discriminate between sound sources from opposing directions (180° apart) in the horizontal (Schuijf and Buwalda 1975) and vertical (Buwalda et al. 1983) planes. In both experiments it was shown that the phase relation between acoustic particle velocity and sound pressure provides information necessary for the discrimination. These and other results support the "phase model" of directional hearing by fishes developed by Schuijf (1975). In essence, this model proposes that the axis of acoustic particle motion is represented by the ensemble of primary afferents responding in a directional manner (see Popper and Fay, Chapter 3), and a decision as to which

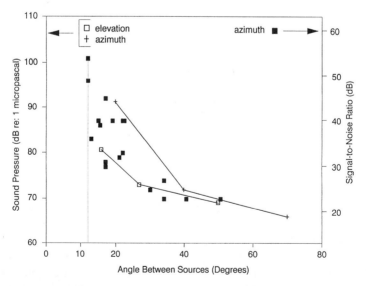

FIGURE 7.13. Minimum detectable differences between two sound source angles (with reference to 0° azimuth and elevation) determined as a function of signal level (and S/N) for *Gadus morhua* (cod) (Chapman and Johnstone 1974, Hawkins and Sand 1977, Buwalda 1981).

end of the axis points to the source is based on computations using the phase relations between particle velocity and pressure.

Although it can be demonstrated that *G. morhua* can discriminate between sources in different locations, including different distances (Schuijf and Hawkins 1983), it is still not clear whether and to what extent other fishes have knowledge about the actual location of sound sources. Only experiments of the first type can demonstrate this.

10.2 Anurans

Frogs show good accuracy in localizing complex sound sources in both the horizontal and the vertical dimensions. Binaural cues are necessary for accurate horizontal localization; occluding one ear with a layer of silicone grease results in circling movements and an inability to localize the source (Feng et al. 1976). These binaural cues might arise from the inherent directional-sensitivity of the frog's ear, which operates as a combined pressure/pressure-gradient receiver (Rheinlaender et al. 1979; Feng and Shofner 1981; Michelsen et al. 1986). The role played by extratympanic input to the inner ear organs in localization tasks has not been systematically examined.

During a typical phonotactic approach toward a nonelevated sound source, females make scanning movements, turn, and then jump toward the source following a zigzag course (Rheinlaender et al. 1979). Scanning movements and the zigzag approach may allow the female to alternate the "leading" ear during the approach, thereby maximizing binaural cues, and to constantly update her position relative to the position of the source. The accuracy of the approach can be quantified in terms of the mean jump angle (deviation from a straight line between the release point and the source) during the approach. Data relating accuracy of localization with interaural distance for four different species are shown in Table 7.1. Given the small interaural distances in species that have been tested, the accuracy of the phonotactic approach is quite good. For example, in female *H. cinerea*, one of the larger animals tested, the most accurate jump angle measured during an approach to a complex sound source was 4.3° (mean from 41 approaches was 16.1°; Rheinlaender et al. 1979). Frogs such as *C. nubicola* and the painted reed frog (*Hyperolius marmoratus*) with smaller interaural distances show somewhat worse acuity in the two-dimensional grid, but they were tested with higher frequency stimuli where localization acuity might be expected to be poorer.

Because the frogs were only localizing one source in these experiments, the data cannot be used to make estimates of a minimum audible angle. Passmore and Telford (1981) found that female *H. marmoratus* could locate a sound source broadcast through one of two speakers 25° apart; they did not test smaller angles of separation. Data from Klump and Gerhardt (1989) suggest that female barking treefrogs (*Hyla gratiosa*) could discriminate

TABLE 7.1. Accuracy of sound localization in anurans.

Species	Interaural distance	Stimulus	Mean jump angle	Reference
C. nubicola (dendrobatid frog)	5 mm	5–6 kHz	23° ± 17° (2D)	Gerhardt and Rheinlaender 1980
H. cinerea (green treefrog)	0.9–1.4 cm	0.9 kHz 0.9 + 2.7 + 3.0 kHz	15.1° ± 13.3° 16.1° ± 14.5° (2D)	Rheinlaender et al. 1979
H. marmoratus (painted reed frog)	8 mm	FM sweep, about 3.0–3.5 kHz	22° ± 29.7° (2D) 43° ± 31.7° (3D)	Passmore et al. 1984
H. versicolor (gray treefrog)	1.6–2.3 cm	1.1 + 2.2 kHz 1.4 + 2.2 kHz 1.0 kHz 2.0 kHz 1.4 kHz	23° (3D) 25° (3D) 24° (3D) 30° (3D) 36° (3D)	Jørgenson and Gerhardt 1991

Mean jump angle is the angular error between the frog's position and the position of the sound source averaged across all jumps made by the animals during all phonotactic approaches. 2D, localization in a two-dimensional grid; 3D, localization in a three-dimensional grid.

between different angles of sound incidence, in particular 45° from 15° and 30°. Only one sound source was presented at a time in these experiments, but they suggest that the minimum audible angle might be as low as 15°. As a comparison, the minimum audible angles for localization of noise bursts by small rodents with interaural distances comparable to that of the treefrogs range from about 4° to 30° (Fay 1988); for the *G. morhua*, maximum acuity measured is about 12° to 20° in azimuth and elevation (Fig. 7.13).

Vertical localization has been examined by measuring phonotactic approaches to complex sound sources in a three-dimensional grid (Gerhardt and Rheinlaender 1982; Passmore et al. 1984; Jørgensen and Gerhardt 1991). Approaches were typically accompanied by lateral head scanning and changes in head position. Only two species have been tested in these conditions (Table 7.1). As expected, *H. versicolor*, the larger animal, was more accurate than *H. marmoratus*, the smaller animal, but it is unclear whether these differences in accuracy are attributable to these size differences or to differences in the spectral content of the sources used. Although the jump error angles for vertical localization are larger than those measured in some mammalian species of comparable size, the ability of frogs to localize sounds in the vertical dimension is impressive given that their lack of external ears precludes pinna filtering effects.

The directionality of the tympanum probably contributes to the frog's ability to localize sound sources in space. The frequency response of the tympanum, as measured by laser vibrometry, changes with elevation of the sound source (Jørgenson and Gerhardt 1991; see Fig. 7.14). Notches in

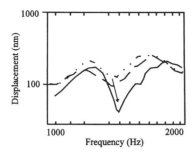

FIGURE 7.14. Frequency response of the external tympanum of *H. versicolor* measured by laser vibrometry Changes in the elevation of the sound source (0°, solid line; 30°, dashed line; 60°, dotted line) at a given azimuth (−60° in this example) produce changes in the position of the peaks and notches in the frequency response. The pattern of changes in the peaks and notches may produce cues for vertical localization. Data are replotted from Jørgenson and Gerhardt (1991).

the transfer function change in frequency with changes in sound source elevation, producing patterns of shifts similar to those seen in the eardrum response of humans (Batteau 1967), cats (Musicant et al. 1990), ferrets (Carlile 1990), and bats (Wotton et al. 1995). Like these mammals, frogs may use the directionally dependent pattern of spectral peaks and notches for localization. It is noted, however, that the frog data are based on a relatively small sample size and need to be replicated using sources at a greater variety of spatial positions.

11. Perception of Complex Sounds

The psychophysical studies on detection and discrimination thresholds reviewed above measure the limits of hearing sensitivity and acuity, and thus help define the sense of hearing in fishes and amphibians in a way that is comparable with psychoacoustic studies on human listeners. However, it is clear that the human sense of hearing is far richer and more complex than is usually revealed in psychoacoustic studies. The most general functions of the human auditory system are to analyze the complex mixture of sounds reaching the ears from multiple sources into component frequencies and temporal events, group those components that arise from each source, and then synthesize a representation of the auditory scene (Bregman 1990) in which the source of each sound is perceived as an object with a particular spatial location. A sense of hearing thus permits individuals to draw the right conclusions about the objects and events in the local world so that they may act accordingly. This general function would seem to be of survival value to all species, including fishes and amphibians. How can these important hearing functions be studied quantitatively in nonhuman animals?

Stimulus generalization methods have been used to approach these and other questions in *C. auratus* (Fay 1970b, 1972, 1992b, 1994a,b, 1995; Fay et al. 1996). Generalization experiments investigate the animal's judgments of sound similarity. An animal is first conditioned to respond to a given simple or complex sound and is then tested for its response to novel sounds that differ from the conditioning sound along various dimensions. After conditioning to a given sound, *C. auratus* tend not to respond or generalize to all sounds; they seem to generalize only to the extent that the conditioning and test sounds have similar features (Fay 1994a). This method reveals what stimulus dimensions are salient or what the animals normally pay attention to. A gradient of response magnitude along a stimulus dimension such as frequency can be interpreted as revealing a parallel perceptual dimension (Guttman 1963), perhaps similar to pitch. Further, the shapes of response gradients have been interpreted as revealing the quantitative relationships between the physical and sensory dimensions (Shepard 1965). Generalization experiments have shown that *C. auratus* behave as if they have perceptual dimensions similar to pure-tone pitch, periodicity pitch, roughness, and timbre as defined in studies on human listeners. In addition, stimulus generalization methods have been used to demonstrate analytic listening in *C. auratus* (Fay 1992b).

The natural vocal behavior of anurans exemplifies the biological relevance of auditory scene analysis for social and reproductive behaviors. In many species, males congregate into choruses and emit advertisement calls to announce their presence at a calling site, both to neighboring males and to females approaching in search of a mate. The advertisement calls of many species, such as *R. catesbeiana* and *H. cinerea*, are complex, multiple harmonic sounds with distinct envelope periodicities. Both the distribution of harmonics in these calls (a spectral feature) and the envelope periodicity (a temporal feature) are important for perceptual segregation of biologically relevant signals from background noise. The perception of both complex spectral and temporal features of these sounds has been examined in field studies of evoked calling by males and selective phonotaxis by females, and the results of these studies can be interpreted as showing that frogs also behave as if they have perceptual dimensions similar to pure tone pitch, periodicity pitch, and timbre (Simmons and Buxbaum 1996).

11.1 Perception of Pure Tones

After conditioning to respond to a pure tone, *C. auratus* respond only to novel test frequencies close to the conditioning frequency. Responses to novel frequencies above and below the conditioning frequency fall to chance levels along substantially symmetrical, monotonic gradients (Fay 1970b, 1992b). These results were interpreted as indications that tone frequency corresponds to a salient perceptual dimension analogous to pure tone pitch as defined by pitch scaling studies on human listeners. Although it is not possible to conclude that *C. auratus* perceive pitch as it is defined in studies on human listeners, the animals certainly behave as if they have the perceptual dimension of pitch in common with humans.

Generalization gradients for pure tone frequency in frogs were measured by Narins (1983) in an experiment examining how small changes in frequency affect the synchronous calling response of *E. coqui*. The frequency tuning curves describing how changes in frequency affect calling were asymmetrical in shape, such that frequencies lower than the center frequency were considered more similar to the standard stimulus than frequencies higher than the center frequency. This asymmetry derives from simultaneous vocal interactions of the target male with neighboring males. Because of this asymmetry, generalization gradients for the frog differ from the more symmetrical gradients measured in *C. auratus* (e.g., Fay 1970b). The data suggest, however, that tone frequency is a salient perceptual dimension for *E. coqui*.

Elepfandt (1986) examined the perceptual dimensions of frequency in *Xenopus laevis* by presenting the animal repeatedly with one wave frequency for a period of time. Responses to this frequency were, in different animals, either positively or negatively rewarded. When this familiar frequency was combined in later tests with the opposite reinforcement, the animal's performance declined. Elepfandt interpreted this effect as showing that *Xenopus* can learn and remember the specific frequency at which it had been previously been trained. He also argued that *Xenopus* possess an ability to identify specific water wave patterns in a manner comparable to that of absolute pitch perception in humans.

11.2 Analytic Listening

The ability to "hear out" or independently analyze the individual frequency components in a multicomponent complex sound is referred to as analytic listening. This capability plays an important role in permitting listeners to determine the individual simultaneous sources making up an auditory scene (Bregman 1990). In a study on analytic listening in *C. auratus* (Fay 1992b), animals were first conditioned to a two-tone complex containing 166- and 724-Hz sinusoids, and then tested for generalization to a set of single tones ranging between 100 and 1500 Hz. The resulting generalization function had two peaks (robust generalization) at the frequencies of the tones in the conditioning complex, separated by a valley (little generalization) at intermediate frequencies. Thus, *C. auratus* acquire independent information about the frequencies of two tones presented simultaneously, and can be said to listen analytically. This demonstration of simultaneous frequency analysis suggests that this fundamental aspect of hearing is a primitive character shared among humans (Hartmann 1988), fishes, and perhaps all vertebrates.

In anurans, changing the amplitude of individual frequency components in a complex, species-specific advertisement call can change the response to these stimuli. Capranica (1965) first demonstrated that the evoked calling response was sensitive to changes in the relative amplitude of individual harmonics in a multiple harmonic stimulus, even when the frequency distribution of harmonics and the relative waveform periodicity remain the same. Similarly, female frogs discriminate by their approach responses a standard synthetic stimulus from one in which one of the dominant harmonics is attenuated by either 6 or 12 dB (Gerhardt 1981a,b). In some cases, this discrimination depends on the absolute levels of the different stimuli. These data suggest that frogs detect changes of single attenuated harmonics in a complex signal, but it is not clear whether this is an example of analytic listening, or of timbre perception (see Section 11.6, below).

11.3 Periodicity and Time Interval Processing

11.3.1 Fishes—Periodicity and Time Interval Processing

For human listeners, complex periodic sounds can evoke a pitch perception corresponding to that of a pure tone at the frequency of repetition. In a generalization study, C. auratus were conditioned to a train of impulses repeated periodically at a particular rate or interpulse interval (IPI), and then tested for response to the same impulse repeated with different IPIs (Fay 1994a, 1995). In general, animals failed to generalize robustly to the same pulse repeated with novel IPIs. For every conditioning IPI, the resulting generalization gradients tended to peak at the conditioning IPI, and declined symmetrically as the novel test IPI deviated from the conditioning IPI.

These results are similar to those from analogous experiments, described above, using pure tone frequency as the stimulus dimension (Fay 1970a, 1992b). The similar generalization gradients of the repeating-pulse experiments suggest that C. auratus have a perceptual dimension that is continuous and monotonic with pulse repetition rate or IPI, and that this dimension has at least some of the properties of periodicity pitch as defined in experiments on human listeners. This conclusion is in accord with the results of earlier experiments in which C. auratus were conditioned to respond to a 1000 Hz tone amplitude modulated at 40 Hz and tested for generalization to the same carrier modulated at different rates (Fay 1972). Varying modulation rate resulted in a substantially symmetrical generalization gradient that peaked at a modulation rate of 40 Hz.

11.3.2 Anurans—Periodicity and Time Interval Processing

The first demonstration that anurans are sensitive to the periodicity of complex vocal signals was provided by Capranica (1965) in an analysis of

the evoked vocal response of *R. catesbeiana* housed in a laboratory ter-
rarium. *R. catesbeiana* vocalized most in response to multiple harmonic
signals with periodicities of 100 Hz (the periodicity of the natural advertisement call), and least to periodicities of 50 Hz (Fig. 7.15). Stimuli modulated
at high rates from 100 to 200 Hz produced more responses than stimuli
modulated at low rates. Even within the high-frequency range of preferred
modulation rates, vocalizing male *R. catesbeiana* distinguish between
stimuli modulated at 100 and 200 Hz, responding most vigorously to calls

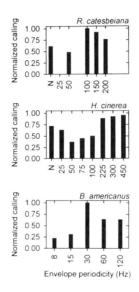

FIGURE 7.15. The influence of waveform periodicity on the evoked calling response
in three species of male anurans. Data are based on the percent of vocal responses
to stimuli with the frequency composition of the advertisement call of each species.
Different degrees of waveform periodicity are then superimposed on this "standard" call. Calling responses are normalized to responses to the stimulus producing
the largest absolute number of responses (100%). Data from *R. Catesbeiana* are
based on laboratory evoked responses collected by Capranica (1965); spectral peaks
in the stimuli are in the regions of 200 and 1400 Hz. Data fom *Hyla cinerea* are based
on field studies of the evoked calling response to playbacks of complex advertisement calls amplitude modulated at different rates and are replotted from Allan and
Simmons (1994). Spectral peaks in these stimuli are at 900, 2700 and 3000 Hz. Data
from the American toad (*Bufo americanus*) are based on field studies to playbacks
of narrow-band noise (center frequency 1400 Hz bandwidth 200 Hz) amplitude
modulated at different rates and are replotted from Rose and Capranica (1984).
Here, percent responses are calculated based on the responses to the standard
stimulus (modulation rate of 30 Hz) presented immediately before the stimuli
modulated at other rates. For all three species, evoked calling is greatest to stimuli
with the periodicity of the natural advertisement call of that species. These data also
suggest that frogs may have a percept similar to that of periodicity pitch.

modulated at the natural periodicity of 100 Hz. In addition, the calling response is more likely to habituate in response to repeated presentations of a 200-Hz periodicity stimulus than of a 100-Hz periodicity stimulus (Hainfeld et al. 1996). Subsequent studies in several different species have confirmed that both male and female anurans can detect and discriminate stimuli based on differences in envelope periodicity, produced either as changes in rate of amplitude modulation or as changes in the internal phase structure of a signal (Gerhardt 1978, 1981b; Rose and Capranica 1984; Simmons et al. 1993; Allan and Simmons 1994; Bodnar 1996; Hainfeld et al. 1996; Simmons and Buxbaum 1996; Fig. 7.15). Male *H. cinerea* show similar sensitivity to modulation of tonal signals and of noise, confirming that the temporal cue associated with the modulation of the envelope, rather than the extra spectral side bands produced by the modulation process, is perceptible to the animal (Allan and Simmons 1994). These data suggest that frogs have a percept similar to that of periodicity pitch.

The data in Figure 7.15 are based on responses to stimuli modulated at depths of 100%, which produces prominent envelope fluctuations. As modulation depth is lowered to 75% or less, calling rate decreases and is no longer significantly influenced by modulation rate (Allan 1992; Allan and Simmons 1994; see also Fay 1982 and Fay and Passow 1982 for similar results in *C. auratus*). These data can be used to construct a temporal modulation transfer function (TMTF) to describe the minimum discriminable depth of modulation at a particular modulation rate. Figure 7.16 shows a TMTF calculated from laboratory evoked calling data for responses to both amplitude modulated (AM) calls and AM noise in *H. cinerea*, using as a criterion the lowest depth of modulation to which the animals responded at a particular modulation rate. When AM calls are used as stimuli, the animals responded at 25% depth of modulation at all AM rates except at 100 Hz. At this rate, no animals responded until depth was increased to 50%. At all rates of AM noise except 100 Hz, threshold was 50%, the lowest depth to which any responses could be evoked. No responses could be evoked to AM rates of 100 Hz until depth of modulation reached 75%. The data in Figure 7.16 show that the resolving power of the auditory system of *H. cinerea* is relatively consistent across frequency, except at an AM rate of 100 Hz, where resolving power is relatively poorer. The estimates for *H. cinerea* are comparable to those for *C. auratus* for responses to modulated noise but not for responses to modulated tones (Fay 1980). In general, the TMTF of *H. cinerea* does not exhibit the low-pass shape typically observed in other vertebrate species (Fay 1988), and shows somewhat less overall sensitivity to modulation. These relatively higher modulation detection thresholds for *H. cinerea* might reflect the relative insensitivity of the evoked calling method as a psychophysical test. They might also reflect a specialization related to the processing of conspecific versus heterospecific vocal signals; for example, the relatively poorer

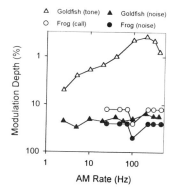

FIGURE 7.16. Temporal modulation transfer functions (TMTF) for *C. auratus* (Fay 1980) and for *H. cinerea* (Allan 1992, Allan and Simmons 1994). Data points show the minimum depth of modulation at which a criterion level of response can be evoked. Data are presented separately for responses to tones, complex calls, and noise. The data for *C. auratus* are based on thresholds measured using conditioning techniques and the data for *H. cinerea* are based on thresholds estimated using the evoked calling response measured in the laboratory. *H. cinerea* and *C. auratus* show similar sensitivity to noise modulated at different depths. The data from *H. cinerea* are based on modulation depths of 25%, 50%, and 75%; smaller changes in depth were not examined.

resolving power at AM rates of 100 Hz for *H. cinerea* occurs at a periodicity matching the advertisement call of a sympatric species of frog, *Hyla squirella*.

11.4 Harmonic Structure Processing

11.4.1 Fishes—Harmonic Structure

Experiments testing sensitivity to harmonic structure as a cue for pitch have produced conflicting results. Fay (1995) reported that *C. auratus* conditioned to respond to a pure tone failed to generalize to harmonic complexes that included the conditioning tone as the fundamental frequency. This indicates that *C. auratus* respond to pure tones and broad-band harmonic complexes as quite different stimuli even though they may share a fundamental frequency component and have the same periodicity. Of course, the animal is quite correct in distinguishing between these two quite different stimuli; the results of this experiment suggest that a sound's spectrum may be more important than temporal periodicity in controlling behavior. On the other hand, animals conditioned to an harmonic complex having a 250-Hz periodicity tended to generalize more to 250-Hz pure tones than to the others making up the conditioning complex (Fay 1994b). This result suggests that the animals' responses were controlled by temporal periodicity. However, it is not clear in this case whether the fish responded to the common periodicity, or listened analytically to "hear out" the 250-Hz component in the harmonic complex.

11.4.2 Anurans—Harmonic Structure

Simmons (1988b) used a reflex modification technique to measure the ability of *H. cinerea* to detect both harmonic and inharmonic complex sounds in noise. In these experiments, sounds were presented singly and aperiodically in a manner that deliberately underemphasized or obscured other species-specific features of the signal besides the change in harmonicity. Thresholds for detection of a harmonic sound in noise could be as much as 5 to 10 dB lower than thresholds for detecting an inharmonic sound in noise. Moreover, the estimated S/N ratio for detection of harmonic complexes with frequencies near 900 and 3000 Hz were lower than those obtained for each tone presented separately (mean of 15 dB for the complex, compared to 21 dB for 900 Hz alone and 23 dB for 3000 Hz alone). These lower thresholds result in a more favorable S/N ratio for extraction of the sound from the background noise. Two subsequent studies using selective phonotaxis (Gerhardt et al. 1990) and evoked calling designs (Simmons et al. 1993) did not find an advantage for detection of harmonic signals. Instead, these studies confirmed the importance of the overall envelope periodicity in controlling the frog's behavior (as also shown physiologically; Schwartz and Simmons 1990). It is possible that responding in those two studies was controlled by other salient species-specific features of the signal, such as call repetition rate, that might have overshadowed the more subtle cue of harmonic structure.

Bodnar (1996) found that female *H. gratiosa* could discriminate between harmonic and inharmonic sounds in a selective phonotaxis design; however, females showed a preference for the inharmonic over the harmonic calls. When frequency modulation is also present in the signals, a preference for harmonic over inharmonic calls occurs. Bodnar showed that mistuning of a four-harmonic signal by as little as 1.1% could be detectable. This degree of sensitivity is greater than that reported by Simmons (1988b), who mistuned a two-harmonic signal by 12%. The signals in Bodnar's experiments contained lower-frequency components, and thus could have provided more cues that made the differences in harmonic structure more detectable. The behavioral data of Bodnar and Simmons (1988b) are consistent with physiological data showing that eighth nerve fibers are sensitive to the harmonic structure of complex signals, and can phase-lock to the pseudoperiod of low-frequency inharmonic signals (Simmons and Ferragamo 1993).

11.5 Phase Processing

11.5.1 Fishes

Changes in the internal phase spectrum of a complex signal can result in changes in the magnitude and periodicity of envelope fluctuation; these changes produce in human listeners a difference in the perceived timbre of the sound. Chronopolous and Fay (1996) conditioned *C. auratus* to an in-phase, three-component complex producing a 100% sinusoidal amplitude modulated signal. Animals were then tested with a variety of amplitude modulated signals; some simply with reduced modulation depth (see also Fay 1972), and others in which the phase of the center component (carrier) was shifted up to 90° [quasi frequency-modulated (QFM) signal]. Generalization decrements in this experiment could be accounted for entirely on the basis of the magnitude of envelope fluctuation, suggesting that possible timbre differences resulting from the phase-shifted carrier did not exert control over behavior.

11.5.2 Anurans

Phase structure does not appear to be easily detectable by frogs, and changes in response apparently due to changes in phase are more readily interpreted by changes in the overall shape of the stimulus envelope (Simmons et al. 1993; Bodnar 1996; Hainfeld et al. 1996). For example, Hainfeld et al. (1996) studied the ability of male *R. catesbeiana* to vocalize to and discriminate complex signals differing in phase spectrum. There were differences in the number and latency of answering calls to stimuli differing in phase; however, these differences appeared only when phase differences produced changes in the shape of the waveform envelope and different intervals between prominent peaks in the internal fine-structure of the signal. These results more likely reflect sensitivity to envelope cues rather than sensitivity to phase per se.

11.6 Complex Pitch and Timbre

C. auratus behave in stimulus generalization experiments as if they are able to distinguish complex periodic stimuli both on the basis of pitch (repetition rate) and timbre (spectral envelope). The results of four experiments of this kind are shown in Figure 7.17 (Fay 1995). In each experiment, animals were conditioned to respond to either a low-frequency impulse ("thud") or a high-frequency impulse ("snap"), repeated slowly (panels A and C), or rapidly (panels B and D). Fish were then tested for generalization to the same impulse (solid lines) and the alternative impulse (dotted lines) repeated at different rates.

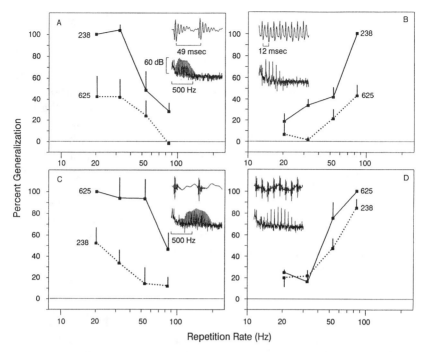

FIGURE 7.17. Similarity judgments (percent generalization) for four groups of *C. auratus* (goldfish) comparing complex periodic sounds on the basis of the repetition rate of the pulse and spectral envelope of the pulse (Fay 1995). In each panel, the center frequency (in Hz) of the pulse used in conditioning is given next to each solid line gradient. Points without standard error bars plotted at 100% generalization indicate the conditioning signal for each group. The center frequency of the alternative pulse used in generalization tests is given next to the dotted line gradient. The insets show the waveform and the spectrum of the conditioning stimulus for each group. The vertical lines at each data point indicate standard errors.

The results show that generalization declines as the repetition rate changes from the conditioning rate (the slopes of the lines in Fig. 7.17), which suggests that the behavior is controlled by a pitch-like dimension. In addition, generalization declines for stimuli with the alternative impulse (separation between curves in each panel of Fig. 7.17), which suggests that the behavior is also controlled by a timbre-like dimension. These two behavioral controls seem independent because changing the repetition rate ("pitch") has the same effect whether the spectral envelope ("source") is the one used in conditioning or is a novel one. These kinds of effects occur no matter what spectral envelope or repetition rate is used in initial conditioning and suggest that *C. auratus* probably perceive the pitch and timbre of complex sounds simultaneously and independently.

This behavior is analogous to the kinds of knowledge humans have about pitches and sources. Imagine listening to a musical instrument (e.g., a flute), playing a given note (e.g., A). We are able to recognize the source, and also recognize a pitch change when another note is played. Now if a saxophone plays the same note, we recognize that the pitch is the same but that the instrument is different. Recognizing and comparing notes is a pitch judgment, and identifying the source instrument is a timbre judgment.

Another aspect of sound source determination has been demonstrated for *C. auratus* in a generalization experiment (Fay et al. 1996). Animals were conditioned to a 6-second-duration 400-Hz pure tone, and then tested for generalization to the tone amplitude modulated by temporally asymmetrical envelopes called "ramps" and "damps." A ramped envelope has a slow rise and a rapid fall, while a damped envelope is the ramp reversed in time. In experiments with human listeners (Patterson 1994a,b), repeated ramps (within a critical range of repetition rates and durations) were judged to contain the sound of a sinusoid, while damps were not. In generalization tests, animals responded more to ramps than to damps, indicating that for both human and fish listeners, ramps have more in common with a sinusoid at the carrier frequency than do damps.

In general, the results of generalization experiments have lead to the conclusion that *C. auratus* "knows" very much about the characteristics of sound sources. So far, these and other experiments have been unable to demonstrate qualitative differences between fish and human listeners in sound source perception and in the sense of hearing.

12. Conclusions

Studies of hearing abilities in fishes and anurans show that auditory perception in these animals shares many similarities with other vertebrates. For example, fishes and frogs in their natural acoustic environment must deal with and solve problems of perceptual segregation and localization of sounds, and may do so using acoustic dimensions that control the perception of complex pitch and timbre for human listeners. The abilities of fishes and anurans to localize sounds in both the horizontal and vertical dimensions is quite accurate given their small effective head widths and lack of pinnae. Level discrimination studies suggest that difference limens are comparable to those measured in many other vertebrates. Frogs generally have a wider range of hearing than fishes, but a more restricted range compared to other vertebrates. Frequency discrimination limens and critical ratio bandwidths for fishes and anurans are broad, but within the upper ends of the range of those measured in other vertebrates. This might reflect a greater reliance on temporal than on spectral coding of sounds in the auditory periphery of both fishes and anurans compared to that of mammals (Fay 1978; Simmons and Buxbaum 1996).

The data so far suggest that hearing in anurans might differ quantitatively in relative acuity and sensitivity from that measured in other vertebrates. These quantitative differences might reflect the biological constraints placed on animals that lead to an emphasis on one aspect of hearing sensitivity over another, but they might also simply reflect differences in the sensitivity of the behavioral techniques used to estimate these various aspects of hearing. We know a great deal about the importance of specific spectral and temporal features of sounds in mediating acoustic communication in anurans (see Zelick, Mann, and Popper, Chapter 9), but we know very little about the absolute or even relative acuity for perception of acoustic features not present in intraspecific vocal signals. This state of affairs is reversed in fishes; we know much about the detection, discrimination, and perception of both simple and complex sounds, but little about the processing of communication sounds. Clearly, more needs to be done to precisely determine the limits of auditory perception in anurans, and to better understand the processing of communication sounds in fishes. Given the differences in the structures of the auditory periphery in fishes and anurans compared to other vertebrates, such behavioral experiments will be useful in testing hypotheses about the mechanical and neural bases underlying auditory perception.

The studies reviewed here suggest that fishes and anuran amphibians share many features of their sense of hearing with humans and other vertebrates. This could mean that the human sense of hearing revealed in detection and discrimination experiments is a general vertebrate character, and further suggests that the essential functions of hearing may be the same in all vertebrates. That is, hearing informs us about the physical characteristics of the sources and reflectors in the nearby environment. These characteristics appear to include their location (azimuth, elevation, and distance), their sizes and natural resonance frequencies (spectral envelope and timbre), and their internal states of excitement (temporal pattern and pitch). The sense of hearing shared among fishes and anurans could be labeled "primitive" only in the sense that it seems shared with most other vertebrate species as well.

Acknowledgments. Preparation of this chapter was supported by National Institute of Health grant NS28565 to AMS and by National Institutes of Health (NIH), National Institute on Deafness and Other Communication Disorders (NIDCD) program project grant to the Parmly Hearing Institute. We thank Susan Allan for permission to use unpublished data (Fig. 7.15), Joshua Schwartz for providing measurements of interaural distances in *Hyla versicolor*, and Mark Sanderson for preparation of some of the figures.

References

Allan SE (1992) The temporal resolving power of the auditory system in the green treefrog (*Hyla cinerea*). Unpublished doctoral dissertation, Brown University.

Allan SE, Simmons AM (1994) Temporal features mediating call recognition in the green treefrog, *Hyla cinerea*: amplitude modulation. Anim Behav 47:1073–1086.

Allen J (1997) OHCs shift the excitation pattern via BM tension. In: Lewis ER, Long GR, Lyon RF, Narins PM, Steele CR, Hecht-Poinar E (eds) Diversity in Auditory Mechanics, World Scientific Publishers, Singapore, pp. 167–175.

Arak A (1983) Sexual selection by male-male competition in natterjack toad choruses. Nature 306:261–262.

Astrup J, Mohl B (1993) Detection of intense ultrasound by the cod, *Gadus morhua*. J Exp Biol 182:71–80.

Batteau DW (1967) The role of the pinna in human localization. Proc R Soc Lond [B] 168:158–180.

Békésy G von (1960) In: Wever EG (ed) Experiments in Hearing. New York: McGraw-Hill.

Boatright-Horowitz SS, Simmons AM (1995) Postmetamorphic changes in auditory sensitivity of the bullfrog midbrain. J Comp Physiol [A] 177:577–590.

Boatright-Horowitz SS, Simmons AM (1997) Transient "deafness" accompanies auditory development during metamorphosis from tadpole to frog. Proc Natl Acad Sci USA 94:14877–14882.

Bodnar DA (1996) The separate and combined effects of harmonic structure, phase, and FM on female preferences in the barking treefrog (*Hyla gratiosa*). J Comp Physiol [A] 178:173–182.

Bregman AS (1990) Auditory Scene Analysis: The Perceptual Organization of Sound. Cambridge, MA: MIT Press.

Brzoska J (1980) Quantitative studies on the elicitation of the electrodermal response by calls and synthetic acoustical stimuli in *Rana lessonae* Camerano, *Rana r. ridibunda* Pallas and the hybrid *Rana esculenta* L. (Anura, Amphibia). Behav Proc 5:113–141.

Brzoska J, Walkowiak W, Schneider H (1977) Acoustic communication in the grassfrog (*Rana t. temporaria L.*); calls, auditory thresholds and behavioral responses. J Comp Physiol [A] 118:173–186.

Buerkle U (1969) Auditory masking and the critical band in Atlantic cod (*Gadus morhua*). J Fish Res Bd Canada 26:1113–1119.

Buwalda RJA (1981) Segregation of directional and nondirectional acoustic information in the cod. In: Tavolga WN, Popper AN, Fay RR (eds) Hearing and Sound Communication in Fishes. New York: Springer-Verlag, pp. 139–172.

Buwalda RJA, Schuijf A, Hawkins AD (1983) Discrimination by the cod of sounds from opposing directions. J Comp Physiol [A] 150:175–184.

Capranica RR (1965) The Evoked Vocal Response of the Bullfrog: A Study of Communication by Sound. Cambridge, MA.: MIT Press.

Capranica RR, Moffat AJM (1983) Neuroethological principles of acoustic communication in anurans. In: Ewert JP, Capranica RR, Ingle DJ (eds) Advances in Vertebrate Neuroethology. New York: Plenum Press, pp. 701–730.

Carlile S (1990) The auditory periphery of the ferret. II: The spectral transformations of the external ear and their implications for sound localization. J Acoust Soc Am 88:2196–2204.

Carr CA, Konishi M (1990) A circuit for detection of interaural time differences in the brainstem of the barn owl. J Neurosci 10:3227–3246.

Chapman CJ (1973) Field studies of hearing in teleost fish. Helgoländer wiss Meeresunters 24:371–390.

Chapman CJ, Hawkins A (1973) A field study of hearing in the cod, Gadus morhua L. J Comp Physiol [A] 85:147–167.

Chapman CJ, Johnstone ADF (1974) Some auditory discrimination experiments on marine fish. J Exp Biol 61:521–528.

Chapman CJ, Sand O (1974) Field studies of hearing in two species of flatfish, Pleuronectes platessa (L.) and Limanda limanda (L.) (family Pleuronectidae). Comp Biochem Physiol 47:371–385.

Chronopolous M, Fay RR (1996) Envelope perception in the goldfish. J Acoust Soc Am 100:2786 (abstr).

Coombs S, Fay RR (1989) The temporal evolution of masking and frequency selectivity in the goldfish (C. auratus). J Acoust Soc Am 86:925–933.

Coombs S, Popper AN (1981) Comparative frequency selectivity in fishes: simultaneously and forward-masked psychophysical tuning curves. J Acoust Soc Am 71:133–141.

Crawford AC, Fettiplace R (1981) An electrical tuning mechanism in turtle cochlear hair cells. J Physiol 312:377–412.

Dijkgraaf S (1952) Uber die Schallwahrnehmung bei Merresfischen. Z Vergl Physiol 34:104–122.

Dijkgraaf S, Verheijen F (1950) Neue Versuch uber das Tonunterscheidungsvermögen der Elritze. Z Vergl Physiol 32:248–265.

Doherty JA, Gerhardt HC (1984) Evolutionary and neurobiological implications of selective phonotaxis in the spring peeper (Hyla crucifer). Anim Behav 32: 875–881.

Dunning DJ, Ross QE, Geoghegan P, Reichle JJ, Menezes JK, Watson JK (1992) Alewives avoid high-frequency sound. North Am J Fish Manag 12:407–416.

Dyson ML, Passmore NI (1988a) The combined effect of intensity and the temporal relationship of stimuli on phonotaxis in female painted reed frogs Hyperolius marmoratus. Anim Behav 36:1555–1556.

Dyson ML, Passmore NI (1988b) Two-choice phonotaxis in Hyperolius marmoratus: the effect of temporal variation in presented stimuli. Anim Behav 36:648–652.

Ehret G, Gerhardt HC (1980) Auditory masking and effects of noise on responses of the green treefrog (Hyla cinerea) to synthetic mating calls. J Comp Physiol [A] 141:13–18.

Elepfandt A (1985) Naturalistic conditioning reveals good learning in a frog (Xenopus laevis). Naturwissenschaften 72:492–493.

Elepfandt A (1986) Wave frequency recognition and absolute pitch for water waves in the clawed frog, Xenopus laevis. J Comp Physiol [A] 158:235–238.

Elepfandt A, Seiler B, Aicher B (1985) Water wave frequency discrimination in the clawed frog, Xenopus laevis. J Comp Physiol [A] 157:255–261.

Fay RR (1970a) Auditory frequency discrimination in the goldfish. J Comp Physiol Psychol 73:175–180.

Fay RR (1970b) Auditory frequency generalization in the goldfish. J Exp Anal Behav 14:353–360.

Fay RR (1972) Perception of amplitude modulated auditory signals by the goldfish. J Acoust Soc Am 52:660–666.

Fay RR (1974a) The masking of tones by noise for the goldfish. J Comp Physiol Psychol 87:708–716.

Fay RR (1974b) Auditory frequency discrimination in vertebrates. J Acoust Soc Am 56:206–209.

Fay RR (1978) Phase-locking in goldfish saccular nerve fibers accounts for frequency discrimination capacities. Nature 275:320–322.

Fay RR (1980) Psychophysics and neurophysiology of temporal factors in hearing by the goldfish: amplitude modulation detection. J Neurophysiol 39:871–881.

Fay RR (1982) Neural mechanisms of an auditory temporal discrimination by the goldfish. J Comp Physiol [A] 147:201–216.

Fay RR (1984) The goldfish ear codes the axis of particle motion in three dimensions. Science 225:951–953.

Fay RR (1985) Sound intensity processing by the goldfish. J Acoust Soc Am 78:1296–1309.

Fay RR (1988) Hearing in Vertebrates: A Psychophysics Databook. Winnetka, IL: Hill-Fay Associates.

Fay RR (1989a) Intensity discrimination of pulsed tones by the goldfish. J Acoust Soc Am 85:500–502.

Fay RR (1989b) Frequency discrimination in the goldfish: effects of roving intensity, sensation level, and the direction of frequency change. J Acoust Soc Am 85:503–505.

Fay RR (1992a) Structure and function in sound discrimination among vertebrates. In: Webster D, Fay R, Popper A (eds) The Evolutionary Biology of Hearing. New York: Springer-Verlag, pp. 229–263.

Fay RR (1992b) Analytic listening by the goldfish. Hear Res 59:101–107.

Fay RR (1994a) Perception of temporal acoustic patterns by the goldfish (*C. auratus*). Hear Res 76:158–172.

Fay RR (1994b) The sense of hearing in fishes: psychophysics and neurophysiology. Sensory Syst 8:222–232.

Fay RR (1995) Perception of spectrally and temporally complex sounds by the goldfish (*C. auratus*). Hear Res 89:146–154.

Fay RR (1997) Frequency selectivity of saccular afferents of the goldfish revealed by revcor analysis. In: Lewis ER, Long GR, Lyon RF, Narins PM, Steele CR, Hecht-Poinar E (eds) Diversity in Auditory Mechanics, World Scientific Publishers, Singapore, pp. 69–75.

Fay RR, Coombs S (1983) Neural mechanisms in sound detection and temporal summation. Hear Res 10:69–92.

Fay RR, Edds-Walton PL (1997) Directional response properties of saccular afferents of the toadfish, Opsanus tau. Hearing Research 111:1–21.

Fay RR, Passow B (1982) Temporal discrimination in the goldfish. J Acoust Soc Am 72:753–760.

Fay RR, Ream TJ (1986) Acoustic response and tuning in saccular nerve fibers of the goldfish (*C. auratus*). J Acoust Soc Am 79:1883–1895.

Fay RR, Ahroon WA, Orawski AT (1978) Auditory masking patterns in the goldfish: psychophysical tuning curves. J Exp Biol 74:83–100.

Fay RR, Yost WA, Coombs S (1983) Repetition noise processing by a vertebrate auditory system. Hear Res 12:31–55.

Fay RR, Chronopolous M, Patterson R (1996) The sound of a sinusoid: perception and neural representations in goldfish (*C. auratus*). Audit Neurosci 2:377–392.

Fellers GM (1979) Aggression, territoriality and mating behaviour in North American treefrogs. Anim Behav 27:107–119.

Feng AS, Shofner WP (1981) Peripheral basis of sound localization in anurans. Acoustic properties of the frog's ear. Hear Res 5:201–216.

Feng AS, Gerhardt HC, Capranica RR (1976) Sound localization behavior of the green treefrog (*Hyla cinerea*) and the barking treefrog (*H. gratiosa*). J Comp Physiol [A] 107:241–252.

Fletcher H (1940) Auditory patterns. Rev Mod Phys 12:47–65.

Freedman EG, Ferragamo M, Simmons AM (1988) Masking patterns in the bullfrog (*Rana catesbeiana*) II: Physiological effects. J Acoust Soc Am 84:2081–2091.

Frisch K von (1936) Über den Gehorsinn der Fische. Biol Rev 11:210–246.

Gerhardt HC (1978) Mating call recognition in the green treefrog (*Hyla cinerea*): the significance of some fine-temporal properties. J Exp Biol 61:229–241.

Gerhardt HC (1981a) Mating call recognition in the green treefrog (*Hyla cinerea*): importance of two frequency bands as a function of sound pressure level. J Comp Physiol [A] 144:9–16.

Gerhardt HC (1981b) Mating call recognition in the barking treefrog (*Hyla gratiosa*): responses to synthetic mating calls and comparisons with the green treefrog (*Hyla cinerea*). J Comp Physiol [A] 144:17–25.

Gerhardt HC (1987) Evolutionary and neurobiological implications of selective phonotaxis in the green treefrog, *Hyla cinerea*. Anim Behav 35:1479–1489.

Gerhardt HC (1991) Female mate choice in treefrogs: static and dynamic acoustic criteria. Anim Behav 42:615–635.

Gerhardt HC, Doherty JA (1988) Acoustic communication in the gray treefrog, *Hyla versicolor*: evolutionary and neurobiological implications. J Comp Physiol [A] 162:261–278.

Gerhardt HC, Klump GM (1988) Masking of acoustic signals by chorus background noise in the green treefrog: a limitation on mate choice. Anim Behav 36:1247–1249.

Gerhardt HC, Rheinlaender J (1980) Accuracy of sound localization in a miniature dendrobatid frog. Naturwissenschaften 67:362–363.

Gerhardt HC, Rheinlaender J (1982) Localization of an elevated sound source by the green tree frog. Science 217:663–664.

Gerhardt HC, Allan S, Schwartz JJ (1990) Female green treefrogs (*Hyla cinerea*) do not selectively respond to signals with a harmonic structure in noise. J Comp Physiol [A] 166:791–794.

Green DM (1973) Minimum integration time. In: Møller A (ed) Basic Mechanisms in Hearing. London: Academic Press.

Green DM (1988) Profile Analysis. Oxford: Oxford University Press.

Guttman N (1963) Laws of behavior and facts of perception In: Koch S (ed) Psychology: A Study of a Science, vol. 5. New York: McGraw-Hill, pp. 114–178.

Hainfeld CA, Boatright-Horowitz SL, Boatright-Horowitz SS, Simmons AM (1996) Discrimination of phase spectra in complex sounds by the bullfrog (*Rana catesbeiana*). J Comp Physiol [A] 178:75–87.

Hartmann WM (1988) Pitch perception and the segregation and integration of auditory entities. In: Edelman GM, Gall WE, Cowan WM (eds) Auditory Function: Neurological Bases of Hearing. New York: Wiley, pp. 623–645.

Hawkins AD (1981) The hearing abilities of fish. In: Tavolga WN, Popper AN, Fay RR (eds) Hearing and Sound Communication in Fishes. New York: Springer-Verlag, pp. 109–137.

Hawkins AD, Chapman CJ (1975) Masked auditory thresholds in the cod, *Gadus morhua* L. J Comp Physiol [A] 103:209–226.

Hawkins AD, Johnstone ADF (1978) The hearing of the Atlantic salmon, *Salmo salar*. J Fish Biol 13:655–673.

Hawkins AD, Sand O (1977) Directional hearing in the median vertical plane by the cod. J Comp Physiol [A] 122:1–8.

Hetherington TE (1987) Timing of development of the middle ear of Anura (Amphibia). Zoomorphology 106:289–300.

Hetherington TE (1992) The effects of body size on functional properties of the middle ear of anuran amphibians. Brain Behav Evol 39:133–142.

Jacobs DW, Tavolga WN (1967) Acoustic intensity limens in the goldfish. Anim Behav 15:324–335.

Jacobs DW, Tavolga WN (1968) Acoustic frequency discrimination in the goldfish. Anim Behav 16:67–71.

Jeffress LA (1948) A place theory of sound localization. J Comp Physiol Psychol 41:35–39.

Jørgenson MB, Gerhardt HC (1991) Directional hearing in the gray tree frog *Hyla versicolor*: eardrum vibrations and phonotaxis. J Comp Physiol [A] 169:177–183.

Kalmijn AJ (1988) Hydrodynamic and acoustic field detection. In: Atema J, Fay RR, Popper AN, Tavolga WN (eds) Sensory Biology of Aquatic Animals. New York: Springer, pp. 83–130.

Karlsen HE (1992a) Infrasound sensitivity in the plaice (*Pleuronectes platessa*). J Exp Biol 171:173–187.

Karlsen HE (1992b) The inner ear is responsible for detection of infrasound in the perch (*Perca fluviatilis*). J Exp Biol 171:163–172.

Klump GM, Gerhardt HC (1987) Use of non-arbitrary acoustic criteria in mate choice by female gray treefrogs. Nature 326:286–288.

Klump GM, Gerhardt HC (1989) Sound localization in the barking treefrog. Naturwissenschaften 76:35–37.

Knudsen FR, Enger PS, Sand O (1992) Awareness reactions and avoidance responses to sound in juvenile Atlantic Salmon, *Salmo salar* L. J Fish Biol 40:523–534.

Knudsen FR, Enger PS, Sand O (1994) Avoidance responses to low frequency sound in downstream migrating Atlantic salmon smolt, *Salmo salar*. J Fish Biol 45:227–233.

Lu Z, Fay RR (1993). Acoustic response properties of single units in the torus semicircularis of the goldfish, *C. auratus*. J Comp Physiol [A] 173:33–48.

Lu Z, Fay RR (1996) Two-tone interaction in primary afferents and midbrain neurons of the goldfish (*C. auratus*). Audit Neurosci 2:257–273.

Lu Z, Popper A, Fay R (1996) Behavioral detection of acoustic particle motion by a teleost fish (*Astronotus ocellatus*): sensitivity and directionality. J Comp Physiol [A] 179:227–234.

Mann D, Lu Z, Popper A (1997) Ultrasound detection by a teleost fish. Nature 389:234.

McCormick C, Popper AN (1984) Auditory sensitivity and psychophysical tuning curves in the elephant nose fish, *Gnathonemus petersii*. J Comp Physiol [A] 155:753–761.

McGill TE (1960) A review of hearing in amphibians and reptiles. Psych Bull 57:165–168.

Megela-Simmons A (1984) Behavioral vocal response thresholds to mating calls in the bullfrog *Rana catesbeiana*. J Acoust Soc Am 76:676–681.

Megela-Simmons A, Moss CF, Daniel KM (1985) Behavioral audiograms of the bullfrog (*Rana catesbeiana*) and the green treefrog (*Hyla cinerea*). J Acoust Soc Am 78:1236–1244.

Michelsen AM, Jørgenson M, Christensen-Dalsgaard J, Capranica RR (1986) Directional hearing of awake, unrestrained treefrogs. Naturwissenschaften 73:682–683.

Moss CF, Simmons AM (1986) Frequency selectivity of hearing in the green treefrog, *Hyla cinerea*. J Comp Physiol [A] 158:257–266.

Musicant AD, Chan JCK, Hind JE (1990) Direction-dependent spectral properties of cat external ear: new data and cross-species comparisons. J Acoust Soc Am 87:757–781.

Narins PM (1982) Effects of masking noise on evoked calling in the Puerto Rican coqui (Anura: Leptodactylidae). J Comp Physiol [A] 147:439–446.

Narins PM (1983) Synchronous vocal response mediated by the amphibian papilla in a neotropical treefrog: behavioural evidence. J Exp Biol 105:95–105.

Narins PM, Capranica RR (1978) Communicative significance of the two-note call of the treefrog *Eleutherodactylus coqui*. J Comp Physiol [A] 127:1–9.

Narins PM, Ehret G, Tautz J (1988) Accessory pathway for sound transfer in a neotropical frog. Proc Natl Acad Sci USA 85:1508–1512.

Nestler JM, Ploskey GR, Pickens J, Menezes J, Schildt C (1992) Responses of blueback herring to high-frequency sound and implications for reducing entrainment at hydropower dams. North Am J Fish Manag 12:667–683.

Offutt G (1973) Structures for detection of acoustic stimuli in the Atlantic codfish *Gadus morhua*. J Acoust Soc Am 56:665–671.

Passmore NI, Telford SR (1981) The effect of chorus organization on mate localization in the painted reed frog (*Hyperolius marmoratus*). Behav Ecol Sociobiol 9:291–293.

Passmore NI, Capranica RR, Telford SR, Bishop PJ (1984) Phonotaxis in the painted reed frog (*Hyperolius marmoratus*). The localization of elevated sound sources. J Comp Physiol [A] 154:189–197.

Patterson RD (1994a) The sound of a sinusoid: spectral models. J Acoust Soc Am 96:1409–1418.

Patterson RD (1994b) The sound of a sinusoid: time-interval models. J Acoust Soc Am 96:1419–1428.

Patterson RD, Moore BCJ (1986) Auditory filters and excitation patterns as representations of frequency resolution. In: Moore BCJ (ed) Frequency Selectivity in Hearing. London: Academic Press.

Pettersen L (1980) Frequency discrimination in the bullhead *Cottus scorpius*, a fish without swimbladder. Master's Thesis, University of Oslo.

Popper AN (1972) Auditory threshold in the goldfish (*C. auratus*) as a function of signal duration. J Acoust Soc Am 52:596–602.

Popper AN, Clarke NL (1979) Non-simultaneous auditory masking in the goldfish *C. auratus*. J Exp Biol 83:145–158.

Rheinlaender J, Gerhardt HC, Yager DD, Capranica RR (1979) Accuracy of phonotaxis by the green treefrog (*Hyla cinerea*). J Comp Physiol [A] 133:247–255.

Rogers PH, Cox M (1988) Underwater sound as a biological stimulus. In: Atema J, Fay RR, Popper AN, Tavolga WN (eds) Sensory Biology of Aquatic Animals. New York: Springer, pp. 131–149.

Rogers P, Lewis PN, Willis MJ, Abrahamson S (1989) Scattered ambient noise as an auditory stimulus for fish. J Acoust Soc Am 85:S35 (abstr).

Rose GJ, Capranica RR (1984) Processing amplitude-modulated sounds by the auditory midbrain of two species of toads: matched temporal filters. J Comp Physiol [A] 154:211–219.

Ryan MJ (1983) Sexual selection and communication in a neotropical frog, *Physalaemus pustulosus*. Evolution 37:261–272.

Sand O, Karlsen HE (1986) Detection of infrasound by the Atlantic cod. J Exp Biol 125:197–204.

Schuijf A (1975) Directional hearing of cod (*Gadus morhua*) under approximate free field conditions. J Comp Physiol [A] 98:307–332.

Schuijf A, Buwalda RJA (1975) On the mechanism of directional hearing in cod (*Gadus morhua*). J Comp Physiol [A] 98:333–344.

Schuijf A, Hawkins AD (1983) Acoustic distance discrimination by the cod. Nature 302:143–144.

Schuijf A, Siemelink M (1974) The ability of cod (*Gadus morhua*) to orient towards a sound source. Experientia 30:773–774.

Schwartz JJ (1993) Male calling behavior, female discrimination and acoustic interference in the neotropical treefrog *Hyla microcephala* under realistic acoustic conditions. Behav Ecol Sociobiol 32:401–414.

Schwartz JJ, Gerhardt HC (1989) Spatially mediated release from auditory masking in an anuran amphibian. J Comp Physiol [A] 166:37–41.

Schwartz JJ, Gerhardt HC (1995) Directionality of the auditory system and call pattern recognition during acoustic interference in the gray tree frog, *Hyla versicolor*. Audit Neurosci 1:195–206.

Schwartz JJ, Simmons AM (1990) Encoding of a spectrally-complex communication sound in the bullfrog's auditory nerve. J Comp Physiol [A] 166:489–499.

Schwartz JJ, Wells KD (1983) The influence of background noise on the behavior of a neotropical treefrog, *Hyla ebraccata*. Herpetologica 39:121–192.

Schwartz JJ, Wells KD (1984) Interspecific acoustic interactions of the neotropical treefrog. *Hyla ebraccata*. Behav Ecol Sociobiol 14:211–224.

Shepard RN (1965) Approximation to uniform gradients of generalization by monotone transformations of scale. In: Mostofsky D (ed) Stimulus Generalization. Stanford, CA: Stanford University Press.

Simmons AM (1988a) Masking patterns in the bullfrog (*Rana catesbeiana*), I: Behavioral effects. J Acoust Soc Am 83:1087–1092.

Simmons AM (1988b) Selectivity for harmonic structure in complex sounds by the green treefrog (*Hyla cinerea*). J Comp Physiol A 162:397–403.

Simmons AM, Buxbaum RC (1996) Neural codes for pitch processing in a unique vertebrate auditory system. In: Moss CF, Shettleworth S (eds) Neuroethological Studies of Cognitive and Perceptual Processes. Boulder, CO: Westview Press, pp. 185–228.

Simmons AM, Ferragamo M (1993) Periodicity extraction in the anuran auditory nerve I. Pitch shift effects. J Comp Physiol [A] 172:57–69.

Simmons AM, Moss CF (1995) Reflex modification: a tool for assessing basic auditory function in anuran amphibians. In: Dooling RJ, Fay RR, Klump G,

Stebbins W (eds) Methods in Comparative Psychoacoustics. Basel, Boston, Berlin: Birkhauser Verlag, pp. 197–208.

Simmons AM, Buxbaum RC, Mirin M (1993) Perception of complex sounds by the green treefrog, *Hyla cinerea*: envelope and fine-structure cues. J Comp Physiol [A] 173:321–327.

Strother WF (1962) Hearing in frogs. J Audit Res 2:279–286.

Tavolga WN (1974) Signal/noise ratio and the critical band in fishes. J Acoust Soc Am 55:1323–1333.

Tavolga WN (1982) Auditory acuity in the sea catfish (*Arius felis*). J Exp Biol 96:367–376.

Viemeister NF, Wakefield GH (1991) Temporal integration and multiple looks. J Acoust Soc Am 90:858–865.

Weiss BA, Strother WF (1965) Hearing in the green treefrog (*Hyla cinerea cinerea*). J Audit Res 5:297–305.

Wever EG (1985) The Amphibian Ear. Princeton, NJ: Princeton University Press.

Wilczynski W, Resler C, Capranica RR (1987) Tympanic and extratympanic sound transmission in the leopard frog. J Comp Physiol [A] 161:659–669.

Witschi E (1949) The larval ear of the frog and its transformation during metamorphosis. Z Naturforsch 4b:230–242.

Wohlfahrt TA (1939) Untersuchungen uber das Tonunterscheidunsvermögen der Elritze. Z Vergl Physiol 26:570–604.

Wotton JM, Haresign T, Simmons JA (1995) Spatially dependent acoustic cues generated by the external ear of the big brown bat, *Eptesicus fuscus*. J Acoust Soc Am 98:1423–1445.

Yost WA, Hill R, Perez-Falcon T (1978) Pitch and pitch discrimination of broadband signals with rippled power spectra. J Acoust Soc Am 63:1166–1173.

Zelick RD, Narins PM (1982) Analysis of acoustically evoked call suppression behaviour in a neotropical treefrog. Anim Behav 30:728–733.

Zelick RD, Narins PM (1983) Intensity discrimination and the precision of call timing in two species of neotropical treefrogs. J Comp Physiol [A] 153:403–412.

8
The Enigmatic Lateral Line System

SHERYL COOMBS AND JOHN C. MONTGOMERY

1. Introduction

Hearing in its broadest sense is the detection, by specialized mechanorecep-tors, of mechanical energy propagated through the environment. In terres-trial vertebrates, this typically means inner ear transduction of air pressure waves radiating out from a sound source, though the detection of substrate vibrations can also be considered as a form of hearing. In aquatic environ-ments, the extended contribution of incompressible flow in the near field of the source adds additional complexities, and both incompressible flow and propagated pressure waves are detected by a range of specialized hair cell mechanosensory systems. Hair cells are generalized mechanical transducers that respond to mechanical deformation of the receptor hairs at their apical surface. One of the interesting stories of hearing in general, and in aquatic vertebrates in particular, is how the structures associated with hair cell organs play a major role in modifying or channeling the environmental stimulus onto the hair cell receptors. Hence the peripheral anatomy deter-mines to a large degree what particular stimulus feature is being encoded at the level of the hair cell.

The particle motions of an acoustic field impart motion to aquatic ani-mals that can be detected by differential motion between the inner-ear otoliths and their underlying hair cells, the so-called direct channel of acoustic stimulation that theoretically exists for all fishes (Popper and Fay, Chapter 3). Conditions leading to differential movement between the animal and the surrounding water result in stimulation of hair cells in the mechanosensory lateral line system. Detection of the pressure component of an underwater acoustic field, the so-called indirect channel, can occur when a compressible air cavity, like a gas-filled swim bladder, converts pressure variations into displacements that can be detected by hair cells of the fish ear (Popper and Fay, Chapter 3) and in rare cases, by the lateral line (e.g., in clupeids; Denton et al. 1979). Likewise, benthic fishes in contact with the substrate are likely to be stimulated by substrate vibrations, which could effect hair cells of both the inner ear and lateral line (Janssen 1990;

Tricas and Highstein 1991). This chapter is divided into four sections that discuss (1) the natural stimuli to the lateral line system and the behaviors that give insight to the multiplicity of uses of this system, (2) the physics of the stimuli and the peripheral anatomy that modifies and channels the stimuli onto the hair cell receptors, (3) our current understanding of how stimuli are encoded by the peripheral and central nervous system, and (4) some of the similarities and differences between lateral line and related auditory systems.

2. Natural Stimuli and Behavior

2.1 Introduction

The lateral line mechanosense has aptly been described as touch at a distance (Hofer 1908; Dijkgraaf 1963). With has sense aquatic vertebrates can feel water movements. So the essential stimulus for the lateral line consists of differential movement between the body surface and the surrounding water. Water movements of interest to aquatic animals range from large-scale ocean currents, tidal streams, and river flows to the minute disturbances created by the movements of planktonic prey (Table 8.1). Like touch, or any other sense for that matter, the lateral line has a multiplicity of uses (Table 8.2) and typically is used in concert with other sensory information. Behavioral studies are required to show just how the lateral line is used, to probe the limits and dimensions of sensory capabilities (see Section 2.6), and to provide insight into the way in which lateral line mechanosensory information is integrated with other senses.

2.2 Water Currents of Inanimate Origin

Large-scale water movements, such as oceanic currents and tides, and river flows, are clearly of great importance to fishes and other aquatic animals (Montgomery et al. 1995). Orientation to currents (i.e., rheotaxis) is important in migration, station holding, and olfactory-initiated search. Recent

---

TABLE 8.1. Common hydrodynamic signals and noises of biotic and abiotic origins.

Low frequency (<10 Hz)
  Steady swimming motions (slow, regular power strokes) of fish and invertebrates
  Ventilatory movements of fish and invertebrates
  Laminar, nonturbulent, slow flows
Broad-band or high frequency (up to at least 100–200 Hz)
  Wakes (shed vortices) behind submerged obstacles in a current
  Wakes behind swimming invertebrates or fish
  Turbulent flow

behavioural evidence shows that the lateral line makes an important contribution to rheotaxis in fish (Montgomery et al. 1997). Visual and tactile stimuli can initiate rheotaxis when the fish is displaced by the current, but when visual and tactile stimuli are absent, particularly at slower flows, the lateral line can provide the necessary information. These behavioural studies also show that the lateral line contribution to this behaviour is mediated by one class of receptors, the superficial neuromasts (described in section 4.2). Earlier behavioural studies had previously shown that station holding adjacent to an obstacle in the stream can also be mediated via the lateral line system (Sutterlin and Waddy, 1975). Though in this case it is the posterior lateral line system which mediates the behaviour, and it may well be that it is the turbulence generated by the obstacle which provides the necessary cues. This study also provides a nice example of the contributions of vision, mechanosensory lateral line, and touch to station holding, and a salutary reminder that behaviors mediated by the lateral line are often only

TABLE 8.2. Lateral line–mediated behaviors and sensory capabilities.

Behavior	Species	References*
Schooling/predator-avoidance	Herring, *Clupea harengus*	6,8
	Sprat, *Sprattus sprattus*	8
	Saithe, *Pollachius virens*	18, 19
	Golden shiner, *Notemigonus crysoleucus*	3
	Tuna, *Euthynnus affinis*	4
	Jacks, *Caranx hippos*	23
	Tadpoles, *Xenopus laevis*	15
Active hydrodynamic imaging	Blind cavefish, *Anoptichthys jordani*	5,10,11,26,27
Passive hydrodynamic imaging	Brook trout, *Salvelinus fontinalis*	24
Courtship communication	Hime salmon, *Oncorhynchus nerka*	21
Subsurface feeding	Mottled sculpin, *Cottus bairdi*	12
	Piper, *Hyporhamphus ihi*	17,22
	Torrentfish, *Cheimarrichythys fosteri*	16
	Blind cavefish, *Typhilchthys subterraneus* *Amblyopsis spelaea, Amblyopsis rosae*	20
	Alewife, *Alosa pseudoharengus*	14
	Axolotl, *Ambystoma mexicanum*	25
Surface wave feeding	Topminnow, *Aplocheilus lineatus*	1,2
	African butterflyfish, *Pantodon buchholzi*	13
	Banded kokopu, *Galaxius fasciatus*	9
	African clawed frog, *Xenopus laevis*	7

* 1. Bleckmann (1988), 2. Bleckmann et al. (1989), 3. Burgess and Shaw (1981), 4. Cahn (1972), 5. Campenhausen et al. (1981), 6. Denton and Gray (1983), 7. Görner (1973, 1976), 8. Gray and Denton (1991), 9. Halstead (1994), 10. Hassan (1986), 11. Hassan et al. (1992), 12. Hoekstra and Janssen (1985, 1986), 13. Hoin-Radkovski et al. (1984), 14. Janssen et al. (1995), 15. Katz et al. (1981), 16. Montgomery and Milton (1993), 17. Montgomery and Saunders (1985), 18. Partridge and Pitcher (1980), 19. Pitcher et al. (1976), 20. Poulson (1963), 21. Satou et al. (1987, 1991, 1994a,b), 22. Saunders and Montgomery (1985), 23. Shaw (1969), 24. Sutterlin and Waddy (1975), 25. Takeuchi and Namba (1989, 1990) and Takeuchi et al. (1991), 26. Teyke (1985, 1988), 27. Weissert and von Campenhausen (1981).

revealed in the absence of other cues. Flow patterns behind stationary obstacles can vary from being relatively laminar to fully turbulent wakes (Fig. 8.1; Vogel 1994). In the case of turbulent flows, it has been suggested that fish such as trout and salmon may exploit oncoming vortices from submerged obstacles like rocks and boulders to boost their swimming efficiency during their arduous upstream migration (Triantafyllou and Triantafyllou 1995). Moreover, it is quite likely that the detection of stream distortions caused by all sorts of irregularities in a stream bed can be used in a general sense to form hydrodynamic images of the surroundings.

2.3 Water Currents Produced by Other Animals

Generally on a much smaller scale are the water currents produced by other animals. Virtually all animals maintain some ventilatory flow, and many sedentary animals filter feed, producing quite significant water currents. Flow velocity at the excurrent siphon of bivalves (Fig. 8.2A) ranges between 6 and 14 cm/s (Price and Schiebe 1978; LaBarbera 1981; Ertman and Jumars 1988). Little if anything is known of the use of these cues by fishes, but given the number of fishes that prey on filter feeding bivalves, including just the exposed siphons (Peterson and Quammen 1982), and crustaceans, it would be surprising if these cues were not used (Montgomery and

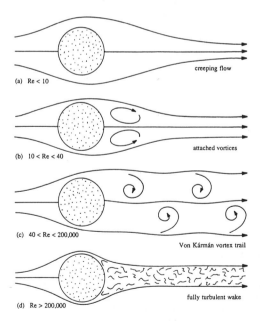

FIGURE 8.1. Flow distortions created behind a stationary cylinder in a stream for different Reynolds numbers. (From Vogel 1994. Copyright © 1994 by Princeton University Press. Reprinted by permission of Princeton University Press.)

Hamilton 1997). Many sedentary animals have rapid withdrawal responses, and the ability to aim a strike without disturbing the prey would seem to be a necessity for feeding on exposed parts, such as the siphons. For fish that dig out buried sedentary animals, an assessment of the location of individual prey, and even perhaps prey density, would be a useful clue as to where to dig (Montgomery and Skipworth 1997).

Animals that move cannot help but produce water disturbances. Even relatively slow crawling will produce some kind of wake, which is a potential lateral line cue for a nearby fish. Axolotl apparently use their lateral lines when feeding on tubifex worms in the lab, for example (Takeuchi and Namba 1989, 1990; Takeuchi et al. 1991). There are now a number of examples of fish feeding on swimming invertebrates such as cladocerans and deleatidium larvae using lateral line cues to locate them (Poulson 1963; Hoekstra and Janssen 1985, 1986; Montgomery and Milton 1993). The last of these studies illustrates an advantage of the fish predator facing upstream. Prey could still be located even if they stopped moving and dropped to the bottom of the artificial stream, presumably by the downstream flow perturbations they produced.

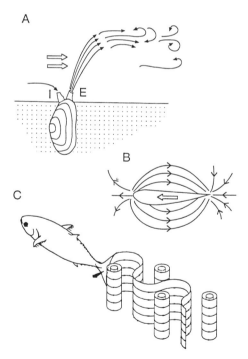

FIGURE 8.2. Examples of water disturbances produced by animals. A: Currents produced by filter-feeding bivalves. B: Dipole flow field created by a gliding fish. C: Vortices shed by carangiform swimming motions of a fish. (From Montomery et al. 1995b, © Chapman & Hall.)

Depending on the circumstances it is appropriate for the fish to employ either a sit-and-wait, or move-stop hunting strategy. There are many examples of sit-and-wait predators, and the move-stop strategy (or saltatory search) is common in many predators (O'Brian et al. 1990). The saltatory search mode is generally thought of in terms of the visual requirements for locating prey, but saltatory movements would also be appropriate to lateral line search. In laboratory studies blind mottled sculpin approach a vibrating target in a series of hops (Hoekstra and Janssen 1986), and a similar search behavior is exhibited by the Antarctic fish *Trematomus bernacchii* (Nototheniidae). This fish has been observed feeding from a sub-ice observation chamber (Janssen et al. 1991). The fish moved up-current a body length at a time, pausing between hops presumably to "feel," either by direct contact or via the lateral line system, for prey animals moving in, or on, the sponge mat that makes up the bottom in this area. Motionlessness of the predator presumably assists the detection of small signals, and reduces the ability of the prey to detect the predator's presence. In effect, motionlessness is the ideal lateral line camouflage.

As prey speed increases, prey become easier to detect, but harder to catch. Piper (*Hyporhamphus ihi*, Hemiramphidae), which feed on demersal zooplankton using lateral line cues, have an anterior extension of the lateral line system that appears to provide lateral line information sufficiently far in advance for the piper to intercept the fast moving plankton (Montgomery and Saunders 1985; Saunders and Montgomery 1985).

When fish are gliding, they produce a dipole flow field (Fig. 8.2B), where water pushed out in front flows around the fish and back in where the body has passed (Hassan 1985, 1989). During active swimming, the oscillatory movement of the tail produces a substantial turbulent wake, with pronounced vortices, that can persist in the water for some time after the fish has passed (Fig. 2C) (Rosen 1959; Magnuson 1978; Blickhan et al. 1992; Stamhuis and Videler 1995). Bleckmann (1993) suggests that this "vortex wake" will produce a potent lateral line stimulus allowing a fish that swims into the wake to detect the fish that has passed. Enger and colleagues (1989) report that the bluegill sunfish has no trouble detecting and capturing a goldfish in a darkened aquarium provided that the lateral line is intact.

Water movements produced by swimming are also used in a form of communication. The coordination of schooling fish is in part maintained by lateral line stimuli (Partridge and Pitcher 1980), and in the case of the sprat, it has been shown that the lateral line is capable of providing the necessary information needed to maintain position in relation to other fish in the school (Denton and Gray 1983; Gray 1984). Communication by lateral line stimulation is also part of the spawning behavior of salmon (Satou et al. 1987, 1991, 1994a,b). Vibrations produced by a quivering of the body are an important component of courtship behavior in these and other fishes, for example, pygmy angelfish (Bauer and Bauer 1981), and it is likely that the lateral line plays an important communication role in synchronizing spawning behavior.

2.4 Self-Induced Currents

As mentioned earlier, the flow field around a gliding fish can provide a lateral line cue to other animals. Extensive studies of blind cave fish conclude that they use their lateral line to monitor their own flow field. Distortions in the flow field caused as the fish swims close to an inanimate, or stationary object provide the fish with its major source of information as to the presence and nature of the object (Hassan 1989). This ability is quite acute and it has been shown that blind cavefish can detect millimeter differences in spacings between vertically oriented rods (Hassan 1986). In this case the lateral line is being used as an active sense, analogous to electrolocation (e.g., Bastian 1986c) or echolocation (e.g., Pollak and Casseday 1989) where motor activities of the animal produce sensory reafference, which provides useful information. Thus, hydrodynamic imaging of the environment by the lateral line is possible in both an active and passive sense.

Self-induced lateral line stimulation may also play a role in optimization of swimming. High-speed video recordings of herring (Rowe et al. 1993) show that head turning during swimming in these fish occurs in the way that is predicted to make swimming more economical and to diminish lateral line stimuli generated by the fish's own movements (Denton and Gray 1993; Lighthill 1993). It has also been suggested that fish may propel themselves by successively generating and destroying tail tip vortices of alternating directions (Ahlborn et al. 1991). The control and monitoring of vortex formation by the lateral line system is proposed as a key feature that enables fish to swim as efficiently and skillfully as they do (Triantafyllou and Triantafyllou 1995).

2.5 Surface Waves

In addition to the "free field" subsurface stimulation described above, a few species of fish [e.g., the topminnow, *Aplocheilus lineatus* (Bleckmann 1988; Bleckmann et al. 1989); the african butterflyfish, *Pantodon bucholzi* (Hoin-Radkovsky et al. 1984); and *Galaxias fasciatus* (Halstead 1994)] and amphibians (e.g., clawed frog *Xenopus laevis*; Görner 1973, 1976) take advantage of surface waves for prey detection and for communication. Insects that fall onto the water surface often become trapped there, forming a potential source of food. In freshwater ecosystems, terrestrial insects can be an important component of the diet of many fish, as dry-fly trout anglers will attest. In most cases prey are taken visually, but a few species of fish can use surface waves to direct them to the prey. Recent field and laboratory studies of surface wave detection by the banded kokopu (*G. fasciatus*, Galaxiidae) illustrate the point that surface wave detection via the lateral line can allow feeding at night in bush-covered streams when visual prey detection would be ineffective (Halstead 1994).

2.6 Behaviorally Measured Sensory Capabilities

A number of behavioral studies on the lateral line system have used naturally occurring behaviors, usually orienting and feeding responses, in combination with easily manipulated and controlled artificial stimuli in the lab, to probe the limits of lateral line sensory capabilities (Coombs 1995). Unconditioned orienting responses have been used to measure threshold levels of detection in fish (Bleckmann 1980; Coombs and Janssen 1990a) and how well fish and amphibians can determine both the distance (Schwartz 1967; Bleckmann et al. 1984, Hoin-Radkovsky et al. 1984) and the angular location (Elepfandt 1982; Tittel 1988; Janssen 1990; Coombs and Conley 1997a,b) of sources of water disturbance. In addition, operant conditioning of the orienting response has been employed to measure frequency discrimination (Bleckmann et al. 1981; Elepfandt et al. 1985) and intensity discrimination (Coombs and Fay 1993) abilities. Many of these abilities, in terms of threshold levels of detection, discrimination, or acuity, compare quite favorably to fish auditory capabilities, as reviewed by Fay (1988). The frequency range of detection for the lateral line system extends from below 10 Hz to a little over 100 Hz. The frequency range of the fish auditory system differs only in the high-frequency cutoff, ranging from as low as 100 Hz in species without swim bladders to 2 to 3 kHz in species with pressure-sensitive adaptations (Popper and Fay, Chapter 3). Threshold levels of detection in terms of best displacement sensitivity are also quite comparable and are in the nanometer range.

3 Physics of Stimuli

The inability to adequately specify and measure the stimulus to the lateral line has probably been the biggest impediment to understanding the function of this still somewhat enigmatic sensory system. It is for this reason that we devote considerable attention to the physics of lateral line stimuli. The biotic and abiotic sources of water disturbances discussed above are examples of those likely to be of behavioral importance to fishes. Embedded in these examples, are three fundamentally different types of signal sources: (1) subsurface vibratory sources (e.g., swimming fish or invertebrate prey), (2) surface wave vibratory sources (e.g., struggling insects at the water surface), and (3) nonvibratory sources in a flow field (e.g., submerged obstacles in a stream). For all of these sources, it is important to note that the primary stimulus to the lateral line is relative movement between the surface of the fish and the adjacent water and that regional variations in flow along the fish are likely to be of particular importance in the fish's ability to recognize and locate biologically significant sources of water disturbance. For most subsurface sources, this means that the effective stimulus to the lateral line is a hydrodynamic phenomenon of incompressible bulk flow close to the source, rather than the propagated sound pressure waves that form at some distance from the source.

3.1 Subsurface Vibratory Sources—Near Fields, Far Fields, and Pressure Gradient Patterns

The effective stimulus and distance limitation of the lateral line system can probably best be understood in terms of a vibrating sphere, or dipole source. Unlike monopole (pulsating bubble) sources, which are characterized by changes in volume, dipole sources are closer in reality to most biologically relevant sources (e.g., moving prey) in that the volume of the source stays nearly constant (van Bergeijk 1964; Kalmijn 1988). That is not to say that dipoles are particularly important sources biologically, but only that their fluid dynamics are tractable, well understood, and, with some simplifying assumptions, more or less applicable to other more biologically realistic and important vibratory sources. For a complete and detailed mathematical description of the stimulus field about a dipole source with an emphasis on fish lateral line and auditory systems, the reader is referred to Kalmijn (1988, 1989). In this subsection, we simply describe the information-bearing dimensions of the dipole stimulus field for the lateral line system while trying to clear up some common misconceptions about what happens in the "near" and "far" fields (defined below) of a dipole source.

As a sinusoidally vibrating sphere is displaced from its starting position, pressure is elevated in front of the source and reduced in the rear (Fig. 8.3A), resulting in nearly incompressible flow that predominates close to the source, the so-called "near field". Further away from the source, the compressions and rarefactions of propagated pressure waves predominate in the far field. This does not mean that there is an absence of pressure in the near field or an absence of water motion in the far field, as is so often misconstrued. Propagated particle motion occurs along the axis of sound propagation during compression and rarefaction of the water. In this case, particles oscillate about a fixed point, transferring their energy to adjacent particles without resulting in any net movement of water. This phenomenon exists in both the near and far fields, but the amplitude of this motion is surpassed by the amplitude of incompressible flow in the near field (Fig. 8.4A). Likewise, incompressible flow extends into the far field, but it

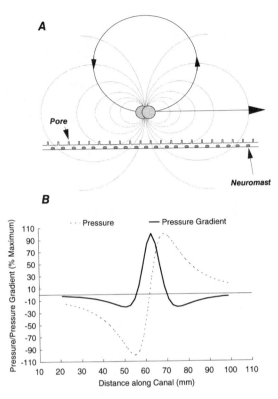

FIGURE 8.3. A: Schematic representation of iso-pressure contours (dashed lines) and flow lines (solid lines with arrows) about a dipole source. Iso-pressure contours are depicted for a single plane that bisects the source along its axis of oscillation, indicated by the large arrowhead to the right of the source. A lateral line trunk canal is modeled as a simple tube with an array of pressure sampling points (canal pores) separated by 2-mm intervals (not to scale). In this example, the modeled canal is confined to a single horizontal plane through the source center and its long axis is parallel to the axis of source oscillation. B: Corresponding plots of pressure (dashed line) and pressure gradient (solid line) distributions across the modeled trunk canal. Note that the maximum pressure gradient is centered at the source, arbitrarily located at x distance $=61$ mm along the modeled canal. (From Coombs et al. 1995.)

FIGURE 8.4. Signal attenuation with distance for a subsurface dipole source (A), a surface wave source (B), and a stationary source in the flow field created by a gliding fish (C). Relative amplitude is represented in arbitrary units along the y-axis in all three graphs, whereas distance is represented in terms of wavelengths in A, as centimeters in B, and as fish lengths in C (100 units equivalent to one fish length). Different functions represent incompressible flow vs. propagated pressure changes in A, different frequencies in B, and stationary sources of different diameters in C. Note that in all three cases the rate of signal attenuation is greatest closest to the source. (A adapted from Coombs and Janssen 1990b, with permission of John Wiley & Sons, Inc.; B from Bleckmann 1985; C from Hassan 1985.)

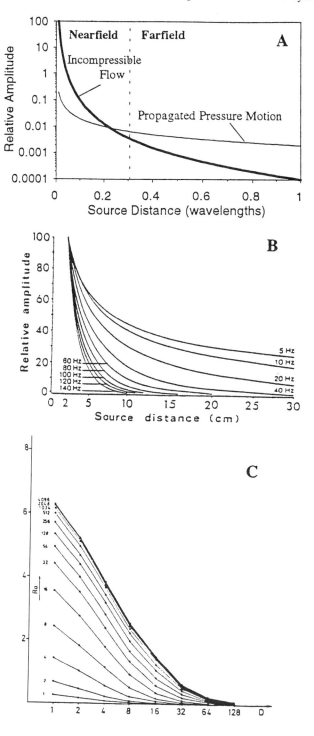

declines so rapidly with distance that its amplitude is far exceeded by the amplitude of propagated particle motion in the far field, which declines more gradually. It is also important to point out that there can be no incompressible (net) flow in the near field without a pressure gradient, and the path of water as it is pushed out in front of the advancing sphere and as it returns to the back of the sphere along the pressure gradient (dashed lines) is shown by the solid lines in Figure 8.3A.

A clear understanding of the distinction between near- and far-field phenomena for underwater sources is important for three reasons: First, the spatial extent of the near field in water is roughly five times that in air and thus likely to be significantly more important to aquatic vertebrates than to terrestrial vertebrates. Second, the extent of the near field is frequency-dependent, being greater for lower frequencies, where both the lateral line and auditory systems of fish operate. Third, stimulation of the lateral line can occur only in the inner regions of the near field (Denton and Gray 1983; Kalmijn 1988, 1989). Although there is no precise boundary between near and far fields, the spatial extent of near-field predominance in water is approximately one third of a wavelength for a dipole "point" source (diameter of source less than about $\frac{1}{10}$ of a wavelength). Thus, at 50 Hz the approximate near field/far field boundary is about 10 m, and at 10 Hz about 50 m. For sources larger than $\frac{1}{10}$ of a wavelength, the near field usually extends a distance of at least two times the largest dimension of the source.

Water motion in both the near and far fields can be described in terms of displacement (d), velocity (μ), or acceleration (a) amplitude. It turns out that pressure gradients, which from Euhler's equation are proportional to acceleration when inertial forces predominate, can be very conveniently measured and modeled for the purposes of predicting lateral line responses to small dipoles (Coombs et al. 1995). But no matter how the amplitude of water motion is expressed, it is proportional to the source amplitude. Thus, flow velocity varies with the source velocity and this quantity should not be confused with the propagation velocity (1500 m/s) of the sound pressure wave. Probably the most important point to be made about flow amplitude in the near field (whether it be expressed as a displacement, velocity, acceleration, or a pressure gradient) is that it falls off very steeply with distance—at the rate of $1/r^3$ for a dipole source (Fig. 8.4A). Since the lateral line system is essentially a spatially distributed array of pressure-gradient (i.e., flow) detectors (Section 4), this means that the lateral line system can operate in the inner regions of the near field only, where significant differences in pressure occur over short distances along the length of the fish (Denton and Gray 1983; Kalmijn 1988, 1989).

Figure 8.3B illustrates the spatial distribution of pressure (solid line) and pressure gradients (dashed line) that would exist across a linear array of pressure sampling points centered on the source and parallel to the axis of vibration for a spatial sampling period of 2 mm and a source distance of 1 cm. The pressure gradient pattern is related to the anatomy and physiology of the lateral line system in Section 4.3, but for the moment it serves to illustrate the kinds of stimulus fields available to the lateral line for sources that are less than approximately $\frac{1}{10}$ of a wavelength in diameter. Note that the field is broad and complex and that pressure gradients differ in both direction and amplitude along the spatial array. These pressure gradient patterns vary with the location of the source relative to the array and thus provide rich information about both source azimuth and distance (Denton and Gray 1983; Gray 1984; Gray and Best 1989; Coombs et al. 1995).

3.2 Surface Wave Sources

Water surface waves are boundary waves caused by moving a solid through the water, by shock, wind, and all similar forces that generate and maintain waves (Bleckmann et al. 1989). As with propagated pressure waves, surface waves transport energy but not mass. In this sense, surface waves can be distinguished from incompressible flow—the primary stimulus to the lateral line for subsurface vibratory sources. The signal characteristics of propagated, concentric surface waves are quite different from those of propagated subsurface pressure waves. The most conspicuous differences are that (1) the propagation velocity of subsurface pressure waves is 1500 m/s at all vibration frequencies, whereas that of surface waves is orders of magnitude slower (cm/s) and frequency-dependent; and (2) the rate of attenuation with distance for surface waves is frequency-dependent (Fig. 8.4B) (Bleckmann et al. 1989), whereas the rates of attenuation for subsurface incompressible flow ($1/r^3$) and propagated pressure waves ($1/r$) are frequency-independent (Fig. 8.4A). The physics of surface waves and the key features of the stimulus that allow the fish to detect direction and distance of the source are very well covered in Bleckmann's (1988, 1993) reviews of the lateral line system and only the essentials will be covered in this section to facilitate comparisons among different surface and subsurface stimulus sources to the lateral line.

When an object falls on the water or moves at its surface, it creates a short disturbance that emanates as a wave packet, composed of a number of different frequencies. Concentric surface waves are strongly attenuated during propagation. Because of geometrical spreading, attenuation occurs mostly in the vicinity of the source, meaning that even for propagated surface waves, the operation of the lateral line will be restricted to short distances. As these waves radiate, high frequencies move to the front of the packet due to their higher propagation velocities, but higher frequencies also attenuate more quickly (Figs. 8.4B and 8.5). By attending to the frequency components of the signal and the rate at which the frequency components sweep from high to low frequency, the fish can obtain information as to the distance of the source. Computer-generated waveforms presented close to the fish, but having the frequency characteristics of signals

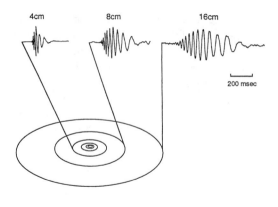

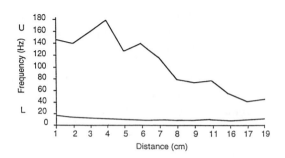

FIGURE 8.5. Distant-dependent changes in the propagation of surface waves. A: Changes in the time waveform of the wave packet at different distances during concentric spreading of the surface wave. B: Lower (L) and upper (U) frequency limits of the surface wave amplitude spectrum as a function of distance. High frequencies attenuate more rapidly with distance, producing a compression of the bandwidth. (From Montgomery et al. 1995b, © Chapman & Hall.)

that have traveled further, are indeed interpreted by the fish as having come from a more distant source (Bleckmann 1993). This kind of cue is completely unavailable for subsurface vibratory sources. This serves as a potent reminder that the physics of surface and subsurface wave propagation are fundamentally different, meaning that the available cues and thus, the lateral line–encoding mechanisms for stimulus features such as source distance may be dramatically different.

3.3 Stationary Objects in Flow Fields: Reynolds Numbers, Boundary Layers, and Vortices

Stationary objects in flow fields produce flow distortions that can be detected by the lateral line. Although there is a plethora of information on the

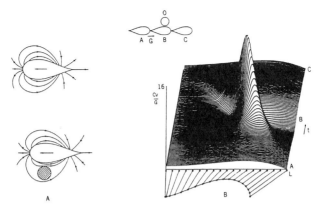

FIGURE 8.6. Active hydrodynamic imaging by gliding fish. A: Distortions in the self-generated flow field due to the presence of an obstacle. B: Flow velocity distribution along the surface of the fish as it glides past the obstacle from position A to position C (inset, upper right). Flow velocity (Cv) is normalized to the gliding velocity (G) of the fish. (From Hassan 1985.)

nature of these distortions in the literature on fluid mechanics (see Vogel 1994 for a very entertaining smorgasbord of biological examples), there is little information about the information-bearing dimensions of these distortions with respect to the lateral line. This is despite the inescapable conclusion that these kinds of sources are likely to be very important in the everyday lives of fish. One notable exception is the body of work by Hassan in connection with active hydrodynamic imaging by characid blind cavefish. He has modeled the flow distribution along a fish-shaped object gliding past and toward a stationary cylinder (Hassan 1985) (Fig. 8.6) and plane surface (Hassan 1992a,b). Although these modeled predictions involve many complicated variables (e.g., fish size and shape), steps, and assumptions, the end results are similar to those modeled for subsurface vibratory sources (Fig. 8.3B) in at least two fundamental ways. One is that the spatial distribution of both flow amplitudes and directions along the fish convey rich information about the source. The other is that flow amplitudes decline steeply with distance from the source such that they approach zero at distances less than about two body lengths away (Fig. 8.4C).

The flow distributions modeled by (Hassan 1985) (Fig. 8.6) assume that flow conditions are laminar, meaning that fluid particles move very nearly parallel to each other in smooth paths. As Figures 8.1 and 8.7 show, however, the disturbances caused by a stationary cylinder in a flow field vary from fairly laminar to fully turbulent, depending on a number of variables, including the size of the cylinder, the flow velocity, and the viscosity and density of the fluid. These variables are all taken into account in a dimension-less number called the *Reynolds number*, which is essentially a measure of the relative contributions of viscous vs. inertial forces (see Vogel 1994 for further detail). High viscosity, low velocities, and small sizes

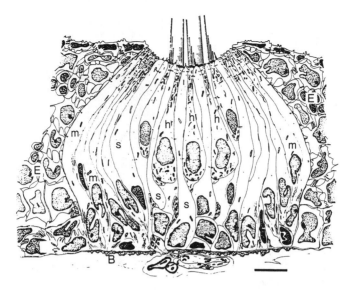

FIGURE 8.7. Schematic cross section through a lateral line neuromast from a teleost fish showing the central hair cells (h) surrounded by supporting cells (s and m). (From Münz 1979.)

are associated with small Reynolds numbers and laminar flow, whereas low viscosity, high velocities, and large sizes produce large Reynolds numbers and turbulent flow. As explained by Vogel (1994), from which much of this subsection is based, the value of this number is that it serves as a general predictive guide as to what happens when solids and fluids move with respect to each other.

A closely related concept is the *boundary layer*, which is simply a velocity gradient at the interface between a solid and a fluid moving relative to one another. If the solid is stationary and the fluid is moving, there must be a small region at the interface where fluid velocity increases from zero at the surface of the solid to 99% of its "free stream" velocity some distance away. This somewhat "fuzzy" region, as Vogel (1994) calls it, is defined as the boundary layer; relatively thick boundary layers are associated with small Reynolds numbers, whereas thinner boundary layers are associated with large Reynolds numbers.

As Figure 8.1A illustrates, flow disturbances due to the presence of a cylinder are fairly laminar at Reynolds numbers below 10. Reynolds numbers between 10 and 2×10^5, on the other hand, produce vortices. Water draining in a bathtub is a perfect example of a vortex, which is characterized by an angular velocity that decreases with distance from its center. At the peripheral edge of the vortex, water height and pressure is highest and decreases more centrally. Thus, pressure gradient information to the lateral line is certainly present in vortices. Bleckmann and colleagues (1991) used laser Doppler anemometry to measure the spectral content of vortices 2 cm behind a small cylinder in a water tunnel (Reynolds number = 5600). The water disturbances measured in the wake had frequency components up to at least 200 Hz (well within the detection range of the lateral line) and were significantly (10–35 dB) greater in amplitude than flow levels in the absence of the cylinder, especially in the 75- to 200-Hz region.

Vortices can also happen under many other circumstances, for example, flow across a furrow, flow turning upstream from a sharp corner, and flow across a sharp edge. Moreover, vortex formation is influenced by the spatial separation of stationary obstacles. It is easy to see how all of these examples could be found in the natural environments (e.g., stream beds, coral reefs, etc.) of fish.

At Reynolds numbers greater than about 2×10^5, the somewhat predictable formation of vortices give way to highly unpredictable and fully turbulent wakes (Fig. 8.1D). In turbulent flow, individual water particles move in a highly irregular fashion, even though the water as a whole may appear to move smoothly in a single direction. At very high Reynolds numbers, incompressible flow dynamics start to yield to compressible dynamics, meaning that propagated sound pressure waves are no longer negligible.

It is important to point out that Reynolds numbers, boundary layers, and vortices are applicable not only to stationary sources in flow fields, as discussed in this section, but also to vibratory sources in stagnant water, and in general to the biomechanics of lateral line end organs (see Section 4.2). For example, vortices can arise in stagnant waters from moving sources like the caudal fin movements of a swimming fish (Fig. 8.2C). In the case of vibratory sources, low-frequency vibrations will be associated with small Reynolds numbers, and thick boundary layers and higher frequency vibrations will be associated with larger Reynolds numbers and thinner boundary layers (see also Kalmijn 1988, 1989).

3.4 Stimulus Measurement

This subsection provides a brief list of the types of methods that have been used in lateral line research for measuring flow fields. In doing so, we wish to demonstrate that the tools and methods for measuring water disturbances are as varied as the disturbances themselves, and that there is no single method suitable for all applications, each method having its own advantages and disadvantages.

Flow visualization techniques (Merzkirch 1987) range from simple observations of dye dispersal in real time to particle streak photography (Breithaupt and Ayers 1996; Stamhuis and Videler 1995; Montgomery and Hamilton 1995), which provides an instantaneous, usually 2-D snapshot of the flow field. Although dye plumes are very useful for getting general information about the overall direction of flow, they have very short lifetimes before they diffuse into the water. In particle streak photography, small, neutrally buoyant particles are suspended in the fluid and illuminated with a plane of light. Their displacement over time ("streak") is then photographed or filmed. From streak lengths at fixed exposure times, flow velocities can be calculated. This technique is noninvasive and particularly useful for depicting complicated, unpredictable fields and overall spatial patterns of flow, but relatively cumbersome for extracting measurements of time-varying flow amplitudes. For surface wave motions, Bleckmann and colleagues (1981) have successfully applied a helium neon laser beam reflected from the water surface onto a position sensitive photodiode (Unbehauen 1980). Hot-film (Coombs et al. 1989b) and laser-Doppler (Bleckmann et al. 1991; Blickhan et al. 1992) anemometry provide relatively inexpensive and expensive tools, respectively, for measuring flow velocities at a single point in space. Of these, the latter is superior in that it is totally noninvasive. Both are useful for measuring temporal changes in flow amplitudes and for extracting amplitude spectra, but neither is very useful for mapping spatial patterns at a given instant in time. Miniature hydrophones have also recently been shown to be useful for mapping pressure gradient patterns about a dipole source (Coombs et al. 1995). In this application, both spatial and temporal patterns are extracted from a single hydrophone responding to the changing locations of a dipole source. Hydrophones respond to oscillating (AC) flows only, however, so they are not very useful for measuring DC flows.

4. Peripheral Processing

The lateral line system is a collection of small, mechanoreceptive patches called neuromasts, distributed on the head and trunk of all fishes and on larval and some postmetamorphic amphibians, such as *Xenopus laevis*, the clawed frog (Northcutt 1989). Each neuromast consists of a centrally located sensory epithelium, composed of hair cells, and surrounding support cells, both of which are overlaid by a gelatinous cupula (Fig. 8.7). Free or superficial neuromasts are found on the skin surface, usually in several specific locations and typically aligned in rows on the body of fishes (Fig. 8.8A) and amphibians (Fig. 8.8B). All neuromasts are superficial on amphibians, but fish additionally have neuromasts that are enclosed in fluid-filled canals just below the skin surface (Figs. 8.8, 8.9 and 8.10). The canal breaks through to the skin/water interface at periodic intervals through a series of tubules and/or pores, such that there is usually one neuromast between every two pores (Figs. 8.3A and 8.9A). The following sections summarize (1) the distribution, orientation, and innervation of superficial and canal neuromasts; (2) the biomechanical and physiological response properties of each type of neuromast; and (3) how pressure gradient patterns are encoded by the peripheral lateral line system.

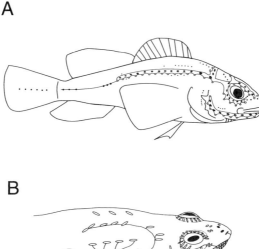

A

B

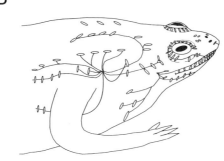

FIGURE 8.8. Distribution of superficial and canal neuromasts in on a teleost fish, the mottled sculpin (A), and of lateral line stitches (rows of superficial neuromasts) on an amphibian, the clawed frog (B, adapted from Shelton 1970, with permission of Company of Biologists.)

Superficial Neuromast

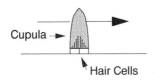

Cupula →

Hair Cells

Canal Neuromast

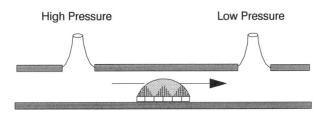

High Pressure   Low Pressure

FIGURE 8.9. Schematic diagram of superficial and canal neuromasts.

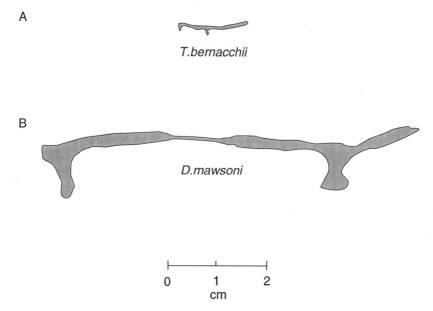

FIGURE 8.10. Plastic casts outlining the actual shapes of lateral line canals on the head of two species of antarctic fish, *Trematomus bernacchii* (A) and the giant antarctic cod, *Dissostichus mawsoni* (B). The projections on each cast represent the tubules connecting the canals to the exterior pore openings. (From Montgomery et al. 1994, with permission of S. Karger AG, Basel.)

4.1 Peripheral Anatomy: Neuromast Distribution, Orientation, and Innervation

In amphibians, short rows of up to 10 or so superficial neuromasts form a stitch (inset, Fig. 8.8B); several stitches in a row form lines along the head and trunk of the animal (Fig. 8.8A) (Harris and Milne 1966; Shelton 1970). In most amphibians, there are three lines on the trunk—the dorsal, medial, and ventral lines—and several on the head, including one above (supraorbital) and one below (infraorbital) the eye, and one along the lower jaw and one on the dorsal surface of the head. In general, lateral line canal distributions on fish follow a very similar pattern, except that superficial, rather than canal, neuromasts make up dorsal and ventral trunk lines (Northcutt 1989). Superimposed on this general pattern, however, is considerable variability in both the distribution and relative abundance of superficial and canal neuromasts among different fish species, especially teleosts (see Coombs et al. 1988 for review). Because all canal neuromasts originate as superficial neuromasts during development and because canal neuromasts are believed to be the most primitive condition, it is quite likely that paedomorphic truncation of canal development is responsible for at least some of this interspecific variability, including the total absence of canals in amphibians and the absence of dorsal and ventral canals on most fish (Northcutt 1989).

In the *Axolotl*, where the development and innervation of the lateral line has been extensively investigated (Northcutt et al. 1994, 1995; Northcutt and Brandle 1995), neuromasts around the eye are innervated by the anterodorsolateral line nerve and those on the ventral head by the anteroventral lateral line nerve. Neuromast lines behind the eye and across the top of the head are innervated by the middle and supratemporal lateral line nerves, respectively, and those on the trunk by branches of the posterior lateral line nerve. Each of these five nerves and the neuromasts they innervate originate from five separate lateral line placodes, which are also separate from the eighth nerve complex and the otic placode that gives rise to the inner ear (Northcutt et al. 1994; Northcutt and Brandle 1995). The situation in fishes is likely to be quite similar with at least three and perhaps as many as seven separate cranial nerves serving the lateral line system (Northcutt 1989; Song and Northcutt 1991a). Since so much of the older literature describes the lateral line system as being innervated by branches of other cranial nerves, such as the glossopharyngeal or facial nerves, it is worth emphasizing here that the lateral line system is innervated by its own set of cranial nerves, each with its own ganglia. Thus, in fish and amphibians with lateral line systems, there are far more than the traditional 12 cranial nerves described in most textbooks.

As the classic work of Flock (1965a,b) on the lateral line system of the burbot has shown, hair cells are both anatomically and physiologically polarized. The stereocilia at the apical end of each hair cell increase in length in a stepwise fashion, leading up to a single, eccentrically placed kinocilium (Fig. 8.7). In the mammalian ear, the kinocilium is absent, but the stepwise arrangement of stereocilia remains and underlies the basic physiological response properties of these cells. Flock and colleagues (Flock and Wersäll 1962; Flock 1965a,b, 1971) provided some of the first experimental evidence that bending of the hairs in the direction of the eccentrically placed kinocilium led to a depolarization of the hair cell and an increase in the firing rate of the innervating afferent fiber, whereas bending of the hairs in the opposite direction led to a hyperpolarizing response and a decrease in the firing rate.

The functional significance of this polarization has yet to be fully understood, but it is quite clear that hair cells in the lateral line system are arranged in one of two directions along the neuromast. This is most easily visualized in canal neuromasts, which are frequently elongated along the axis of the canal (Flock 1965; Coombs et al. 1988). In these neuromasts, the hair cell epithelium is also elongated along the canal axis, which defines the polarization pattern of the hair cells. Adjacent hair cells along the sensory strip typically have opposite orientations of their ciliary bundles such that one of them will respond best to water flow in one direction along the canal and the other to water flow in the opposite direction. Each of these hair cells are, in turn, innervated by separate afferent fibers, meaning that there are separate channels for encoding the direction of water movement inside the canal. The situation is a bit more complex for superficial neuromast lines in amphibians and teleosts in that the orientation of the hair cell epithelium is either parallel or perpendicular to the stitch or line axis (Harris and Milne 1966; Shelton 1970; Blaxter et al. 1983; Janssen et al. 1987; Harvey et al. 1992; Coombs and Montgomery 1994). Nevertheless, all neuromasts that have been examined in both amphibians and fish appear to have a single axis of hair cell orientation and to be innervated by a minimum of two afferent fibers—one for each of two hair cell polarities.

Given that neuromasts are spatially arrayed along the heads and bodies of fish and amphibians, it is reasonable to ask if neuromasts at different locations are innervated by separate groups of afferent fibers. Unfortunately, there are very few detailed anatomical data bearing on this question, but in the few cases where such data exist, this appears to be primarily the case. In the African cichlid fish, *Saratherodon niloticus*, individual trunk canal neuromasts, one per scale, may be innervated by as many as 20 different afferent fibers, but fewer than 4% of these appear to innervate adjacent canal neuromasts (Münz 1985). Similarly, although fibers branch to innervate several superficial neuromasts in a single trunk scale row, they do not innervate superficial neuromast rows on adjacent scales. Finally, fibers innervating canal neuromasts do not innervate nearby superficial neuromasts, and fibers innervating superficial neuromasts do not innervate canal neuromasts. In the African clawed frog, *Xenopus laevis*, the story appears to be somewhat similar with the majority of neuromast stitches being innervated by afferent fibers that do not contact neuromasts in adjacent stitches (Mohr and Görner, 1996). Neither of these studies, however, rules out the possibility that there may be a small percentage of fibers that integrate information across neuromasts at different locations.

4.2 Peripheral Biomechanics and Physiology

Although the anatomy of the peripheral lateral line system varies in a number of different dimensions, many of which may affect function (see Coombs et al. 1988 for review), one of the most obvious and perhaps most significant dimensions of variability is the absence or presence of canals. The biomechanics of canal and superficial neuromasts are essentially the same in that the cupulae of both, being of nearly the same density as the surrounding fluid, are driven primarily by viscous forces (Harris and Milne 1966; Flock 1971; von Netten and Kroese 1987, 1989; Kalmijn 1988, 1989; Denton and Gray 1988). That is, water flowing past the cupula causes it to move by virtue of friction coupling with the cupula surface. This means that the cupula response is largely proportional to the velocity of water flowing past it, and for this reason superficial neuromasts are often described as being sensitive to water velocity. This proportionality has basically been confirmed physiologically in the phase and amplitude responses of primary afferent fibers innervating superficial neuromasts in both fish (Coombs and Janssen 1990a; Kroese and Schellart 1992; Montgomery and Coombs 1992) and amphibians (Görner 1963; Kroese et al. 1978).

Because Denton and Dray (1982, 1983) have shown that flow velocity inside the canal is proportional to the net acceleration between the fish and the surrounding water, however, canal neuromasts are often described as being sensitive to water acceleration. Another way of thinking about canal biomechanics is that there has to be an external pressure gradient across the canal pores in order for fluid to flow inside the canal. Since there is usually one neuromast between every two canal pores, the response of each neuromast will be proportional to the pressure gradient across the two adjacent pores, as illustrated in Figure 8.3. No matter how one thinks about the effective stimulus, however, the bottom line is that there has to be relative movement between the cupula and surrounding water in order for the cupula to move and for the underlying hair cell cilia to be displaced; this is true for both superficial and canal neuromasts.

The major difference between the overall function of these two sub-classes of lateral line end organs lies in the filtering properties of canals and the subsequent frequency response of the neuromast. These filtering properties are illustrated in Figure 8.11A with frequency response functions from fibers innervating both superficial and canal neuromasts in an antarctic fish, *Dissosticus mawsoni* (Montgomery et al. 1994). To generate these functions, the responses of single fibers to a small, sinusoidally vibrating sphere (dipole source) were recorded for different vibration frequencies. As the frequency was increased, the pk-pk displacement of the source was decreased in order to maintain a constant pk-pk velocity across frequencies. As these results show, and discussed above, the response of superficial neuromast fibers is relatively constant and thus proportional to velocity up to around 20 to 30 Hz (solid line, Fig. 8.11A). Thus, in the velocity frame of reference, superficial neuromasts behave as low-pass filters with the low-pass filter residing in the biomechanical properties of the cupula and the fact that viscous forces dominate over inertial forces.

In contrast, the response function from canal neuromast fibers shows a general reduction in responsiveness at frequencies below 30 Hz and some amplification at or above 30 Hz relative to superficial neuromast responses (dashed line, Fig. 8.11A). The reduction at low frequencies is due primarily to the high-pass filtering effects of canals (Denton and Gray 1988). The high-pass nature of the canal comes from its increased internal surface area/volume ratio, which results in the formation of thick boundary layers (see Section 3.3) inside the canal at low frequencies. A neuromast cupula submerged in a thick boundary layer means that the flow velocity along the cupula is significantly reduced relative to the free stream velocity beyond

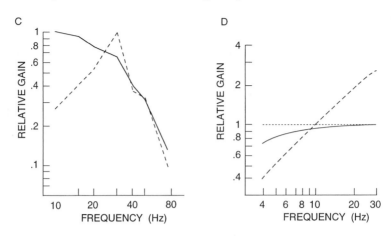

FIGURE 8.11. A: Frequency response functions (gain vs. frequency, constant velocity stimulus) for afferent fibers innervating superficial (solid line) and canal (dashed line) neuromasts in *D. mawsoni*. B: Modeled responses of water motion in the absence of a canal (horizontal line), inside a wide-bore canal of uniform width (solid line) and inside a wide-bore canal with a narrow section (dashed line) in response to a constant velocity stimulus. (From Montgomery et al. 1994, with permission of S. Karger AG, Basel.)

the boundary layer. As the frequency increases, however, and the boundary layer becomes thinner, the cupula is exposed to more of the free stream velocity. The slight amplification at higher frequencies is due to the narrow constriction of the canal in the vicinity of the neuromast (Fig. 8.10B), which causes an increase in the flow velocity, an effect that can be explained by the law of continuity (Vogel 1994). Figure 8.11B summarizes the filtering effects predicted for neuromasts in the absence of canals and in the presence of uniformly wide and non–uniformly wide canals.

While the frequency response of both superficial and canal neuromast fibers below about 30 Hz or so can be understood in terms of cupular and canal biomechanics, the steep decline in response at frequencies above this for both canal and superficial neuromasts is most likely due to later-stage filters residing at or beyond the hair cell (Coombs and Montgomery 1992). It is important to point out that we have illustrated the frequency response properties of canal neuromasts with data obtained from a cold-adapted antarctic fish, for which the high-frequency cutoff (around 30–50 Hz) may be temperature-limited (Coombs and Montgomery 1992). In temperate-water fish like the rainbow trout, *Salmo gairdneri* (Kroese and Schellart 1992) or the mottled sculpin, *Cottus bairdi* (Coombs and Janssen 1990a), the bandwidth of detection may extend much higher for canal neuromasts—up to about 100 Hz. In amphibians, like the clawed frog, *Xenopus laevis* (Kroese et al. 1978), the upper-frequency limit of superficial neuromasts (20–40 Hz) is about the same as that reported for superficial neuromasts in both cold-adapted (Montgomery and Coombs 1992; Coombs and Montgomery 1994; Montgomery et al. 1994) and temperature water fishes (Coombs and Janssen 1990a; Kroese and Schellart 1992). Nevertheless, in all cases it can be seen that the lateral line system is essentially a low-frequency system relative to the auditory system of most vertebrates, seldom responding to frequencies above about 150 Hz. Despite this low-frequency limitation, behavioral and physiological threshold curves from a number of different fish and amphibian species indicate that in terms of best displacement sensitivity, lateral line hair cells are as sensitive as auditory hair cells—responding to displacements in the nanometer range (Kroese and van Netten 1989).

4.3 Pressure-Gradient and Lateral-Line Excitation Patterns

The spatial distribution of lateral line end organs on the head and body of fish and amphibians suggest that this sensory system is designed to encode spatial patterns of activity across end organs. Figure 8.3 illustrates the pressure-gradient and, thus, excitation pattern that would exist across an array of canal neuromasts near a small (6 mm in diameter), 50-Hz dipole source. In this example, a lateral line canal is modeled as a series of sensors (neuromasts) with pressure sampling points, or pores, on either side; pore spacing (2 mm) is based on actual interpore distances measured on the trunk canal of goldfish and mottled sculpin. The pressure gradient pattern that exists along the sensors (solid line of Fig. 8.3B) is simply derived by computing the pressure difference that exists across each consecutive pair of pores. It turns out that the changes in the pressure gradient direction and amplitude that make up this pattern are faithfully encoded by single posterior lateral line nerves in both species (Coombs et al. 1995; Coombs and Conley 1997b). Pressure gradient patterns modeled similarly for a source moving at constant velocities are also encoded by lateral line nerves (Montgomery and Coombs 1998). This means that it is theoretically possible to predict lateral line excitation patterns if one knows (1) the course of lateral line canals on the animal's body, (2) the pressure sampling period (interpore spacing), and (3) the pressure distributions about the source. Whether these patterns are preserved and used by the central nervous system remains an open question at this stage of lateral line research.

5. Central Processing

The primary ascending projection of mechanosensory afferent neurons is to the medial octavolateralis nucleus (MON) located in the dorsolateral wall of the hindbrain. The MON is turn projects to its opposite partner (the contralateral MON), two secondary brain-stem nuclei, and to the midbrain. The details of the final portion of the ascending pathway are less well known, but information from the midbrain is relayed to the thalamus and from there to the telencephalon (Striedter 1991). It is reasonably clear that lateral line and auditory pathways in the CNS are largely parallel but separate—at least up to the level of the diencephalon. For a complete description of what is known about lateral line pathways in the CNS, the reader is referred to McCormick (1989). In this section, we focus on what is known about signal processing in the CNS—specifically with respect to the hindbrain and midbrain, for which we have the most information.

5.1 Hindbrain Processing

Lateral line signal processing in the hindbrain occurs primarily in the MON, although other brain-stem structures—the Mauthner cells and the octavolateralis efferent nucleus—are involved as well. The structure of the goldfish MON has been described in detail by New et al. (1996), and the distinctive similarities between the MON and other hindbrain nuclei processing electrosensory and auditory input is reviewed in Montgomery et al. (1995a). The key features of the organization of the MON are (1) a superficial molecular layer of parallel fibers derived from cerebellar granule cells, (2) a principal cell layer of large multipolar projection neurons that extend their apical dendrites into the overlying molecular layer, and (3) deeper layers where primary afferents and interneurons synapse with the ventral dendrites of the overlying principal cells (Fig. 8.12). All of these features are shared by hindbrain nuclei of allied senses: the electrosensory dorsal octavolateralis nucleus (DON) of most nonteleost fishes, the electrosensory lateral line lobe (ELL) of at least two independently evolved teleost lineages, and the dorsal cochlear nucleus (DCN) of the mammalian auditory system. A portion of the descending octaval nucleus of some teleosts may also have the same basic organization (McCormick and Bradford 1994; McCormick, Chapter 5). In the large gap between fish octavolateralis systems and the mammalian auditory system, first-order auditory nuclei with these features either do not exist or have not been closely studied in amphibians, birds, and reptiles.

The similarities in structure of these nuclei are matched by similarities in function (Montgomery et al. 1995a). In general terms, the receptive fields of principal cells and their selectivity for spatial (or frequency) characteristics of the stimuli are determined by afferent and interneuron inputs to the cell body and ventral dendrites. This generality has been demonstrated for DON, ELL, DCN and to some extent, the MON (Coombs et al. 1998). Due to the complex physical nature of lateral line stimuli (see Section 4), the receptive field characteristics of MON principal cells are less clear. The second major generality generated by comparative study (Montgomery et al. 1995a) is that dynamic signal conditioning occurs within the molecular

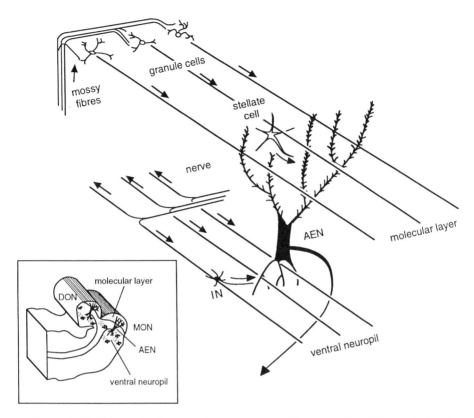

FIGURE 8.12. Schematic diagram of synaptic interactions occuring at the ventral and dorsal surface of an MON principal cell or ascending efferent neuron (AEN). Direct excitatory input from peripheral nerve fibers and indirect inhibitory input from small interneurons (IN) impinges on the basal dendritic surface of the AEN in the ventral neuropil. Descending input from cerebellar granule cells and stellate interneurons occurs in the molecular layer at the apical dendritic surface of the AEN. Inset at the lower left shows how these cells are situated in adjacent electrosensory (dorsal octavolateralis nucleus, DON) and mechanosensory (medial octavolateralis nucleus, MON) first-order nuclei in the brain stem of some amphibians and cartilaginous fish. (Reprinted from Montgomery and Bodznick 1994, with kind permission from Elsevier Science Ireland Ltd., Bay 15K, Shannon Industrial Estate, Co. Clare, Ireland.)

layer, which provides a substrate for adaptive, context-specific modification of principal cell responses. One aspect of this dynamic signal conditioning is the removal of reafference or sensory inputs associated with the animal's own activity (Bell et al. 1996). With respect to the MON, the finding means that principal cells learn to ignore stimuli associated with the animal's own movement (Fig. 8.13). If a vibration stimulus is triggered by opercular closure during ventilation, then after some minutes of coupling it ceases to be an effective stimulus for principal cells of the MON. A similar stimulus

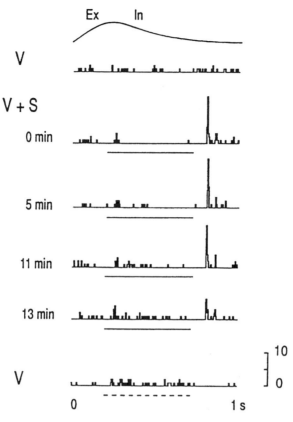

FIGURE 8.13. Second-order mechanosensory lateral line neuron in the MON of the scorpion fish illustrates adaptive cancellation of the response to a stimulus burst (S) (70 Hz vibrating ball 8 mm away from fish) coupled to the fish's ventilation (V). Top trace indicates exhalation (Ex) and inhalation (In) ventilation phases. Peristimulus time histogram records of evoked spikes, which were taken in the order presented, show what happens to the response over time. Initial presentation of the stimulus (solid line below histogram) causes an inhibition of activity followed by a excitatory burst at stimulus offset. Both the inhibitory and excitatory response decline over time. Note that after 13 minutes when the stimulus is abruptly turned off (dashed line), there is an apparent cancellation signal—the inverse of that first evoked by the stimulus plus ventilation. (Reprinted from Montgomery and Bodznick 1994, with kind permission from Elsevier Science Ireland Ltd., Bay 15K, Shannon Industrial Estate, Co. Clare, Ireland.)

uncoupled or uncorrelated with ventilation retains its efficacy. The basis of the adaptive filter is thought to be synaptic plasticity in the molecular layer which allows the formation of a cancellation signal that nulls the sensory reafference driven by movement (Montgomery and Bodznick 1994).

Given the spatial distribution and separate innervation, by location, of end organs in the lateral line system, an important question is whether terminals of peripheral fibers form a somatotopic map in the CNS. There is considerable evidence to show that there is a crude rostral/caudal map in the MON of different fish species (Claas and Münz 1981; McCormick 1981, 1983; Finger and Tong 1984; New and Northcutt 1984; Blubaum-Gronau and Münz 1987; DeRosa and Fine 1988; Puzdrowski 1989; Song and Northcutt 1991b; New and Singh 1994). That is, anterior lateral line nerves innervating head neuromasts terminate in a separate part of the nucleus, usually the ventromedial portion of the nucleus, whereas posterior lateral line nerves innervating trunk neuromasts tend to terminate in the dorsolateral portion. The precision of this map remains an open question, however, and a number of investigators using HRP tract tracing techniques have reported considerable overlap in the terminal fields of primary afferents. One interpretation of these results is that there is no precise somatotopic map in the CNS, but that the periphery is mapped in the CNS in some other way. A distinct separation of fiber terminals in the MON of the skate (Bodznick and Schmidt 1984) and the mottled sculpin (New, personal communication) into two groups suggests a possible substrate for preserving polarity differences of opposing hair cell populations at the brainstem level. Thus, phase maps, instead of or in addition to somatotopic maps, may help to explain some of the overlap that has been observed.

Golgi studies on the goldfish indicate that there are several cell types in the MON (New et al. 1996), including two distinct populations of principal output (crest) cells very similar to basilar and non-basilar pyramidal cells in the gymnotid ELL (Maler 1979). Recent neurophysiolgical studies on the responses of goldfish primary lateral line afferents and MON principal cells to different locations of a small, 50 Hz dipole source indicate that the receptive fields of some, but not all MON cells, can be modeled with either an excitatory center/inhibitory surround or an inhibitory center/excitatory surround organization of primary afferents onto principal cells (Coombs et al. 1998), as has been demonstrated for basilar and non-basilar pyramidal cells in the gymnotid ELL (Bastian 1981; Maler et al. 1981).

Studies on the responses of catfish (*Ancistrus*) and goldfish MON cells to a target moving past the fish also indicate that there are a variety of response types in the MON (Müller et al. 1995; Bleckmann et al. 1996). These studies show that most of these MON cells appear to be sensitive to target movement, with the average firing rate response increasing with increasing target velocity (above about 2 cm/s). Some MON cells show direction-specific responses consisting of an increase in firing rate preceded and followed by firing rate decreases when the target moves in one direction and a decrease in firing rate surrounded by firing rate increases when the target moves in the opposite direction.

In addition to the MON, there are other brainstem sites where lateral line information is processed. Primary afferent fibers also project to the large, bilaterally paired Mauthner cells, located along the midline of the brainstem beneath the fourth ventricle (Zottoli 1978). The projection of lateral line fibers to the Mauthner cell, which mediates escape behavior, is clear evidence of the importance of the lateral line in this behavior. However, Mauthner cells receive additional inputs from auditory and visual systems, which appear to dominate in the sense that an input from one of these two senses alone can excite the Mauthner cell beyond threshold, whereas this has not yet been demonstrated for the lateral line system (Zottoli and Danielson 1989). Thus, the precise role that the lateral line system plays in the escape response is presently unclear.

There are also secondary hindbrain nuclei, nucleus Praeminentialis and the perilemniscal nucleus, that receive projections from the MON (see Montgomery et al. 1995a for review). The nucleus Praeminentialis, in turn, provides descending inputs to the ipsilateral MON. Little is known about the contribution of the secondary hindbrain nuclei to information processing in the mechanosensory lateral line, but in electrosense these form part of a feedback system that acts as a gain control mechanism (Bastian 1986a,b).

Descending control of the lateral line system at the periphery is also thought to be mediated by efferent fibers, whose cell bodies reside in a single hindbrain nucleus, the octavolateralis efferent nucleus, in close association with the branchiomotor column and facial motor nucleus (see Roberts and Meredith 1989 for review of this system). Efferent fibers from this nucleus synapse onto the hair cells and afferent fibers of the auditory, vestibular, and lateral line systems and thus are capable of modulating primary afferent activity. Efferent effects on the lateral line system are generally inhibitory (Russell and Roberts 1972; Flock and Russell 1973), although excitatory effects have been reported for other octavolateralis systems (e.g., Boyle and Highstein 1990). It is currently postulated that the efferent system functions as part of a feedback or feed-forward regulatory system that controls hair cell sensitivity. In the former case, control is mediated by lateral line sensory reafference, and in the latter case by motor or sensory relay nuclei in the CNS. An example of the latter was provided by an elegant experiment by Tricas and Highstein (1991) in which responses from primary lateral line afferent and efferent fibers were recorded in free swimming toadfish, *Opsanus tau*, when the efferent system was activated with visual stimuli. These investigators showed that lateral line activity evoked by respiratory gill movements was significantly reduced when the efferent system was activated by stroboscopic illumination or presentation of natural prey, and that this reduction was not due to a reduction in the amplitude of gill movements. As these investigators discuss, activation of the efferent system appears to be closely associated with arousal behaviors and may serve to enhance the signal-to-noise processing capabilities of the sensory system when biologically relevant signal sources, like predators or prey, are at hand.

5.2 *Midbrain Processing*

The primary ascending output of the MON is to the torus semicircularis of the midbrain. Very little is known about lateral line signal processing per se in the torus, primarily because stimuli in physiological studies of the torus have been insufficiently defined to discriminate between auditory and lateral line cells. However, at least three types of acousticolateralis cells have been identified in the torus of the teleost midbrain on the basis of frequency response characteristics (Schellart and Kroese 1989). Of these, two types (low-frequency cells responding to <125 Hz and broad-band cells responding to frequencies both below and above 125 Hz) are potential candidates for receiving lateral line input. On the basis of HRP tract-tracing experiments, it is likely that information from auditory and lateral line systems are kept separate in the torus for at least some species, although there is clearly physiological and anatomical evidence for some overlap as well (McCormick 1989; Schellart and Kroese 1989).

Perhaps the best-known account with respect to midbrain processing lies in the tectum, which receives projections from the torus and is implicated in the representation of the position of sensory targets in space. In the clawed toad (*Xenopus laevis*), lateral line units recorded in layer 6 of the tectum respond to surface waves and show a sharp tuning for stimulus direction (Claas et al. 1989). The lateral line units are arranged topographically according to their receptive fields and form a map of directions on the water surface around the animal. As in the midbrain auditory maps of the barn owl (Konishi 1993), the lateral line midbrain map is not a simple topological projection from the sensory periphery, but is a computed map. The computational algorithm required to build up the map and its implementation by the nervous system are as yet unknown for the lateral line system.

6. Concluding Remarks

The notion of the lateral line system as a mere accessory auditory structure in fish and aquatic amphibians has long since been abandoned for a variety of reasons (Dijkgraaf 1963; Coombs et al. 1989a), including innervation by cranial nerves that are not part of the eighth nerve complex (Northcutt 1989), pathways in the CNS that are separate from auditory pathways (McCormick and Bradford Jr. 1988; Will 1988, 1989; McCormick 1989) and developmental origins from separate epidermal placodes that do not give rise to the inner ear (Northcutt 1986; Northcutt et al. 1994). Table 8.3 summarizes these and other functional differences between the lateral line and auditory systems of fish. Aside from the obvious feature that links these two systems together—the hair cell—perhaps the only remaining shared characteristic is that at frequencies below 200 Hz and at source distances less than one to two body lengths, both the lateral line and the ear will be

TABLE 8.3. Fish lateral line and auditory systems compared.

	Lateral line system	Auditory system	
Receptor organs	Superficial and canal neuromasts	Otolithic ear	Otolithic ear and air cavity
Receptor distribution	Dispersed on body surface	Clustered in cranial cavity	
Innervation	3 to 5 separate cranial nerves	Eighth nerve complex	
Effective stimulus	Differential movement between fish and surrounding water	Whole-body acceleration	Compression of air cavity
Stimulus encoding	Pressure gradient patterns	Acceleration	pressure fluctuations
Distance range	1 to 2 body lengths	10 body lengths	100 body lengths
Frequency range	<1 Hz to 200 Hz	<1 Hz to 500 Hz	<1 Hz to 2000 Hz

stimulated by many moving sources. For many fish, especially those without specialized connections between the ear and swim bladder (see Popper and Fay, Chapter 3), much of hearing is confined to these low frequencies. So the important question is, What does the lateral line system buy the fish that the ear doesn't? The simplest answer is that the spatial distribution of end organs and the biomechanical response properties of the lateral line system means that information in pressure gradient patterns, like that shown in Figure 8.3, is available through the lateral line system only. That is, the lateral line system is capable of extracting and encoding the spatial nonuniformities in the stimulus field, whereas the otolithic ear of the fish, responding only to acceleration of the whole animal (Popper and Fay, Chapter 3), extracts information about the amplitude and direction of acceleration (and perhaps phase, if the swim bladder is involved) at a single point in space (Kalmijn 1988, 1989). This is not to say that one system is superior to the other or that both haven't, for example, developed equally useful but different mechanisms for encoding source location and distance. It does, however, mean that the fine details present in pressure gradient patterns very close to the source are primarily the domain of the lateral line. Thus, unlike most vertebrate ears, information-processing "channels" in the lateral line system appear to be spatially, rather than spectrally, distributed.

Spectral tuning of sorts, however, does occur in the lateral line system. Superficial neuromasts respond best to frequencies below 30 Hz or so, whereas canal neuromasts respond best to frequencies above this point. Based on this kind of information, one might speculate that superficial neuromasts would be best at detecting low-frequency signals, such as those generated by the fish's own steady swimming movements (Table 8.1), but that the usefulness of superficial neuromasts in detecting exogenous signal sources, such as swimming prey, would be compromised by low frequency noise whenever the fish moves or finds itself in moving water (Montgomery et al. 1994). In this context, the primary function of the canal would be to improve the signal-to-noise ratio at higher frequencies—perhaps for detecting wakes behind swimming prey or submerged obstacles (Table 8.1).

Given that we can firmly establish significant functional differences between the lateral line and ear of aquatic vertebrates, why should students of the auditory system be interested in the lateral line and vice versa? In other words, why does this chapter appear in a series of volumes on vertebrate hearing? The answer is partly historical, in that the two systems have been linked as part of a single acoustico or octavolateralis system before the distinctions between them were clearly understood (e.g., van Bergeijk 1966, 1967). The answer is also partly practical, in that the superficial location of lateral line organs on the body of the fish make sensory organs and receptor cells readily assessable and visible. A prime example is the classic work of (Flock 1965a,b) on the lateral line system of the burbot, which established the anatomical and physiological basis of directional responses in hair cells. More recent examples include the usefulness of the lateral line system in studies of hair cell regeneration (Corwin et al. 1989; Jones and Corwin 1996), cell division (Lewis 1986; Winklebauer 1989; Winklebauer and Harsen 1983a,b, 1985a,b), sensory system development (Smith et al. 1990; Collazo et al. 1994; Northcutt et al. 1994, 1995; Northcutt and Brandle 1995), and hair cell diversity (Song and Popper 1994, 1995).

Finally, there may be other answers in store. One possibility that holds promise is the close resemblance between the organization of the lateral line brainstem nucleus and the DCN in terrestrial mammals, as summarized in Section 6 of this chapter and reviewed by Montgomery and colleagues (1995a). One of the major functions of this nucleus, as revealed primarily be research on the mechano- and electrosensory lateral lines (Bell et al. 1996), seems to be the cancellation of self-induced noise, such as that generated when fish ventilate by moving water through their mouth and over their gills. It is interesting to note that self-induced stimulation of the lateral line can be both a blessing and a curse. It forms the basis of hydrodynamic imaging in the blind cave fish, and possibly swimming optimization in herring and tuna, but it must also potentially mask the reception of external signals. With respect to the latter, one can recognize a cascade of behaviors, structures, and filters that would work together to reduce the masking problem: sit-and-wait predation strategies, modified slow movement, displacement of lateral line organs away from noise-generating fins (Dijkgraaf 1963), the enclosure of neuromasts in canals, efferent modulation at the periphery, and, finally, adaptive filtering by brainstem circuits. The problem of self-induced noise, which seems so obvious for the lateral line system and which has been such an active area of research, has received little or no attention for the closely related auditory system. Yet, as Montgomery and colleagues (1995a) point out, self-stimulation of the ear, through chewing, breathing, and even heart beats (Lewis and Henry 1992; Veluti et al. 1994) is a problem that even terrestrial vertebrates have to deal with. Thus, the lessons learned about the functions of the lateral line MON may very well provide clues about the functional significance of the similarly organized, but very enigmatic, DCN of the mammalian auditory system.

References

Ahlborn B, Harper DG, Blacke RW, Alborn D, Cam M (1991) Fish without footprints. J Theor Biol 148:521–533.

Bastian J (1981) Electrolocation II. The effects of moving objects and other electrical stimuli on the activities of two categores of posterior lateral line lobe cells in Apternotus albifrons. J Comp Physiol 144:481–494.

Bastian J (1986a) Gain control in the electrosensory system: a role for the descending projections to the electrosensory lateral line lobe. J Comp Physiol 158:505–515.

Bastian J (1986b) Gain control in the electrosensory system mediated by descending inputs to the electrosensory lateral line lobe. J Neurosci 6:553–562.

Bastian J (1986c) Electrolocation: behavior, anatomy, and physiology. In: Bullock TH, Heiligenberg W (eds) Electroreception. New York: Wiley, pp. 577–612.

Bauer JA Jr, Bauer SE (1981) Reproductive biology of pygmy angelfish of the genus Centropyge (Pomacanthidae). Bull Mar Sci 3:495–513.

Bell CC, Bodznick D, Montgomery JC, Bastian J (1996) Generation and subtraction of sensory expectations within cerebellar-like structures. Brain Behav Evol 1997;50(Suppl 1):17–31.

Bergeijk WA van (1964) Directional and nondirectional hearing in fish. In: Tavolga WN (ed) Marine Bio-Acoustics. Oxford: Pergamon Press, pp. 281–299.

Bergeijk WA van (1966) The evolution of the sense of hearing in vertebrates. Am Zool 6:371–377.

Bergeijk, WA van (1967) The evolution of vertebrate hearing. In: Neff WD (ed) Contributions to Sensory Physiology, New York: Academic, pp. 1–48.

Blaxter JHS, Gray JAB, Best ACG (1983) Structure and development of the free neuromasts and lateral line system of the herring. J Mar Biol Assoc UK 63:247–260.

Bleckmann H (1980) Reaction time and stimulus frequency in prey localization in the surface-feeding fish Aplocheilus lineatus. J Comp Physiol [A] 140:163–172.

Bleckmann H (1985) Perception of water surface waves: how surface waves are used for prey identification, prey localization and intraspecific communication. Prog Sensory Physiol 5:147–166.

Bleckmann H (1988) Prey identification and prey localization in surface-feeding fish and fishing spiders. In: Atema J, Fay RR, Popper AN, Tavolga W (eds) Sensory Biology of Aquatic Animals. New York: Springer-Verlag, pp. 619–641.

Bleckmann H (1993) Role of the lateral line in fish behaviour. In: Pitcher TJ (ed) Behavior of Teleost Fishes, 2nd ed. New York: Chapman & Hall, pp. 201–246.

Bleckmann H, Waldner I, Schwartz E (1981) Frequency discrimination of the surface-feeding fish *Aplochelius lineatus*—a prerequisite for prey localization? J Comp Physiol [A] 143:485–490.

Bleckmann H, Muller U, Hoin-Radkovski I (1984) Determination of source distance by the surface-feeding fishes *Aplocheilus lineatus* (Cyprinodontidae) and *Pantodon bucholzi* (Pantodontidae). In: Varju D, Schnitzler U (eds) Orientation and Localization in Engineering and Biology Heidelberg, New York: Springer, pp. 66–68.

Bleckmann H, Tittel G, Blubaum-Gronau E (1989) Lateral line system of surface-feeding fish: anatomy, physiology and behavior. In: Coombs S, Görner P, Münz H (eds) The Mechanosensory Lateral Line: Neurobiology and Evolution. New York, Berlin: Springer-Verlag, pp. 501–526.

Bleckmann H, Breithaupt T, Blickhan R, Tautz J (1991) The time course and frequency content of hydrodynamic events caused by moving fish, frogs, and crustaceans. J Comp Physiol [A] 168:749–757.

Bleckmann H, Mogdans J, Fleck A (1996) Integration of hydrodynamic information in the hindbrain of fishes. Mar Fresh Behav Physiol 27:77–94.

Blickhan R, Krick C, Zehren D, Nachtigall W (1992) Generation of a vortex chain in the wake of a subundulatory swimmer. Naturwissenschaften 79:220–221.

Blubaum-Gronau E, Münz H (1987) Topologische Reprasentation primarer Afferenzen einzelner Seitenlinienabschnitte beim Schmetterlingsfisch *Pantodon buchholzi* (in German). Verh Dtsch Zool Ges 80:268–269.

Bodznick D, Schmidt AW (1984) Somatotopy within the medullary electrosensory nucleus of the little skate, *Raja erinacea*. J Comp Neurol 225:581–590.

Boyle R, Highstein SM (1990) Efferent vestibular system in the toadfish: action upon horizontal canal afferents. J Neurosci 10:1570–1582.

Breithaupt T, Ayers J (1996) Visualization and quantitative analysis of biological flow fields using suspended particles. In: Lenz PH, Hartline DK, Purcell JE, Macmillan DL (eds) Zooplankton: Sensory Ecology and Physiology. Amsterdam: Gordon and Bheach, pp. 117–129.

Burgess JW, Shaw E (1981) Effects of acoustico-lateralis denervation in a facultative schooling fish: a nearest-neighbor matrix analysis. Behav Neural Biol 33:488–497.

Cahn PH (1972) Sensory factors in the side-to-side spacing and positional orientation of the tuna, *Euthynnus affinis*, during schooling. Fish Bull 70:197–204.

Campenhausen von C, Reiss I, Weissert R (1981) Detection of stationary objects by the blind cave fish *Anoptichthys jordani* (Characidae). J Comp Physiol [A] 143:369–374.

Claas B, Münz H (1981) Projection of lateral line afferents in a teleost's brain. Neurosci Lett 23:287–290.

Claas B, Münz H, Zittlau KE (1989) Direction coding in central parts of the lateral line system. In: Coombs S, Görner P, Münz H (eds) The Mechanosensory Lateral Line: Neurobiology and Evolution. New York, Berlin: Springer-Verlag, pp. 409–419.

Collazo A, Fraser SE, Mabee PM (1994) A dual embryonic origin for vertebrate mechanoreceptors. Science 264:426–430.

Coombs S (1995) Natural orienting behaviors for measuring lateral line function. In: Klump GM, Dooling RJ, Fay RR, Stebbins WC (eds) Methods in Comparative Psychoacoustics. Basel: Birkhauser Verlag, pp. 237–248.

Coombs S, Conley RA (1997a) Dipole source localization by mottled sculpin. I. Approach strategies. J Comp Physiol A 180:387–399.

Coombs S, Conley RA (1997b) Dipole source localization by mottled sculpin. II. The role of lateral line excitation patterns. J Comp Physiol A 180:401–415.

Coombs S, Fay RR (1993) Source level discrimination by the lateral line system of the mottled sculpin. J Acoust Soc Am 93:2116–2123.

Coombs S, Janssen J (1990a) Behavioral and neurophysiological assessment of lateral line sensitivity in the mottled sculpin, *Cottus bairdi.* J Comp Physiol [A] 167:557–567.

Coombs S, Janssen J (1990b) Water flow detection by the mechanosensory lateral line. In: Stebbins WC, Berkley MA (eds) Comparative Perception: Complex Signals, vol. 2. New York: Wiley, pp. 89–124.

Coombs S, Montgomery JC (1992) Fibers innervating different parts of the lateral line system of the Antarctic fish, *Trematomus bernacchii,* have similar neural responses despite large variations in peripheral morphology. Brain Behav Evol 40:217–233.

Coombs S, Montgomery J (1994) Function and evolution of superficial neuromasts in an antarctic notothenioid fish. Brain Behav Evol 44:287–298.

Coombs S, Janssen J, Webb JF (1988) Diversity of lateral line systems: evolutionary and functional considerations. In: Atema J, Fay RR, Popper AN, Tavolga WN (eds) Sensory Biology of Aquatic Animals. New York: Springer-Verlag, pp. 553–593.

Coombs S, Görner P, Münz H (1989a) A brief overview of the mechanosensory lateral line system and the contributions to this volume. In: Coombs S, Göner P, Münz H (eds) The Mechanosensory Lateral Line: Neurobiology and Evolution. New York, Berlin: Springer-Verlag, pp. 3–5.

Coombs S, Fay RR, Janssen J (1989b) Hot-film anemometry for measuring lateral line stimuli. J Acoust Soc Am 85:2185–2193.

Coombs S, Janssen J, Montgomery J (1991) Functional and evolutionary implications of peripheral diversity in lateral line systems. In: Webster D, Popper AM, Fay RR (eds) The Evolution of Hearing. New York: Springer-Verlag, pp. 267–288.

Coombs S, Hastings M, Finneran J (1995) Modeling and measuring lateral line excitation patterns to changing dipole source locations. J Comp Physiol [A] 178:359–371.

Coombs, S, Mogdans J, Halstead M, Montgomery J (In Press). Transformations of peripheral inputs by the first order brainstem nucleus of the lateral line system. J Comp Physiol A ••

Corwin JT, Balak KJ, Borden PC (1989) Cellular events underlying the renegenerative replacement of lateral line sensory epithelia in amphibians. In: Coombs S, Görner P, Münz H (eds) The Mechanosensory Lateral Line: Neurobiology and Evolution. New York, Berlin: Springer-Verlag, pp. 161–186.

Denton EJ, Gray JAB (1982) The rigidity of fish and patterns of lateral line stimulation. Nature 297:679–681.

Denton EJ, Gray JAB (1983) Mechanical factors in the excitation of clupeid lateral lines. Proc R Soc Lond [B] 218:1–26.

Denton EJ, Gray JAB (1988) Mechanical factors in the excitation of the lateral line of fishes. In: Atema J, Fay RR, Popper AN, Tavolga WN (eds) Sensory Biology of Aquatic Animals. New York: Springer-Verlag, pp. 595–617.

Denton EJ, Gray JAB (1993) Stimulation of the acoustico-lateralis system of clupeid fish by external sources and their own movements. Philos Trans R Soc Lond [B] 341:113–127.

Denton EJ, Gray JAB, Blaxter JHS (1979) The mechanics of the clupeid acoustico-lateralis system: frequency responses. J Mar Biol Assoc UK 59:27–47.

DeRosa F, Fine ML (1988) Primary connections of the anterior and posterior lateral line nerves in the oyster toadfish. Brain Behav Evol 31:312–317.

Dijkgraaf S (1963) The functioning and significance of the lateral-line organs. Biol Rev 38:51–105.

Elepfandt A (1982) Accuracy of taxis response to water waves in the clawed toad (*Xenopus laevis* Daudin) with intact or with lesioned lateral line system. J Comp Physiol [A] 148:535–545.

Elepfandt A, Seiler B, Aicher B (1985) Water wave frequency discrimination in the clawed frog, *Xenopus laevis*. J Comp Physiol [A] 157:255–261.

Enger PS, Kalmijn AJ, Sand O (1989) Behavioral identification of lateral line and inner ear function. In: Coombs S, Görner P, Münz H (eds) The Mechanosensory Lateral Line: Neurobiology and Evolution. New York, Berlin: Springer-Verlag, pp. 575–587.

Ertman SC, Jumars PA (1988) Effects of bivalve siphon currents on the settlement of inert particles and larvae. J Mar Res 46:797–813.

Fay RR (1988) Hearing in Vertebrates: A Psychophysics Databook. Winnetka: Hill-Fay Associates.

Finger TE, Tong SL (1984) Central organization of eighth nerve and mechanosensory lateral line systems in the brainstem of ictalurid catfish. J Comp Neurol 229:129–151.

Flock A (1965a) Electron microscopic and electrophysiological studies on the lateral line canal organ. Acta Otolaryngol Suppl S199:7–90.

Flock A (1965b) Transducing mechanisms in the lateral line canal organ receptors. Cold Spring Harb Symp Quant Biol 30:133–145.

Flock A (1971) Sensory transduction in hair cells. In: Lowenstein WR (ed) Handbook of Sensory Physiology, vol. 1: Principles of Receptor Physiology. Berlin, Heidelberg: Springer-Verlag, pp. 396–441.

Flock A, Russel IJ (1973) The postsynaptic action of efferent fibres in the lateral line organ of the burbot Lota lota. J Physiol 235:591–605.

Flock A, Wersäll J (1962) A study of the orientation of the sensory hairs of the receptor cells in the lateral line organs of fish, with special reference to the function of the receptors. J Cell Biol 15:19–27.

Görner P (1963) Untersuchungen zur Morpholgie und Elektrophysiologie des Seitenlinienorgans vom Krallenfrosch (*Xenopus laevis* Daudin). Z Vergl Physiol 47:316–338.

Görner P (1973) The importance of the lateral line system for the perception of surface waves in the clawed toad, *Xenopus laevis* Daudin. Experientia 29:295–296.

Görner P (1976) Source localization with the labyrinth and lateral-line in the clawed toad (*Xenopus laevis*). In: Schuijf A, Hawkins A (eds) Sound Reception in Fish. Amsterdam, New York: Elsevier, pp. 171–183.

Gray J (1984) Interaction of sound pressure and particle acceleration in the excitation of the lateral-line neuromasts of sprats. Proc R Soc Lond [B] 220:299–325.

Gray JAB, Best ACG (1989) Patterns of excitation of the lateral line of the ruffe. J Mar Biol Assoc UK 69:289–306.

Gray JAB, Denton EJ (1991) Fast pressure pulses and communication between fish. J Mar Biol Assoc UK 71:83–106.

Halstead MDB (1994) Detection and location of prey by the New Zealand freshwater Galaxiid, Galaxias fasciatus (Pisces: Salmoniformes). Master's thesis, University of Auckland.

Harris GG, Milne DC (1966) Input-output characteristics of the lateral-line sense organs of *Xenopus laevis*. J Acoust Soc Am 40:32–42.

Harvey R, Blaxter JHS, Hoyt RD (1992) Development of superficial and lateral line neuromasts in larvae and juveniles of plaice (*Pleuronectes platessa*) and sole (*Solea solea*). J Mar Biol Asoc UK 72:651–668.

Hassan ES (1985) Mathematical analysis of the stimulus for the lateral line organ. Biol Cybern 52:23–36.

Hassan ES (1986) On the discrimination of spatial intervals by the blind cave fish (*Anoptichthys jordani*). J Comp Physiol 159:701–710.

Hassan ES (1989) Hydrodynamic imaging of the surroundings by the lateral line of the blind cave fish *Anoptichthys jordani*. In: Coombs S, Görner P, Münz H (eds) The Mechanosensory Lateral Line: Neurobiology and Evolution. New York, Berlin: Springer-Verlag, pp. 217–227.

Hassan ES (1992a) Mathematical description of the stimuli to the lateral line system of fish derived from a three-dimensional flow field analysis. I. The case of moving in open water and of gliding towards a plane surface. Biol Cybern 66:443–452.

Hassan ES (1992b) Mathematical description of the stimuli to the lateral line system of fish derived from a three-dimensional flow field analysis. II. The case of gliding alongside or above a plane surface. Biol Cybern 66:453–461.

Hassan ES, Abdel-Latif H, Biebricher R (1992) Studies on the effects of Ca^{++} and Co^{++} on the swimming behavior of the blind Mexican cave fish. J Comp Physiol [A] 171:413–419.

Hoekstra D, Janssen J (1985) Non-visual feeding behavior of the mottled sculpin, *Cottus bairdi*, in Lake Michigan. Environ Biol Fish 12:111–117.

Hoekstra D, Janssen J (1986) Lateral line receptivity in the mottled sculpin (*Cottus bairdi*). Copeia 1986:91–96.

Hofer B (1908) Studien uber die hautsinnesorgane der fische. I. Die funktion der seitenorgane bei den fischen. Ber Kgl Bayer Biol Versuchsstation Munchen 1:115–164.

Hoin-Radkovsky I, Bleckmann H, Schwartz E (1984) Determination of source distance in the surface-feeding fish *Pantodon buchholzi* Pantodontidae. Anim Behav 32:840–851.

Janssen J (1990) Localization of substrate vibrations by the mottled sculpin (*Cottus bairdi*). Copeia 1990:349–355.

Janssen J, Coombs S, Hoekstra D, Platt C (1987) Anatomy and differential growth of the lateral line system of the mottled sculpin *Cottus bairdi* (Scorpaeniformes: Cottidae). Brain Behav Evol 30:210–229.

Janssen J, Sideleva V, Montgomery J (1991) Under-ice observations of fish behavior at McMurdo Sound. Antarctic J US 26:174–175.

Janssen J, Jones WR, Whang A, Oshel PE (1995) Use of the lateral line in particulate feeding in the dark by juvenile alewife (*Alosa pseudoharengus*). Can J Fish Aquat Sci 52:358–363.

Jones JE, Corwin JT (1996) Regeneration of sensory cells after laser ablation in the lateral line system: hair cell lineage and macrophage behavior revealed by time-lapse video microscopy. J Neurosci 16:649–662.

Kalmijn AJ (1988) Hydrodynamic and acoustic field detection. In: Atema J, Fay RR, Popper AN, Tavolga WN (eds) Sensory Biology of Aquatic Animals. New York: Springer-Verlag, pp. 83–130.

Kalmijn AJ (1989) Functional evolution of lateral line and inner-ear sensory systems. In: Coombs S, Görner P, Münz H (eds) The Mechanosensory Lateral Line: Neurobiology and Evolution. New York, Berlin: Springer-Verlag, pp. 187–215.

Katz LC, Potel MJ, Wassersug RJ (1981) Structure and mechanisms of schooling in tadpoles of the clawed frog, *Xenopus laevis*. Anim Behav 29:20–33.

Konishi M (1993) Listening with two ears. Sci Am 268:34–41.

Kroese ABA, Netten SM van (1989) Sensory transduction in lateral line hair cells. In: Coombs S, Görner P, Münz H (eds) The Mechanosensory Lateral Line: Neurobiology and Evolution. New York, Berlin: Springer-Verlag, pp. 265–284.

Kroese ABA, Schellart N (1992) Velocity- and acceleration-sensitive units in the trunk lateral line of the trout. J Neurophysiol 68:2212–2221.

Kroese ABA, van der Zalm JM, Van den Bercken J (1978) Frequency response of the lateral-line organ of *Xenopus laevis*. Pfluegers Arch 375:167–175.

LaBarbera M (1981) Water flow patterns in and around three species of articulate brachiopods. J Exp Mar Biol Ecol 55:185–206.

Lewis J (1986) Programs of cell division in the lateral line system. TINS 4:135–138.

Lewis ER, Henry KR (1992) Modulation of cochlear nerve spike rate by cardiac activity in the gerbil. Hear Res 63:7–11.

Lighthill J (1993) Estimates of pressure differences across the head of a swimming clupeid fish. Philos Trans R Soc Lond [B] 341:129–140.

Magnuson JJ (1978) Locomotion by scrombrid fishes: hydromechanics, morphology and behavior. In: Hoar WS, Randall DJ (eds) Fish Physiology: Locomotion. New York: Academic Press, pp. 239–313.

Maler L (1979) The posterior lateral line lobe of certain gymnotoid fish: quantitative light microscopy. J Comp Neurol 183:323–363.

Maler L, Sas E, Rogers J (1981) The cytology of the posterior lateral line lobe of high frequency electric fish (Gymnotidae): dendritic differentiation and synaptic specificity in a simple cortex. J Comp Neurol 158:87–141.

McCormick CA (1981) Central projections of the lateral line and eighth nerves in the bowfin, *Amia calva*. J Comp Neurol 197:1–15.

McCormick CA (1983) Central connections of the octavolateralis nerves in the pike cichlid, *Crenicichla lepidota*. Brain Res 265:177–185.

McCormick CA (1989) Central lateral line mechanosensory pathways in bony fish. In: Coombs S, Görner P, Münz H (eds) The Mechanosensory Lateral Line: Neurobiology and Evolution. New York, Berlin: Springer-Verlag, pp. 341–364.

McCormick CA, Bradford MR Jr (1988) Central connections of the octavolateralis system: evolutionary considerations. In: Atema J, Fay RR, Popper AN, Tavolga WN (eds) Sensory Biology of Aquatic Animals. New York: Springer-Verlag, pp. 733–756.

McCormick CA, Bradford MR Jr (1994) Organization of inner ear endorgan projections in the goldfish, *Carassius auratuos*. Brain Behav Evol 43:189–205.

Merzkirch W (1987) Flow Visualization. Orlando, FL: Academic Press.

Mohr C, Görner P (1996) Innervation patterns of the lateral line stitches of the clawed toad *Xenopus laevis Daudin* and their reorganization during metamorphosis. Brain Behav Evol 48:55–69.

Montgomery JC, Bodznick D (1994) An adaptive filter that cancels self-induced noise in the electrosensory and lateral line mechanosensory systems of fish. Neurosci Lett 174:145–148.

Montgomery JC, Coombs S (1992) Physiological characterization of lateral line function in the Antarctic fish (*Trematomus bernacchii*). Brain Behav Evol 40:209–216.

Montgomery JC, Coombs S (1998) Peripheral encoding of moving sources by the lateral line system of a sit-and-wait predator. J Exp Biol 201(1):91–102.

Montgomery JC, Hamilton AR (1997) The sensory biology of prey capture in the dwarf scorpion fish (*Scorpaena papillosus*). Mar Fresh Behav Physio 30:209–223.

Montgomery JC, Milton RC (1993) Use of the lateral line for feeding in the torrentfish (*Cheimarrichthys fosteri*). N Z J Zool 20:121–125.

Montgomery JC, Saunders AJ (1985) Functional morphology of the piper *Hyporhamphus ihi* with reference to the role of the lateral line in feeding. Proc R Soc Lond [B] 224:197–208.

Montgomery JC, Baker CF, Carton AG (1997) The lateral line can mediate rheotaxis in fish. Nature 389:960–963.

Montgomery JC, Coombs S, Janssen J (1994) Form and function relationships in lateral line systems: comparative data from six species of antarctic notothenioid fish. Brain Behav Evol 44:299–306.

Montgomery JC, Skipworth E (1997) Detection of weak water jets by the short-tailed stingray *Dasyatis brevicaudatus* (Pisces: Dasyatididae). Copeia 1997:881–883.

Montgomery JC, Coombs S, Conley RA, Bodznick D (1995a) Hindbrain sensory processing in lateral line, electrosensory and auditory systems: a comparative overview of anatomical and functional similarities. Audit Neurosci 1:207–231.

Montgomery JC, Coombs S, Halstead M (1995b) Biology of the mechanosensory lateral line. Rev Fish Biol Fisheries 5:399–416.

Müller HM, Fleck A, Bleckmann H (1996) The responses of central octavolateralis cells to moving sources. J Comp Physiol [A] 179:455–471.

Münz H (1985) Single unit activity in the peripheral lateral line system of the cichlid fish *Sarotherodon niloticus* L. J Comp Physiol [A] 157:555–568.

Netten SM von, Kroese ABA (1987) Laser interferometric measurements on the dynamic behaviour of the cupula in the fish lateral line. Hear Res 29:55–61.

Netten SM von, Kroese ABA (1989) Dynamic behavior and micromechanical properties of the cupula. In: Coombs S, Görner P, Münz H (eds) The Mechanosensory Lateral Line: Neurobiology and Evolution. New York, Berlin: Springer-Verlag, pp. 247–263.

New J, Northcutt RG (1984) Central projections of the lateral line nerves in the shovelnose sturgeon. J Comp Neurol 225:129–140.

New JG, Singh S (1994) Central topography of anterior lateral line nerve projections in the channel catfish, Ictalurus *punctatus*. Brain Behav Evol 43:34–50.

New JG, Coombs S, McCormick CA, Oshel PE (1996) Cytoarchitecture of the medial octavolateralis nucleus in the goldfish, *Carassius auratus*. J Comp Neurol 364:1–13.

Northcutt RG (1986) Evolution of the octavolateralis system: evaluation and heuristic value of phylogenetic hypotheses. In: Ruben RJ, Van De Water TR, Rubel EW (eds) The Biology of Change in Otolaryngology (Proc./9th ARO Medwinter Research Mtg., 1986). Amsterdam, New York: Elsevier, pp. 3–14.

Northcutt RG (1989) The phylogenetic distribution and innervation of craniate mechanoreceptive lateral lines. In: Coombs S, Görner P, Münz H (eds) The Mechanosensory Lateral Line: Neurobiology and Evolution. New York, Berlin: Springer-Verlag, pp. 17–78.

Northcutt RG, Brandle K (1995) Development of branchiomeric and lateral line nerves in the axolotl. J Comp Neurol 355:427–454.

Northcutt RG, Catania KC, Criley BB (1994) Development of lateral line organs in the axolotyl. J Comp Neurol 340:480–514.

Northcutt RG, Brandle K, Fritzsch B (1995) Electroreceptors and mechanosensory lateral line organs arise from single placodes in Axolotls. Dev Biol 168:358–373.

O'Brian WJ, Browman HI, Evans BI (1990) Search strategies and foraging of animals. Am Sci 78:152–160.

Partridge B, Pitcher TJ (1980) The sensory basis of fish schools: relative roles of lateral line and vision. J Comp Physiol [A] 135:315–325.

Peterson CH, Quammen ML (1982) Siphon nipping: its importance to small fishes and its impact on growth of the bivalve Protothaca staminea (Conrad). J Exp Mar Biol Ecol 63:249–268.

Pitcher TJ, Partridge BL, Wardle CS (1976) A blind fish can school. Science 194:963–965.

Pollak GD, Casseday JH (1989) The Neural Basis of Echolocation in Bats. Heidelberg: Springer-Verlag.

Poulson T (1963) Cave adaptation in amblyopsid fishes. Am Midl Nat 70:257–290.

Price RE, Schiebe MA (1978) Measurements of velocity from excurrent siphons of freshwater clams. Nautilus 92:67–69.

Puzdrowski RL (1989) Peripheral distribution and central projections of the lateral-line nerves in goldfish, *Carassius auratus*. Brain Behav Evol 34:110–131.

Roberts BL, Meredith GE (1989) The efferent system. In: Coombs S, Görner P, Münz H (eds) The Mechanosensory Lateral Line: Neurobiology and Evolution. New York, Berlin: Springer-Verlag, pp. 445–460.

Rosen MW (1959) Water flow about swimming fish. U.S. Naval Ord Test Stn Techn Proc 2298:1–94.

Rowe DM, Denton EJ, Batty RS (1993) Head turning in herring and some other fish. Philos Trans R Soc Lond [B] 341:141–148.

Russell IJ, Roberts BL (1972) Inhibition of spontaneous lateral line activity by efferent nerve stimulation. J Exp Biol 57:77–82.

Satou M, Takeuchi H, Takei K, Hasegawa T, Okumoto N, Ueda K (1987) Involvement of vibrational and visual cues in eleciting spawning behaviour in male Hime salmon (landlocked red salmon, *Oncorhynchus nerka*). Anim Behav 35:1556–1584.

Satou M, Shiraishi A, Matsushima T, Okumoto N (1991) Vibrational communication during spawning behavior in the Hime salmon (landlocked red salmon, *Oncorhynchus nerka*). J Comp Physiol [A] 168:417–428.

Satou M, Takeuchi HA, Takei K, Hasegawa T, Matsushima T, Okumoto N (1994a) Characterization of vibrational and visual signals which elicit spawning behavior in the male hime salmon (landlocked red salmon, *Oncorhynchus nerka*). J Comp Physiol [A] 174:527–537.

Satou M, Takeuchi HA, Tanabe M, Kitamura S, Okumoto N, Iwata M, Nishii J (1994b) Behavioral and electrophysiological evidences that the lateral line is involved in the inter-sexual vibrational communcation of the hime salmon (landlocked red salmon, *Oncorhynchus nerka*). J Comp Physiol [A] 174:539–549.

Saunders AJ, Montgomery JC (1985) Field and laboratory studies of the feeding behavior of the piper *Hyporhamphus ihi* with reference to the role of the lateral line in feeding. Proc R Soc Lond [B] 224:209–221.

Schellart NM, Kroese ABA (1989) Interrelationship of acousticolateral and visual systems in the teleost midbrain. In: Coombs S, Görner P, Münz H (eds) The Mechanosensory Lateral Line: Neurobiology and Evolution. New York, Berlin: Springer-Verlag, pp. 421–444.

Schwartz E (1967) Analysis of surface-wave perception in some teleosts. In: Cahn PH (ed) Lateral Line Detectors. Bloomington, IN: Indiana University Press, pp. 123–134.

Shaw E (1969) The duration of schooling among fish separated and those not separated by barriers. Am Mus Novit 2373:1–13.

Shelton PMJ (1970) The lateral line system at metamorphosis in *Xenopus laevis* (Daudin). J Embryol Exp Morphol 24:511–524.

Smith SC, Lannoo MJ, Armstrong JB (1990) Development of the mechanoreceptive lateral-line system in the axolotl: placode specification, guidance of migration, and the origin of neuromast polarity. Anat Embryol 182:171–180.

Song J, Nortcutt RG (1991a) Morphology, distribution and innervation of the lateral-line receptors of the Florida gar, *Lepisosteus platyrhincus*. Brain Behav Evol 37:10–37.

Song J, Northcutt RG (1991b) The primary projections of the lateral-line nerves of the Florida gar, *Lepisosteus platyrhincus*. Brain Behav Evol 37:38–63.

Song J, Popper AN (1994) Two types of hair cells in the mechanosensory lateral line receptors: evidence based on ototoxicity sensitivity. Abstracts of the 18th Midwinter Meeting of the Association of Res Otolaryngology, vol. 75.

Song J, Popper AN (1995) Differences in postembryonic growth and ultrastructure of sensory and supporting cells in canal and superficial neuromasts of the lateral line. Abstracts of the 18th Midwinter Meeting of the Association of Res Otolaryngology, vol. 44.

Stamhuis EJ, Videler JJ (1995) Quantitative flow analysis around aquatic animals using laser sheet particle image velocimetry. J Exp Biol 198:283–294.

Striedter GF (1991) Auditory, electrosensory and mechanosensory lateral line pathways through the forebrain in channel catfishes. J Comp Neural 312:311–331.

Sutterlin AM, Waddy S (1975) Possible role of the posterior lateral line in obstacle entrainment by brook trout (*Salvelinus fontinalis*). J Fish Res Bd Canada 32:2441–2446.

Takeuchi H, Namba H (1989) Feeding behavior and lateral line system in axolotl, *Ambystoma mexicanum*. Zool Sci 6:1216.

Takeuchi H, Namba H (1990) Feeding behavior and lateral line system in axolotl, *Ambystoma mexicanum*. II. Effects of blocking the mechanosensory lateral line system by cobalt ions. Zool Sci 7:1177.

Takeuchi H, Nakamura S, Nagai T (1991) Feeding behavior of the eyeless mutant in the axolotl, *Ambystoma mexicanum*. Zool Sci 8:1190.

Teyke T (1985) Collision with and avoidance of obstacles by blind cave fish *Anoptichthys jordani* (Characidae). J Comp Physiol [A] 157:837–843.

Teyke T (1988) Flow field, swimming velocity and boundary layer: parameters which affect the stimulus for the lateral line organ in blind fish. J Comp Physiol [A] 163:53–61.

Tittel G (1988) Untersuchungen zum Beutefangverhalten des Streifenhechtlings Aplocheilus lineatus (Cu. & Val.). Ein Modell zur Richtungsdetermination unter Berucksichtigung von ontogenetischen Prozessen des peripheren Seitenliniensystems (Ph.D. Thesis), Giessen, W. Germany: Justus-Liebig-Universitat Giessen.

Triantafyllou MS, Triantafyllou GS (1995) An efficient swimming machine. Sci Am 64–70.

Tricas TC, Highstein SM (1991) Action of the octavolateralis efferent system upon the lateral line of free-swimming toadfish, *Opsanus tau*. J Comp Physiol [A] 169:25–37.

Unbehauen H (1980) Morphologische und elektrophysiologische Untersuchungen zur Wirkung von Wasserwellen auf das Seitenlinienorgan des Streifenhechtlings (Aplocheilus lineatus) (Ph.D. Thesis), Giessen, W. Germany: Justus-Liebig-Universitat Giessen.

Velluti RA, Pena JL, Pedmonte M, Narins PM (1994) Internally-generated sound stimulates cochlear nucleus units. Hear Res 72:19–22.

Vogel S (1994) Life in Moving Fluids: the Physical Biology of Flow, 2nd ed. Princeton, NJ: Princeton University Press.

Weissert R, Campenhausen von C (1981) Discrimination between stationary objects by the blind cave fish *Anoptichthys jordani* (Characidae). J Comp Physiol 143:375–381.

Will U (1988) Organization and projections of the area octavolateralis in amphibians. In: Fritzsch B, Ryan M, Wilczynski W, Hetherington T (eds) The Evolution of the Amphibian Auditory System. Chichester, NY: Wiley, pp. 185–208.

Will U (1989) Central mechanosensory lateral line system in amphibians. In: Coombs S, Görner P, Münz H (eds) The Mechanosensory Lateral Line: Neurobiology and Evolution. New York, Berlin: Springer-Verlag, pp. 365–386.

Winklbauer R (1989) Development of the lateral line system in Xenopus. Progress in Neurobiology 32:181–206.

Winklbauer R, Hausen P (1983) Development of the lateral line system in Xenopus laevis. I. Normal development and cell movement in the supraorbital system. J Embryol Exp Morph 76:265–281.

Winklbauer R, Hausen P (1983) Development of the lateral line system in Xenopus laevis. II. Cell multiplication and organ formation in the supraorbital system. J Embryol Exp Morphol 76:283–296.

Winklbauer R, Hausen P (1985) Development of the lateral line system in Xenopus laevis. IV. Pattern formation in the supraorbital system. J Embryol Exp Morphol 88:193–207.

Winklbauer R, Hausen P (1985) Development of the lateral line system in Xenopus laevis. III. Development of the supraorbital system in triploid embryos and larvae. J Embryol Exp Morphol 88:183–192.

Zotolli SJ (1978) Comparative morphology of the Mauthner cell in fish and amphibians. In: Faber D, Korn H (eds) Neurobiology of the Mauthner Cell. New York: Raven Press, pp. 13–45.

Zottoli SJ, Danielson PD (1989) Lateral line afferent and efferent systems of the goldfish with special reference to the Mauthner cell. In: Coombs S, Görner P, Münz H (eds) The Mechanosensory Lateral Line: Neurobiology and Evolution. New York, Berlin: Springer-Verlag, pp. 461–480.

9
Acoustic Communication in Fishes and Frogs

Randy Zelick, David A. Mann, and Arthur N. Popper

1. Introduction

Many fish and amphibian species use sounds for communication in a wide range of behavioral and environmental contexts. The behaviors most often associated with acoustic communication in both groups include territorial behavior, mate finding, courtship, and aggression. Unlike most other communication channels (e.g., chemical, visual, touch), sound provides a means for long-distance communication as well as for communication in areas where there is poor visibility. Both fishes and frogs tend to use fairly broad-band pulsed sounds, although in both groups there are species known to use narrow bands of noise or even relatively pure tones.

Although there are a number of similarities in the uses and features of sound in both fishes and amphibians, direct comparisons between the groups is difficult for several reasons. Of these, the most significant is the substantial difference in what we know about acoustic communication in the two groups. The basis for this difference arises not from the potential breadth of communications involving sounds in the two groups, but more from the difficulties in studying acoustic communication in an aquatic versus a terrestrial environment. An appropriate analogy here might be that the difficulties in studying fish bioacoustics are only paralled by the difficulties in studying bioacoustics of marine mammals, while amphibian bioacoustics parallels the study of bird communication.

The problems in studying fish bioacoustics arise from the difficulty of finding and seeing the subjects of study. While it has been known for some time that the marine environment is quite noisy (reviewed in Tavolga 1971; also see Section 2.1), it still requires a good deal of equipment to record from fishes, and divers or underwater video systems to observe behavior. Even if these problems can be overcome, hydrophones (underwater microphones) and underwater observers are poor at localizing sounds in water, and so it is not easy to tell exactly which animal in a group (or even spread over a reef) is a sound producer. Not until very recently, with the limited availability of very expensive equipment, has it become possible to determine the source of a sound in a natural underwater environment.

Given the relative ease in studying the acoustics of amphibians and the difficulties in studying acoustics of fishes, there is a far greater database for amphibians than for fishes, not only in the absolute numbers of species surveyed and the amount of information available for each species, but also in the fraction of each group that has been studied to determine if it uses acoustic communication. Moreover, we have a good sense of sounds, strategies of behavior, detection mechanisms, and central processing of sounds by amphibians. In fishes we know most about sounds, sound production mechanisms, and detection capabilities (see Popper and Fay, Chapter 3; Lewis and Narins, Chapter 4; Feng and Schellart, Chapter 6; and Fay and Megela Simmons, Chapter 7), but little about acoustic processing in the CNS (see Feng and Schellart, Chapter 6) and the behavioral contexts in which sounds are used.

While it might have been appropriate to put fishes and amphibians in separate chapters, it is of some interest to treat the two groups together to provide some comparison and contrast between the groups, and to use the information for guidance as to where future studies in the each group might go, based on what we know of the other group. Accordingly, we have divided the chapter into sections on fishes and amphibians. The information is then brought together in a summary section in which future research directions are considered.

2. Fishes

2.1 History of Studies of Fish Sound Production

Early work (1800s–1940s) on fish sound production focused on identifying sonic fishes and the mechanisms of sound production. Many of these fish sounds were associated with reproduction. Darwin (1874) hypothesized, "In this, the lowest class of the Vertebrata, as with so many insects and spiders, sound-producing instruments have, at least in some cases, been developed through sexual selection, as a means for bringing the sexes together" (p. 367).

Research on fish sound production has followed two lines of inquiry: the behavioral ecology of fish sound production and the physiology of sound production mechanisms. The behavioral ecology studies have continued from the earliest studies of identifying sound-producing fishes, but recent work has extended questions of identification with attempts to develop an understanding of variation in fish sounds and the timing of fish sound production. Physiological studies of sound production have gone from simply understanding the physiological control mechanisms of sound production to understanding the bases of sex and species differences in sonic behavior.

Previous reviews of fish sound production have focused on different paths of inquiry, and are important to consult. Tavolga (1977) reviewed and catalogued early scientific studies on fish sound production. Myrberg (1981) reviewed fish sound production in the light of communication theory, and also in the context of data on the hearing abilities of fishes (Myrberg 1981; Hawkins and Myrberg 1983). This section provides a review of all aspects of fish sound production, giving the reader several avenues into the literature on fish sound production and identifying important questions that remain unresolved.

2.2 The Acoustic Structure of Fish Sounds and Mechanisms of Sound Production

2.2.1 Sonic Muscles and Swim Bladder Sounds

The air-filled swim bladder, which is located in the abdominal cavity and used to regulate buoyancy in most fishes, is an integral part of sound production in many species because of its compressibility (it is also used in hearing; see Popper and Fay, Chapter 3; Fay and Megela Simmons, Chapter 7). Muscles on the swim bladder, or muscles connected to bones around the swim bladder, are used by many fishes to produce sound. These muscles are either intrinsic (in which the muscle is entirely on the swim bladder) or extrinsic (in which one end inserts on bone and the other end either on bone or on the swim bladder) (Schneider 1967) (Fig. 9.1). Contractions of these muscles lead to rapid changes in the volume of the swim bladder, which in turn produces sound.

The sounds of most fishes with sonic muscles, including all deep-sea fishes, have not been described. Even for species whose sounds have been described, there are often many closely related species that have not been studied. For example, of 69 species of toadfishes (Batrachoididae), the sounds of only four species are known (*Opsanus tau, O. beta, O. phobetron,* and *Porichthys notatus*) (Tavolga 1958b; Fish and Mowbray 1970; Ibara et al. 1983). Of 209 species of cusk-eels (Ophidiidae), only the striped cusk eel (*Ophidion marginatum*) has been studied acoustically (Mann et al. 1997). Of 285 rattails (Macrouridae), none has been studied. Still, despite this lack of data, generalizations can be made from the species that have been recorded.

Fishes with sonic muscles on the swim bladder produce pulsed or tonal sounds depending on the pattern of muscle contraction (Fig. 9.2). One species of toadfish, *P. notatus* (midshipman) produces tonal sounds, referred to as boatwhistles, that range in frequency from about 80 to 180 Hz, and that can last for several minutes to hours (Ibara et al. 1983; Bass and

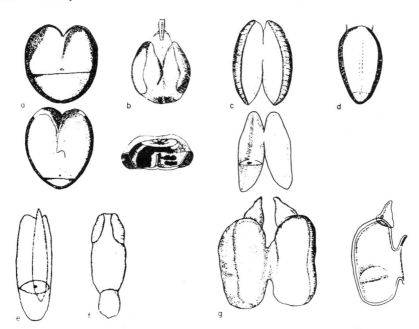

FIGURE 9.1. Sound-producing systems using intrinsic muscles and diaphragms in the swim bladder from a variety of species. (A) *Opsanus tau*, redrawn from Tower (1908); (B) *Porichthys notatus*, redrawn from Green (1924) (From Schneider 1967); (C) *Prionotus carolinus*, redrawn from Tower (1908); (D) *Trigla lineata*, redrawn from Rauther (1945); (E) *Sebasticus marmoratus*, redraw from Dôtu (1951); (F) *Zeus faber*, redrawn from DuFossé (1874); (G) *Dactylopterus volitans*, redrawn from DuFossé (1874). (From Schnieider 1967)

Baker 1990). The dominant frequency of the sound produced by toadfishes (*O. tau* and *P. notatus*) and Scorpaeniformes (searobins and sculpins) (*Prionotus carolinus*, *Myoxocephalus scorpius*, and *Leptocottus amatus*) depends on the rate of sonic motor neuron discharge (which determines the rate of muscle contraction) and is a linear function of water temperature (Fine 1978; Bass and Baker 1991). Toadfishes and cods can also produce grunting sounds which often grade into tonal sounds (Tavolga 1958b; Hawkins and Rasmussen 1978). The neurophysiology of sound production in toadfishes and Scorpaeniformes has been relatively well studied and will be discussed later.

2.2.2 Stridulatory Sounds

Many sonic fishes do not possess swim bladder muscles, but use stridulation of bones to produce sounds. Grunts (Haemulidae) produce sounds by grinding their pharyngeal jaws (Burkenroad 1930). Croaking gouramis (*Trichopsis vittatus*) strum tendons attached to the fourth and fifth pectoral

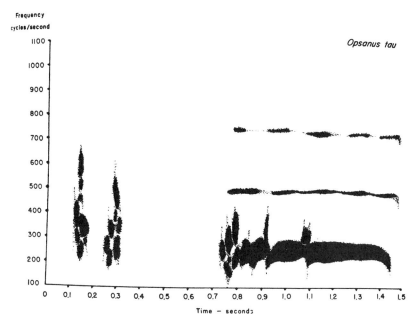

Frequency
cycles/second

Opsanus tau

FIGURE 9.2. Sounds of the Oyster toadfish, *Opsanus tau*. Sound spectrogram on the left is a double grunt, and the sound on the right is a long hoot. Fish were recorded in Florida. (From Tavolga 1958b, with permission.)

fin rays over two raised areas on the second and third fin rays (Daugherty and Marshall 1976).

Damselfishes and cichlids have been hypothesized to grind their pharyngeal jaws to produce courtship and aggressive sounds (Myrberg et al. 1965; Chen and Mok 1988). Surprisingly, this hypothesis has never been experimentally tested, even though damselfishes are among the best-studied sonic fishes (e.g., Schneider 1964; Spanier 1979; Chen and Mok 1988; Myrberg et al. 1993; Lobel and Mann 1995). It seems that sound production by pharyngeal stridulation has been transformed from a reasonable hypothesis to accepted fact through repeated citation of the hypothesis.

Stridulatory sounds are typically pulsed, broad-band sounds, similar to sounds produced by fishes with extrinsic sonic muscles (Fig. 9.3). The pulses are of short duration (10–50 ms), and the pulse number and repetition rate varies between species (e.g., Spanier 1979). The dominant spectral components of the sound have been correlated with fish size; larger fish produce lower frequency sounds (Myrberg et al. 1993; Lobel and Mann 1995). This relationship is likely due to the resonance properties of the swim bladder, since the resonance of a sphere is inversely related to its volume (Urick 1983). It must be noted that swim bladders are not isolated spheres of air, and that there is a wide variety of shapes and amounts of tissue around the swim bladder, resulting in significant damping of the swim bladder response

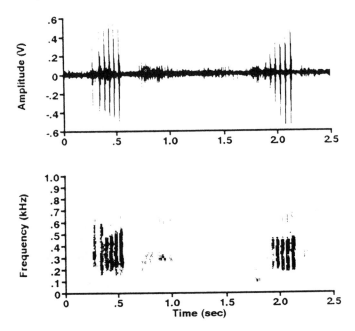

FIGURE 9.3. Oscillogram (top) and spectrogram (bottom) of two courtship calls produce by a single male pomacentrid, *Dascyllus albisella*. (From Lobel and Mann 1995, with permission.)

and changes in resonance frequency of the bubble from that of an unrestricted bubble of the same size (Clay and Medwin 1977). The ability of a fish that uses stridulation to modulate the resonant frequency of its swim bladder has not been investigated.

2.2.3 Other Sonic Mechanisms

Vibration of sonic muscles on the swim bladder and stridulation are the most common mechanisms of sound production, but some unique mechanisms of sound production have also been described and proposed, and others are likely to exist. The triggerfish (Balistidae) drums its pectoral fin spine against evaginations of the swim bladder that lie near the body wall (Salmon et al. 1968). The characid, *Glandulocauda inequalis*, has been hypothesized to use air gulped at the surface to produce sounds by vibrating the gill rakers (Nelson 1964). Sound production in the goby *Bathygobius soporator*, which has no swim bladder, is associated with forced ejection of water through the gills (Tavolga 1958a).

Just as there are many fishes with sonic muscles whose sounds have never been recorded, there are many fishes where the sounds are known but for which the mechanism of production has not been demonstrated (e.g., hamletfish, cichlids, and trunkfishes). It is relatively easy to dissect a fish and search for sonic muscles, and even stimulate these muscles to produce sounds. When no sonic muscles are found, pharyngeal stridulation is often proposed but rarely confirmed experimentally.

2.2.4 Frequency Modulation and Sound Pressure Level

Frequency modulation is extremely rare in fishes. Some toadfish sounds show frequency modulation as the rate of contraction changes at the beginning or end of boatwhistles while the hamletfish (*Hypoplectrus* spp.) produces a sound with a frequency sweep during spawning (Lobel 1991). The rarity of frequency modulation is likely due to the limitations of changing the contraction rate of sonic muscles (as pointed out above, toadfish contraction rates only vary between 80 and 180 Hz, and depend on temperature), and the resonance properties of the swim bladder.

Very few measurements of sound pressure levels of fish sounds have been made. Based on existing data, fishes with sonic muscles on the swim bladder (e.g., toadfish 140 dB re: 1 μPa) (Tavolga 1971), produce louder sounds than fishes that use stridulatory mechanisms (e.g., damselfish 110 dB re: 1 μPa) (Myrberg and Riggio 1985). Choruses of croakers and drums (Sciaenidae) have been measured as loud as 90 dB re: 1 μPa (50 dB above background noise), but the source levels were unknown, as was the distance between the source and the recording device (Fish and Cummings 1972).

2.3 Unintentional Sounds

Many fishes that do not produce sounds during courtship and aggression do produce "incidental" sounds during feeding and swimming (hydrodynamic sounds) (Tavolga 1977). While these sounds are often referred to as unintentional, even unintentionally produced sounds could provide information to other fishes about the location of food supplies or spawning fishes (Moulton 1960; Lobel 1991). Little effort has been spent investigating the behavioral implications of these sounds, and they will not be discussed further. But one should recognize their potential importance for communication and signaling.

2.4 Neurophysiology of Sound Production

Physiological and neurophysiological research on fish sound production has focused on fishes with sonic muscles, especially toadfishes, including the oyster toadfish (*O. tau*) and the midshipman (*P. notatus*), and Scorpaeniformes, including the sea robin (*Prionotus carolinus*) and Pacific staghorn sculpin (*Leptocottus armatus*) (reviewed by Bass 1990, 1992, 1993). In these fishes, each pulse of a sound is produced by contraction of the sonic muscle (Packard 1960; Bass and Baker 1991); tones are produced by rapid contraction of the sonic muscles. The sonic muscle has extremely fast response properties. Investigations by Skoglund (1961) demonstrated that the contraction time of sound-producing muscles in *O. tau* is on the order of 5 ms, with a relaxation time of 8 ms, and that this rate is broadly seen among fishes. Gainer and colleagues (1965) demonstrated similar contraction and relaxation rates in sonic muscles of the squirrelfish *Holocentrus*, and that nonsonic muscles in the same species had contraction and relaxation times of 12 and 25 ms, respectively.

The drumming muscles of these fishes are innervated ipsilaterally by nerves from the sonic motor nucleus (SMN) (Fig. 9.4) and the nerve firing pattern determines the sonic muscle contraction pattern (Bass 1985; Fine and Mosca 1989; Bass and Baker 1991). The SMN lies on the midline at the junction of the medulla and spinal cord in toadfishes, and bilaterally along the ventrolateral margin of the caudal medulla and rostral spinal cord in Scorpaeniformes (Bass 1985). In toadfishes the sonic muscles on both sides of the swim bladder contract simultaneously due to simultaneous discharge of the SMN triggered by bilateral inputs from pacemaker neurons (Bass and Baker 1991). In the sea robin, however, the muscles contract out of phase with each other, possibly through the action of local pacemaker neurons (Bass and Baker 1991).

A wide variety of neural systems are used to generate sound in fishes that stridulate their pectoral fins, such as the croaking gourami (*Trichopsis vittatus*) and Pimelodid catfishes. The SMN in the croaking gourami is composed of pectoral motoneurons, and there is no difference in the innervation of the pectoral fin compared with the Siamese fighting fish (*Betta splendens*), a non–sound-producing fish in the same family (Ladich and Fine 1992). Pimelodid catfishes have three sound-production mechanisms: a stridulatory mechanism composed of the pectoral girdle and the first pectoral fin ray, a swim bladder with extrinsic muscles, and a tensor tripodis muscle that inserts on the swim bladder (Ladich and Fine 1994). Each of these sonic motor systems is innervated by different sonic motor nuclei, thus demonstrating that fish can evolve multiple sonic motor nuclei.

Toadfish (*O. tau* and *P. notatus*) males make both a pulsed, broad-band agonistic sound (grunts) and a tonal boatwhistle, while females produce only the agonistic sound (*O. tau*) or are silent (*P. notatus*). Several studies have investigated the neurophysiological bases for these differences in sound production between genders. Males have larger neurons in the SMN than females in both species, although dendrite diameters are the same in both sexes of *O. tau*, but larger in males in *P. notatus* (Fine et al. 1984; Bass and Baker 1991; Fine and Mosca 1995). Bass and his colleagues also found that some males of *P. notatus* have a SMN like that of females, and that these males do not vocalize to attract females (Bass and Marchaterre 1989; Bass 1993). These "satellite" males intrude on spawning pairs of females and vocalizing males, behave like females, and fertilize the female's eggs

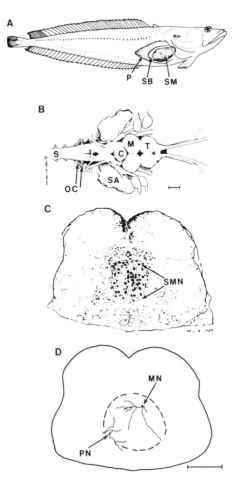

FIGURE 9.4. The vocal-sonic motor system of *Porichthys notatus*, the plainfin midshipman. A: Drawing showing the relative position of the swim bladder (SB) and the attached sound-producing muscles (SM). These are overlain by the pectoral fin (P). B: Drawing of a top view of the brain of the midshipman showing the position of the ventral occipital roots (OC) that carry sonic motor axons. Rostral is to the right. C: Transverse section of the brain at about the level of OC in B, showing the position of the fused sonic motor nuclei (SMN) that lie along the midline of the brain in the region of the caudal medulla and the rostral region of the spinal cord (S). D: An outline drawing of C, showing the relative positions of the pacemaker (PN) and sonic motor (MN) neurons. These neurons are reconstructed from serial sections. Only the ipsilateral branching of the pacemaker neuron is shown. C, cerebellum; M, midbrain; SA, saccular otolith; T, telencephalon. Scale bars 1 mm in B and 500 μm in C and D. (From Bass 1992, with permission.)

while she lays them. Thus, work with the neurophysiology of sound production has led to discoveries about the spawning behavior and tactics of these fishes. Much work remains to be done, extending neurophysiological studies to fishes that produce sounds using stridulation and linking species-specific variability of sound production with neurophysiology.

2.5 Behavioral Context of Sound Production

2.5.1 Standard Behavioral Repertoire

Fishes produce sounds during aggression, defense, territorial advertisement, courtship, and mating. The role of many fish sounds in reproduction was evident in early studies where sounds were often associated with breeding seasons (Smith 1905). Several studies have found that the rate of sound production increases with the timing of reproduction, especially in fishes that form spawning aggregations, such as croakers (Sciaenidae) (Brawn 1961; Saucier and Baltz 1993; Mann and Lobel 1995). This relationship is largely due to increases in courtship calling rather than production of spawning sounds. Few fishes are known to produce sounds associated with spawning. Damselfish (*Dascyllus albisella*) and several freshwater goby males (*Padogobius martensii* and *Knipowitschia punctatissima*) produce sounds during female egg-laying, and the hamletfishes (*Hypolplectrus* spp.) produce sounds during gamete release (Lobel 1991; Lobel and Mann 1995; Lugli et al. 1995). Other fishes, like the drum and weakfish, do not produce sounds during spawning (Guest 1978; Connaughton and Taylor 1996).

Only a limited number of studies have investigated the differences between sounds used in different behavioral contexts. Toadfish (*O. tau*) produce a boatwhistle as an advertisement and courtship signal but produce a pulsed sound in aggressive or defensive situations (Gray and Winn 1961). Damselfishes (*Stegastes partitus* and *Dascyllus albisella*), which produce only pulsed sounds, produce a multiple-pulse courtship sound, but a single-pulse aggressive sound (Myrberg 1972; Mann 1995). *Dascyllus albisella* also produces multiple-pulse aggressive sounds that are similar to courtship sounds, but with faster pulse rates (Mann 1995). The electric fish, *Pollimyrus isidori*, produces grunts, growls (bursts of pulses), and moans (tonal sounds) during courtship, and hoots (tonal sounds with rapid frequency modulation) and single-pulse pops (Fig. 9.5) (Crawford et al. 1986, 1997a,b).

Evidence for the use of sounds as territorial advertisement is indirect, and difficult to demonstrate because some signals serve both courtship and territorial advertisement functions. Damselfish (*D. albisella*) males produce advertisement sounds outside of the reproductive periods (Mann and Lobel 1995), but at a lower rate, suggesting that the sounds also function for territorial advertisement. Playback experiments in which nonresident dam-

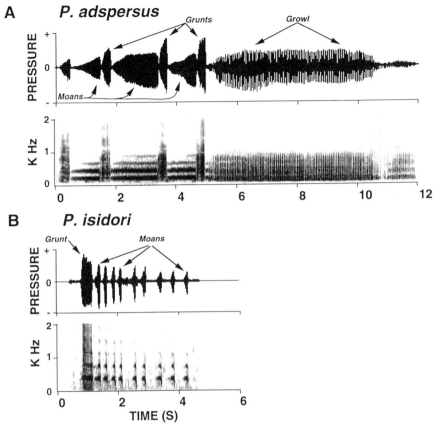

FIGURE 9.5. Sounds produced by two species of African mormyrids, (A) *Pollimyrus adspersus* and (B) *P. isidori*. Both species produce two types of sound, a grunt and a moan. In each case, the top figure is the acoustic waveform of the signal and the bottom is the corresponding sonogram. These sounds are produced by males and are species-specific. (From Crawford et al. 1997a, with permission. Copyright 1997 Acoustical Society of America.)

selfish (*Stegastes partitus*) sounds were played from a given territory elicited greater sonic responses from neighboring conspecifics than playbacks of resident fish sounds (Myrberg and Riggio 1985). This shows that damselfish produce sounds in response to changes in the sounds from neighboring territories.

Crepuscular peaks in sound production are common in sonic fishes and are analogous to the dawn chorus in birds. Breder (1968) found a peak in the calling of the sea catfish, *Galeichthys felis* (Ariidae) at dusk, and that most boatwhistling of the toadfish, *O. beta*, occurred at sunset. Winn and colleagues (1964) and Steinberg and colleagues (1965) found that sound production by the squirrelfish, *Holocentrus rufus* (Holocentridae), peaked at both dawn and dusk. The dawn peak may have been associated with territorial interactions as these nocturnal fish returned to cover. Damselfishes also have dawn and dusk peaks of sound production (Steinberg et al. 1965; Myrberg 1972; Mann and Lobel 1995). Overall, though, the biological importance of crepuscular peaks in sound production is not known. In birds they are thought to be related to optimal allocation of time devoted to singing versus foraging, and to environmental variables, such as wind, that influence sound propagation (Henwood and Fabrick 1979; Kacelnik and Krebs 1983).

2.5.2 Echolocation in Catfishes

Perhaps the most unique use of sound among fishes, is the presence of a low-frequency form of echolocation in the sea catfish, *Arius felis* (= *Galeicthyes felis*). *Arius* produces pulsed burst grunts using a special "elastic-spring" mechanism (Tavolga 1962). Tavolga (1971, 1976) noticed that these sounds, which have most of their energy around 100 Hz, bear some resemblance to the low-frequency sounds often used by visually impaired humans when they are trying to navigate around obstacles. In a series of experiments, Tavolga (1971, 1976) demonstrated that *Arius* can use its sounds to navigate around objects and he suggested that the sounds are used for low-frequency echolocation so that the fishes can move around obstacles in the dark.

While it is not known if other species use similar mechanisms, it is possible that deep-sea fishes, many of which are known to have sound-producing muscles (Marshall 1967), use sounds in a similar way to *Arius*.

2.6 *Importance of Fish Sounds*

The temporal patterning of fish sounds, especially the pattern of pulsing in pulsed sounds, has long been suspected of being their most important feature (Winn 1964). However, there has been little quantification of either intraspecific or interspecific variation in fish sounds in combination with studies on fish hearing abilities. Thus, little is known about what variation is detectable.

Myrberg and Riggio (1985) performed the only study on individual recognition of fish sounds with the bicolor damselfish, *Stegastes partitus*. Males of this species respond to playbacks of conspecific sounds by producing their own sounds and also respond more to playbacks of the sounds of nonresident fish from the territories of its nearest neighbors than playbacks of the sounds made by the resident. They hypothesized that this difference was likely due to frequency differences in the sounds produced by males of different sizes rather than to differences in pulse rates.

Recognition of species-specific sounds has been studied in four species of damselfish (*Stegastes* spp.) (Spanier 1979). The sounds of these species vary in the number of pulses and pulse rate. Spanier measured the response of males to playbacks of their own and other species sounds and found that fish responded most to sounds of their own species. However they did respond, albeit to a lesser degree, to the sounds of other species. The number of pulses in a sound and pulse rate were the most important features governing the response. A similar study with two species of African mormyrid fishes, *Pollimyrus adspersus* and *P. isidori* (Crawford et al. 1997a), demonstrated species discrimination based on several acoustic parameters including pulse repetition rate for grunt sounds and fundamental frequency peaks in groans (Fig. 9.5).

Although many fish sounds have been categorized as courtship sounds, their role in mate choice has not been well demonstrated. The dominant frequency of courtship sounds of damselfishes, like many anurans, is related to male size, in which larger males produce lower frequency sounds (Myrberg et al. 1993; Lobel and Mann 1995). Note that this relationship is not likely to hold for fishes with intrinsic sonic muscles in which the rate of contraction is related to the frequency of the sound, rather than the size of the swim bladder. Playback experiments performed in the field with the bicolor damselfish (*Stegastes partitus*) showed that females would respond to the playbacks of male courtship sounds by moving toward the loudspeaker (Myrberg et al. 1986). When sounds of two different males were used that differed in dominant frequency (710 and 780 Hz), the females went toward the speaker broadcasting the lower frequency more often (15 times versus 2 times). In experiments where white noise or the sounds of another damselfish (*Stegastes variabilis*) were broadcast, female bicolor damselfish never went toward the speaker producing these sounds. These experiments show that the bicolor damselfish can distinguish potential mates based on sound, and they may choose mates based on the dominant frequency of the sound. However, whether they actually do this is open to debate, since in damselfishes, in which both female choice and male sounds have been studied, there is little or no relationship between male size and male reproductive success (Petersen 1995).

Recent work with damselfishes suggests that the rate of courtship is used by females in selecting mates, and may be an indicator of male energy reserves and the likelihood the male will successfully defend a brood of eggs (Knapp and Kovach 1991; Karino 1995; Knapp 1995; Mann and Lobel 1995). If this is extended to other fishes, it would suggest that the repetition rate or duration of the sound is the most important feature for mate choice, rather than the structure of the sound (pulse number, pulse rate, dominant frequency), which may only be important for species-specific recognition [but see Ryan and Rand (1993), who suggest that species recognition and sexual selection are just ends of a single continuum of preference].

2.7 Sound Propagation in Water

The propagation of animal sounds can be greatly affected in air, where wind, temperature gradients, the ground, and foliage can restrict or enhance the distance over which signals can be used for communication (Marten and Marler 1977; Wiley and Richards 1978; Wells and Schwartz 1982). Propagation of fish sounds has only been studied in very shallow water (relative to the sound wavelength). In such environments, theory indicates low frequencies that have long wavelengths do not propagate (Rogers and Cox 1988). Propagation does occur at higher frequencies in shallow water (when the wavelength is sufficiently small), which may have selected for high-frequency hearing in otophysans (e.g., goldfish, catfish, and relatives, which have a series of bones, the Weberian ossicles, connecting the swim bladder and inner ear), which commonly live in shallow water.

Playbacks of the tonal boatwhistle of the toadfish in 1-m-deep water lost 18 dB over 5 m distance, hypothetically restricting communication to several meters (Fine and Lenhardt 1983). Grunt sounds produced by squirrelfishes in 5-m-deep water attenuated 10 dB over 40 cm, to about 25 dB signal-to-noise ratio (S/N) (Horch and Salmon 1973). Other work on sound propagation in shallow water shows that the high reflectivity of the air-water interface could degrade signal qualities over short distances (Forrest et al. 1993). This has been observed in a field study on the propagation of damselfish (*Dascyllus albisella*) sounds in 7-m-deep water (Mann and Lobel 1997). Characteristics of the call, including pulse duration, dominant frequency, and amplitude changes of pulses within calls, were greatly affected by propagation over distances as short as 2 m. Pulse period was least affected by propagation, and varied little over 11 to 12 m.

The range of detectability of fish sounds is likely to be mediated by a combination of attenuation from propagation and the level of background noise, which can be quite high in marine environments (Knudsen et al. 1948). The S/N of damselfish sounds were 17 to 25 dB at 1 to 2 m from the sonic fish, and decreased to 5 to 10 dB at 11 to 12 m (Mann and Lobel 1997). Based on sound detection data from other fishes, S/N is 15 to 20 dB at detection thresholds for pure tones (Fay 1988). Taken together, these data suggest that damselfish sounds are used for short-range communication.

2.8 Hearing Capabilities of Fishes

Since fish sounds are either tonal or pulsed, it is intellectually easy to interpret the importance of variation in sounds in light of discrimination abilities of fishes, especially as compared to birds (see Fay and Megela Simmons, Chapter 7). For tonal sounds where the dominant frequency reflects the rate of muscle contraction, it may be advantageous for females to be able to discriminate sounds of different frequencies. For pulsed sounds where the pulse rate varies both intra- and interspecifically, it might be advantageous to discriminate sounds of different pulse rates. Unfortunately, there are few psychophysical data on either frequency discrimination or pulse rate discrimination for fishes that produce these sounds; they are generally only available for the goldfish, *Carassius auratus* (a fish not known to produce sounds).

Fishes other than otophysans, such as *Gobius niger* (Gobiidae), *Corvina nigra* (Sciaenidae), and *Sagrus annularis* (Sparidae), can discriminate tones differing in frequency of about 10% (Fay 1988 and references within). Otophysan fishes, like the goldfish, can discriminate tones differing in frequency of about 4% to 5% (Jacobs and Tavolga 1968; Fay 1970). Toadfish boatwhistles are typically 160 Hz (the seasonal range is about 140 to 220 Hz). If toadfish discriminatory abilities are similar to that of other non-otophysans, then they should be able to discriminate 160 Hz from 180 Hz. The typical variation in toadfish sounds has not been well characterized, but standard deviations are about 40 Hz (Fine 1978). Thus, their discriminatory abilities may allow them to detect some natural variation, but fine differences (<20 Hz) may go undetected.

Fay and colleagues have performed a number of studies on the abilities of the goldfish to detect amplitude modulation (AM) and changes in the rate of AM, which can be used as a first-order attempt to interpret the importance of variation in pulsed sounds (Fay 1972, 1980; Fay and Passow 1982; also see Fay and Megela Simmons, Chapter 7). Goldfish are better at detecting changes in the rate of AM when the carrier signal is a tone, as opposed to noise. This was hypothesized to result from error in phase-locking to the noise envelope produced by saccular neurons (Fay 1982). The sounds produced by damselfishes and other fishes are more similar to band-limited noise than to tones. If the discrimination abilities of damselfishes are similar to the goldfish, then they might not be able to detect most of the intraspecific variation and much of the interspecific variation in AM rates that are found in their sounds.

However, data on discrimination abilities using the damselfish show that they can distinguish sounds with differences in AM rates from about 23 to 32 Hz (Myrberg et al. 1978; Spanier 1979), which suggests that damselfishes are at least twice as sensitive as goldfish to changes in the rate of AM (Fay 1982). To directly compare these fishes, however, the discrimination abilities need to be measured using the same methods.

2.9 Costs of Communication

Physiological data on the amount of energy invested in sound production are lacking. Energy expenditure is probably high for fishes like toadfishes that contract sonic muscles for long periods of time, and for fishes like damselfishes that produce sounds during an energy-demanding behavioral swimming display.

Cetacean stomach contents reveal that they prey on sound-producing fishes, such as cod and haddock (Recchia and Read 1989). It is not known, however, whether the prey fishes are targeted passively when they are producing sounds, actively with echolocation, or both.

Damselfish males that perform an elaborate swimming display along with sound production have higher mortality than females (Shpigel and Fishelson 1986). However, it is not clear that this is due to predators targeting sound-producing fish, as opposed to targeting fishes performing swimming displays.

2.10 Conclusions

It is curious that many people are still surprised to learn that fishes produce sounds. This may be because most people only encounter fish when they are suitably seasoned and garnished or in home fish tanks, instead of when the fish are in their normal habitats. Clearly, many fishes make sounds. Fishes are the most diverse vertebrates and produce sounds with a wide array of mechanisms. Despite the myriad mechanisms of sound production, the types of sounds currently known to be produced by fishes are quite limited compared to many other vertebrates.

The study of sounds produced by fishes has lagged behind that of insects, frogs, and birds. Much remains to be learned about which fishes produce sounds, how they produce sounds, and how these sounds are used. Perhaps most will be gained through the interface of the study of hearing abilities of fishes and variation in the sounds they produce. Ultimately, this will lead us to a better understanding of how sound production may have influenced the evolution of hearing capabilities, and how hearing capabilities may have constrained the evolution of sound production.

More specifically, while there is a significant number of known sound-producing fishes, there are likely many more to be recorded, and many that may have unique sound production mechanisms. Studies aimed at examining sounds produced by close relatives (Fig. 9.5) (e.g., Crawford et al. 1997a) are greatly needed to study the evolution of sound production, like those that have been done with frogs (e.g., Cocroft and Ryan 1995). Likewise the few studies on sexual selection and mate choice need to be expanded to determine their role in the evolution of sound production. As a part of these studies, more care needs to be taken to measure SPLs and S/N in natural environments.

There are no data on the energetics of sound production in fishes, and only anecdotal data on other costs of sound production. These types of data will be important in understanding the patterns of sound production, and the role of sound in mate choice. Some fish produce sounds in choruses, like croakers, but nothing is known about the patterning of calls in such a chorus. Do fish time their calls like many chorus frogs?

We are beginning to see the fruits of neuroethological studies on sound production in fishes. They are giving us tools to understand the evolution of brain morphologies that are intimately linked with the evolution of behavior. These studies need to be continued and expanded to further comparative work with other sound-producing fishes.

Sound does not act alone in many of these fishes. For example, damselfishes perform an exaggerated swimming movement during sound production. How is information from the different senses integrated? Tavolga (1956), a pioneer in studies of sound production in fishes, has given us insight into how the visual, chemical, and acoustic senses are tightly integrated in the courtship of the goby *Bathygobius*. We would do well to follow his example, if we are to truly understand communication in fishes.

3. Frogs

Historically, the study of frog acoustic communication has been pushed forward by interest from several domains. Most long-standing has been an interest in the evolution of amphibians both because they are the first terrestrial vertebrates to have evolved, and because the dramatic vocal behavior of most frogs invites the study of the evolution of mating systems. Also, frogs have been exploited by neuroethologists trying to understand how relatively simple vertebrate brains and auditory end organs process complex acoustic information. This in turn has had a positive feedback effect, encouraging continued work examining the vocally mediated social behavior of frogs.

3.1 Sound Production

While the other groups of amphibians (salamanders and the amphisbeiana) lack vocal cords and are essentially mute, frogs and toads are prodigious sound producers. In many species, the sound level of an individual advertisement-calling male is on the order of 90 to 110 dB re: 20 μPA measured at a distance of 1 m. It is common, furthermore, for many temperate-zone anurans to assemble into relatively dense aggregations during the breeding season. The collective advertisement calls of, not uncommonly, some hundreds of males at the same breeding pond can produce a chorus sound level in excess of 120 dB re: 20 μPa.

3.2 Terrestrial Sound Production

The dynamics of vocal production and the wide variety of sounds that anurans can produce has been studied systematically in only a few species, primarily toads of the genus *Bufo* (Martin 1972) and frogs of the genus *Rana* (Paulsen 1967). Most of the variation in frog calls of different species comes from (1) the harmonic structure of the sound, and (2) various styles of amplitude modulation of the sound. The rate at which the vocal cords open and close depends on the activation air pressure and the masses of cartilaginous deposits on their lateral portions (Paulsen 1967; Martin 1971). That rate corresponds to the fundamental frequency in the case of quasi-harmonic calls and to the dominant or carrier frequency in amplitude-modulated and tonal calls. In some species of *Rana* and *Hyla*, the calls in their vocal repertoire can range from vowel-like harmonic-rich signals to explosive, noisy consonant-like sounds (Bogert 1960; Capranica 1968). The calls of many species of toads and treefrogs are characterized by their species-specific amplitude-modulated trill rate. Indeed, nearly all frogs use amplitude modulation as the most important feature to convey information, although a few frogs have frequency modulations in their calls (Ryan 1983a; Passmore 1985; Matsui et al. 1993). The rate of amplitude modulation corresponds to the rate of vibration of the arytenoid cartilages, which superimpose their temporal cycles on the sound produced by the vocal cords (Martin 1972). In addition, modulation of the contraction amplitude and rate of body wall musculature modulates air flow through the larynx and so the temporal acoustic pattern. And in some genera, for example *Eleutherodactylus* of Central and South America, the calls can be exceptionally pure, whistle-like tones (Narins and Capranica 1978). The vocal tract configuration and mechanics by which frogs can produce such remarkable pure tones is poorly understood. Whistles presumably are generated by precise sinusoidal vibrational cycles of the vocal cords with the arytenoid cartilages held far apart, but how anurans can achieve such tonal purity is unclear. Further studies of sound production in anurans are needed to elucidate how these animals can produce such a variety of sounds, rivaling the variety of notes in bird songs.

The majority of anurans possess very similar vocal structures. The paired lungs communicate with the buccopharyngeal cavities through a well-developed vocal tract controlled by 16 pairs of muscles (Martin and Gans 1972). Within the tract is a larynx with intrinsic vocal cords that serve as the primary sound source (Fig. 9.6). The larynx skeleton consists of the ring-shaped cricotracheal cartilage, upon which are situated the flexible arytenoid cartilages (Martin 1971). In general the larynx of the female is anatomically similar to that of the male except that it is smaller. It is attached to the hyoid apparatus by muscle and connective tissue and opens into the pharynx through the slit-like glottal opening between the thickened cords (Trewavas 1933). The tension on the vocal cords and their opening

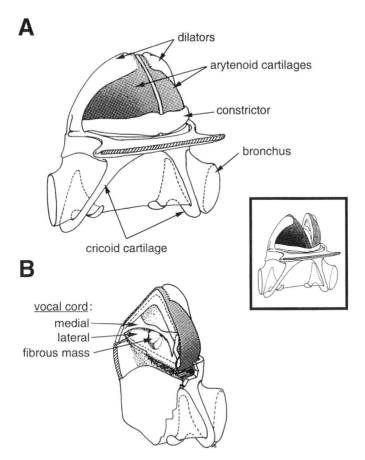

FIGURE 9.6. Larynx from a toad of the genus *Bufo*. A: Air expelled from the lungs enters the bronchus. The dilator laryngis and constrictor laryngis externa cause the arytenoid cartilages to open (inset) and air moves up into the buccal cavity and vocal sac. Rapid modulation of the arytenoid cartilage position amplitude modulates the sound frequencies made by the vocal cords, shown in B. (Redrawn from Martin 1971, with permission. © 1971 John Wiley & Sons.)

and closure are under active control by four pairs of laryngeal muscles, which are innervated by the vagus cranial nerve (Schneider 1988).

The buccal cavity has a simple shape and communicates with the external environment for breathing through the pair of nares at the tip of the upper jaw. The nares can be quickly opened or closed by the action of striated throat and jaw muscles. Since the mouth is normally kept tightly closed by the tonic contraction of specialized musculature serving principally the lower jaw, air intake and expulsion occur through the nares (Fig. 9.7; Gans 1973).

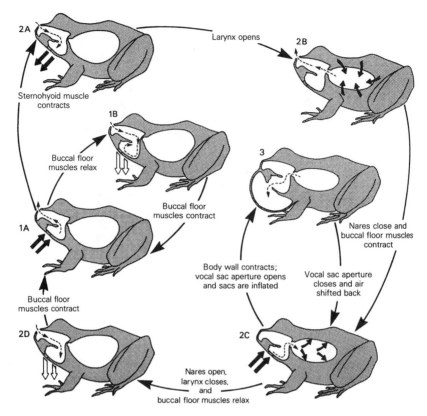

Figure 9.7. Ventilation cycle and vocalization. Buccal pumping with glottis and vocal sac aperture closed, and nares open (oscillatory cycle of buccal floor muscle contraction and relaxation) mixes air in buccal cavity (1A–1B). This oscillation may be significant for olfactory sniffing. A larger inspiration of fresh air stored in the buccal space (2A) starts the cycle of bringing air to the lungs. Stale air is expired by abdominal contraction with the glottis/larynx open and nares open (2B). Contraction of buccal floor muscles with glottis open and nares closed forces fresh air from buccal space into lungs (2C). The glottis is then closed so that fresh air may be drawn in through the nares using elastic relaxation of the buccal floor (unfilled arrows; 2D) without evacuating the lungs. Vocalization involves a detour in the cycle in which the body wall contracts forcing air out of the lungs, but rather than open the nares, the vocal sac aperture is opened (3). The body wall contraction during vocalization is typically much more vigorous than that during exhalation (2B) causing vibration of the vocal cords. (Modified from Gans 1973, with permission.)

In only a few species is the mouth opened during vocalization of any kind (Amiet 1989). Rather it is the vocal sac, a most distinctive feature of nearly all male anurans, that allows sound to be efficiently radiated into the environment. Air expulsed from the lung enters the vocal sac through one opening, or, more commonly, two small bilateral openings, in the floor of the mouth. This subgular vocal sac may itself be a single sac, or there may be paired sacs (Liu 1935). The vocal sac and the paired openings are absent in the female. For each vocalization, then, body wall muscular contraction drives air through the vocal tract and into the buccal and vocal sac cavities. A notable exception is the male fire-bellied toad, *Bombina*, in which sound is produced during airflow in the opposite direction, namely during the inspiration phase, as air flows from the buccal cavity back into the lungs (Schneider 1988). In either case, the flow of air causes the vocal cords to vibrate, and simultaneously the edges of the overlying arytenoid cartilages open and close rhythmically.

Although suspected to be so, the vocal sac is apparently not a cavity resonator, as shown by clever experiments in which the air in the frog's vocal tract was replaced with a mixture of air and helium (Rand and Dudley 1993). This treatment did not change the frequency of the call produced, as would be expected with a resonant system.

3.3 Aquatic Sound Production

There are a number of frog species that, as adults, have secondarily re-turned to the water and are considered aquatic rather than terrestrial anurans. The common African clawed frog (*Xenopus laevis*) is an example of this family, the Pipidae. These frogs have inherited the sound-producing mechanism suitable for making airborne vocalizations from their terrestrial ancestors, but the mechanism has been modified to efficiently make under-water calls. Both the cricoid and arytenoid cartilages of the larynx are modified to form a relatively large structure with calcified and thus hard-surfaced disks that are allowed to strike each other at high velocity and produce quite loud clicks (>105 dB re: 20 µPa at 1 m; Yager 1992). As in other (terrestrial) anurans, only the male gives advertisement calls and has the elaborate enlarged larynx. In both sexes, the vocal cords have been lost.

Interestingly, there is at least one species of terrestrial frog (*Rana subaquavocalis*) that produces calls underwater (Platz 1993). Furthermore, the white-lipped frog from Puerto Rico (*Leptodactylus albilabris*), while producing airborne vocalizations, simultaneously drives the substrate as it inflates its vocal sac (Lewis and Narins 1985; Lopez et al. 1988). In this way males generate seismic signals as vertically polarized surface (Rayleigh) waves, which other white-lipped frogs can detect at least 3 m away.

3.4 Neural Control of Sound Production

The neural correlates of frog calling have been studied over many years by Robert Schmidt (see, for example, Schmidt 1974), using a variety of tech-niques including transection, lesioning, and neural recording of intact animal and isolated brain-stem preparations. Additional data (Wetzel et al. 1985) using horseradish peroxidase to identify particular nuclei in the vocal control pathway have confirmed much of the lesion results and the following picture has emerged: In the hindbrain, motor neurons from cranial nerve nuclei IX and X send axons ipsilaterally to the laryngeal and glottal muscles. Both left and right nuclei IX and X are interconnected and receive ipsilateral projections from two nuclei in the reticular formation of the hindbrain, collectively termed the nucleus reticularis inferior. One of the reticular formation nuclei is probably the same as the "inspiratory phase generator" (Wetzel et al. 1985; Schmidt 1992). The reticular formation nuclei send to and receive ipsilateral projections from the dorsal tegmental area of the medulla (DTAM), which is the same as Schmidt's pretrigeminal nucleus. This nucleus is also connected to the contralateral nucleus. Finally, there are three telencephalic and diencephalic nuclei that send descending projections to the DTAM. These are the ventral striatum, a portion of the thalamus, and the anterior preoptic area. The DTAM and reticular and cranial nerve nuclei all concentrate androgens (Kelley et al. 1975) and no doubt play a role in the seasonal modulation of calling activity.

At least the thalamus and possibly other diencephalic and telencephalic nuclei involved in vocal motor control receive ascending auditory input. Indeed pathways from the thalamus to the DTAM pass through the midbrain torus semicircularis, a major auditory processing center (see McCormick, Chapter 5). It is at these sensorimotor interfaces that much work is needed with the expectation of very interesting results (Walkowiak and Luksch 1994).

3.5 Hormone/Sex Factors

A variety of reproductive behaviors in frogs, including advertisement calling, are influenced by hormones (Schmidt 1982; Aitkin and Capranica 1984; Wetzel et al. 1985). The vocal sac and advertisement call are characteristic of males and thus conspicuous sexual dimorphisms relative to acoustic communication. In *Xenopus laevis*, the brain vocal control pathway, laryngeal muscles, and larynx itself are all under androgen hormonal control, and all are larger in males (Sassoon and Kelley 1986, Kelley et al. 1988). Various measures of increased activity in brain nuclei controlling vocalization are larger in other species of frogs as well (Schmidt 1982; Aitkin and Capranica 1984).

Only male frogs make advertisement calls, but female frogs may also vocalize. The most conspicuous of these is the release call, produced when a nonreceptive female is clasped by a male frog intent on mating. It is this female vocalization and female mating behavior in general that has received the greater share of attention with regard to hormonal influences on frogs. In female frogs, arginine vasotocin (AVT) is a key hormone controlling egg laying and other reproductive behaviors, and when circulating levels of AVT are high, corresponding with a female who is ready to mate, the release call is inhibited (Diakow 1978). The AVT may act indirectly because it causes release of both prolactin and at least one type of prostaglandin (Diakow and Nemiroff 1981; Boyd 1992). Prolactin itself will inhibit release calling, and prostaglandin modifies activity in some nuclei that control vocalization. Interestingly, prostaglandin may also have a role in suppressing male advertisement vocalization (Schmidt and Kemnitz 1989).

The male advertisement call is given only during the appropriate season and depends on seasonal fluctuation in circulating androgens. Female frogs obviously have the ability to vocalize. Is the lack of advertisement calling in females due to the basic difference that females lack androgens at the appropriate time of year? Female *Xenopus* produce a click-train release call, which is modified when adult ovariectomized females are treated with androgens (Hannigan and Kelley 1986). This treatment does not lead to production of a male advertisement call, however. As is the case for many other neural systems, there is a critical developmental window when androgens are needed to masculinize the vocal system. If juvenile female *Xenopus* are implanted with testes (after gonadectomy), they produce upon maturity advertisement calls indistinguishable from males (Watson and Kelley 1992). One component of this ability is the induction of male-type laryngeal muscle, which is in turn controlled by a gene whose expression is androgen-dependent (Catz et al. 1992).

3.6 Plasticity, Diversity, and Information Content

There is no evidence the frogs learn any aspect of acoustic signaling, and it is common to associate such genetically programmed behavior with extreme stereotypy. Variations from the stereotyped call are thought to be due simply to unavoidable environmental factors such as temperature (see Section 3.1) or a lack of natural selection to maintain a particular vocalization parameter constant, perhaps because it is less important in conveying information. This view underestimates the flexibility to convey meaning, which may exist in many anuran vocal communication systems, and has been documented in several. For example, Taigen and Wells (1985) found that male *Hyla versicolor* produce advertisement calls of varying duration. A longer call is more attractive to females, but producing long calls over a prolonged breeding bout is very costly from an energetic standpoint. Thus males give long calls only when they are competing with other nearby males of the same species. Similar behavior has been seen in other species, and the important point is that to at least a certain extent frogs may adjust their "programmed" calls to particular circumstances. In addition to duration, call rate and call complexity may change according to context (Wells and

Schwartz 1984). Even call dominant frequency may be modulated. Cricket frogs (*Acris crepitans blanchardi*) may lower their dominant frequency when they hear another nearby cricket frog. The lowering of frequency seems to signal the resident's ability or willingness to engage in a territorial fight (Wagner 1992). Indeed, although the advertisement call is the most conspicuous type of vocalization, most frogs produce a variety of other sounds that have different meanings and the most common type of nonadvertisement call is that given to maintain a territory (in many species the advertisement call plays the dual role of advertisement and territory maintenance/aggression). Acoustic defense of a resource is common in frogs, and like advertisement calls, specific aggressive vocalizations may be modulated according to context (Narins and Capranica 1978; Schwartz and Wells 1984; Wells and Bard 1987). Furthermore, a given frog may switch between advertisement and aggressive calls to suit the immediate situation. A review of the large variety of call types and their functions may be found in Wells (1977).

In general, it is male frogs who make conspicuous advertisement calls, but this is not strictly true. Female carpenter frogs (*Rana virgatipes*) produce a vocal response to the acoustic advertisement of male carpenter frogs (Given 1993a). The relatively stealthful existence of most female frogs has surely led to an underestimate of the number of species in which females make vocalizations of communicative significance.

3.7 Radiation Pattern and Habitat Effects

Despite several theoretical studies of environmental influences on sound propagation (for a recent review see Forrest 1994), only a few investigations have focused on specific problems relative to anuran acoustic communication. There are two issues here. First, has natural selection operated on advertisement calls of frogs that are adapted to particular microenvironments? Acoustic production could, in principle, be optimized for that particular microenvironment. For different subspecies of the frog *Acris crepitans*, different call structures characteristic of the subspecies do propagate farther in their relative preferred habitats (open vs. forest), leading to the suggestion that selection has indeed adapted the calls for maximum transmission (Ryan and Wilczynski 1991; Ryan et al. 1991).

Ryan and Sullivan (1989) found that the temporal structures of the advertisement calls of two toads (*Bufo valliceps* and *Bufo woodhousii*) were affected differently by environmental propagation in their natural habitat. If there are reflective surfaces in the environment, the receiver will encounter an acoustic signal that is temporally degraded due to multipath distortion. The most sensitive parameter of the advertisement call to this distortion is amplitude modulation percent. Depending on the modulation rate and pulse duration, calls of different species may differentially drop below the frog's detection threshold for amplitude modulation, and potentially then for species recognition, even though they are audible signals.

The second issue is whether a given calling frog selects a location to improve its broadcast. The radiation pattern of the advertisement call must depend, to some extent, on the physical features of the immediate calling site, but in fact there is little evidence that frogs choose a site for its acoustic qualities. Field measurements of calling frogs show that individuals of some species produce a uniform sound field, whereas individuals of other species produce directional fields, with major lobes 5 to 8 dB greater than the minor lobes (Gerhardt 1975). More information on radiation pattern and anuran acoustic active space would be welcome.

3.8 Ecological and Evolutionary Aspects

3.8.1 Species Isolation

The use of frogs to examine principles of evolutionary biology was begun rigorously in the 1950s starting with the work of W. Frank Blair and C.M. Bogert (see, for example, Blair 1958; Bogert 1960). The main focus was to look at frog calls as devices of speciation. Here is the idea: Different species of frogs have different advertisement calls. Indeed, each is referred to as "the species-specific advertisement call." It is inefficient, for several obvious reasons, to mate with the wrong species. Thus natural selection should favor female frogs who can discriminate one species call from that of another, especially if there is a chance of encountering, during the mating season or a time of day, a particular wrong species. Also, there is variation in the advertisement calls of individuals in a given species, and for some related species many spectral or temporal parameters of the call may overlap. Thus ecological theory would predict that in zones of sympatry, where populations of similar species interact, the calls would diverge, and selection would favor those male frogs from both populations with the most different calls and the females best able to discriminate the difference. This divergence is called character displacement. Conversely, where two populations are not likely to encounter each other, there is little selection pressure to have divergence of call parameters. Indeed, when many different species of frogs were analyzed, character displacement in call structure was observed in sympatric zones (Duellman 1967) and, in general, the advertisement call is more attractive to conspecifics than heterospecifics.

Just what is the extent of genetic hybridization in any two frog populations, and what is the effectiveness of one or more mating call parameters in preventing hybridization? Gerhardt has examined this issue extensively in North American hylid frogs. Females of different species are not always very selective about the species of male with which they will breed (Fig. 9.8). For example, *H. andersonii* and *H. cinerea* will hybridize, and in fact

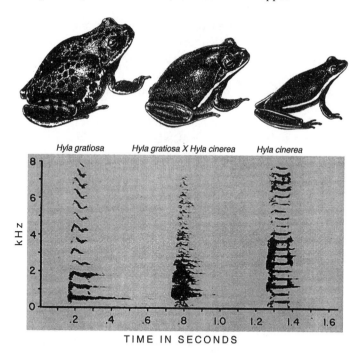

FIGURE 9.8. Sound spectrograms (sonograms) of two North American Hylids and the spectrogram of a hybrid individual. Note the rich harmonic structure in this call, which, interestingly, extends well beyond the hearing range of either species. For approximately 40 years the spectrograph, which produces this output of frequency composition vs. time and with darkness proportional to intensity, was the standard tool for analyzing animal vocalizations. (From Bogert 1960. © 1960 American Institute of Biological Sciences.)

female *H. andersonii* will respond to the calls of *H. cinerea* (Gerhardt 1974). On the other hand *H. crucifer* and *Pseudacris ornata* also can hybridize, but reject each other's advertisement calls, and thus appear to have more robust acoustically mediated behavioral isolation. Naturally occurring viable hybrids of *H. chrysoscelis* and *H. versicolor* can be found, but they are quite rare (probably much less than 0.5%) even though mismatings may occur about 7% to 10% of the time (Schlefer et al. 1986; Gerhardt et al. 1994a). The interpretation of these results is that hybrids are not particularly fit, genetically, and there should be strong natural selection against mismatings. Why then is the acoustic isolating mechanism not better? Female frogs may not be given a choice insofar as male frogs often attempt to mate with any small animal that moves nearby. Thus while females may be phonotactically attracted only to the species-specific mating call, in a mixed-species chorus encountering a male of the wrong species could result in unavoidable mismating. If this is common, females could in theory avoid heterospecific males by diverting their trajectory when they hear a conspecific nearby. This notion has recently been tested (Gerhardt et al. 1994a) with three sympatric species of North American hylids. The female hylids did not, in fact, avoid the heterospecific vocalizations. Perhaps the limit of auditory selectivity for the relevant call features has been reached, and these hylids must accept a certain amount of gametic wastage.

3.8.2 Sexual Selection

Despite the imperfect ability of some frogs to select the calls of their own species, much attention has been paid to investigating if female frogs can perform the more subtle task of acoustically judging the quality of a mate of her own species.

Males of nearly all species of frogs fertilize eggs externally, and parental care is certainly not the rule in amphibians. Compared with birds or mammals, this makes it difficult to judge mating success. Persistent, careful field work has been successful, however, in showing that indeed some males in a population are more successful at mating than their neighbors (Whitney and Krebs 1975; Licht 1976; Wells 1977). Assuming there is female choice, what male characteristics might a female frog be interested in? Size of the calling male is one obvious feature, because a large male has presumably been able to acquire more food resources, live longer, or both, and thus such a male has desirable genes. A number of studies have examined the notion that males of a larger size are more successful at mating (e.g., Howard 1978; Davies and Halliday 1979), and from these studies we conclude that indeed size is a factor. Is body size something that a male frog can communicate acoustically to a prospective mate? In both Bufonids (Gerhardt 1975) and Hylids (Fellers 1979) females choose the louder of two otherwise equal advertisement calls and it is the larger males that produce the louder calls. Furthermore, in many species the dominant frequency of the advertisement call is related to body size such that larger males produce lower frequency calls (Loftus-Hills and Littlejohn 1971; Morris 1991) and females can select larger males on the basis of dominant frequency (Ryan 1983b; Morris 1991). On the bases of call intensity or dominant frequency, then, females could assess male size.

In other anuran species, however, size is not correlated with mating success (Sullivan 1982; Passmore et al. 1992; Cherry 1993). In Woodhouse's toad (*Bufo woodhousei*), painted reed frogs (*Hyperolius marmoratus*), and the raucous toad (*Bufo rangeri*), females are instead interested in the rate at which calls are given, but call rate in these species is not correlated with male size. Indeed, raucous toad males that consistently call at higher rates lose weight faster than other males and are thus unlikely be the largest specimens! What other desirable characteristic might a male communicate with increased call rate?

It has recently been hypothesized that male frogs are selected to have traits that take advantage of female preferences regardless of whether the traits convey any accurate information about the male (Ryan and Keddy-Hector 1992; Ryan and Rand 1993). This idea comes from studies that show that females prefer an extreme variant on a relevant call parameter. For example, the North American cricket frog *Acris crepitans* prefers artificial calls lower in dominant frequency than any natural call (Ryan et al. 1992). Alternatively, females may prefer calling that demonstrates extreme prowess (i.e., the highest rate or longest duration).

3.9 Chorusing Behavior

3.9.1 Leks and Satellite Behavior

Female frogs are often not assessing isolated vocalizations. Rather it is a conspicuous feature, especially of temperate zone frogs, to find male frogs calling in a dense temporary assemblage called a chorus (Fig. 9.9). In some cases frog choruses may be considered leks (Ryan 1991; Bourne 1992), a particular type of mating system in which (1) there is an arena (lek) to which females are attracted, (2) the females choose males at the lek, but (3) the males control no resources of interest to the female at the lek (Bradbury 1981). A fourth criterion is the absence of male parental care. Indeed, in most frog species the only contribution males make to reproductive success is to provide gametes, and the other conditions for lekking are probably satisfied in many frog choruses as well. The significance of lekking in frogs is not clear, except that it is now known to occur in all vertebrate classes, but only sporadically. If frogs are found to satisfy the criteria for lekking on a widespread basis, more information as to the value of this mating system will accrue. Frog acoustic behavior as a model system to investigate the evolutionary significance of leks has already yielded rewards, as it has been shown that female choice can be based on a "direct" or immediate improvement rather than a genetic improvement of fitness (Bourne 1993).

It is clear that one of the main determinants of male reproductive success is the amount of time a male frog spends calling in a chorus (Ritke and Semlitsch 1991; Murphy 1994a,b; Townsend and Stewart 1994). For a male frog, probability favors attracting a mate if the male has a greater presence in the chorus. Thus we can define a supermale frog as having the lowest dominant frequency (see above), the highest call rate, the longest call duration, the loudest call, and the greatest persistence over days in the chorus. All of this effort comes at a great energetic cost, however, as will see below. It is interesting that for the last feature, persistence would work to a male's benefit if females arrive at the chorus at unpredictable, infrequent intervals. This advantage accrues because the probabilistic nature of encountering females simply favors males who are more commonly advertising. It is also possible that a female could instead make an assessment over a relatively prolonged period (days). There is a small amount of data

FIGURE 9.9. A variety of different frogs species may aggregate to form a mixed-chorus. Each has a preferred calling perch. Clockwise from upper right: *Hyla femoralis, Hyla squirella, Hyla cinerea, Bufo quercicus, Microhyla carolinensis, Acris gryllus, Bufo terrestris, Rana pipiens, Hyla gratiosa*. (From Bogert 1960. © 1960 American Institute of Biological Sciences.)

suggesting this is true (Sullivan 1990; Dyson et al. 1994) and more data on this topic would be welcome.

Not all the males in a chorus give advertisement calls. Most often, it is smaller males that do not call, yet they will still attempt to mate with females moving toward calling males. This sort of sexual parasitism is termed satellite behavior (Miyamoto and Cane 1980; Perrill et al. 1982). Both satellite and regular calling males are successful at mating (Forester and Lykens 1986; Ovaska and Hunte 1992). Driving satellite behavior is the advantage of not competing for calling sites, and/or reducing energetic costs associated with calling. Although the energetic argument is sensible, as yet there is no evidence that satellite males are physiologically weaker than other males who do call (Lance and Wells 1993). Lucas and colleagues (1996) have modeled the advantage to becoming a satellite male using the following hypothetical calling pattern: In their first year males do not call, in the second year they participate in the chorus, and in their third year die of old age. For first-year males, with less energy reserve, the optimum strategy is to call initially then switch to satellite behavior. Second-year males should continue to call and not become satellites. An interesting further effect that emerges from the mathematical model is alternating nights of calling, such that the chorus may be present one night and not a subsequent night, without prolonging the entire seasonal length of the chorus. It is interesting that the results of the model fit well the observed dynamic behavior of many temperate frog choruses.

3.9.2 Temporal Aspects of Chorus Life

When motivated, a male frog in isolation commonly produces advertisement calls at a more or less regular rate. Typical values range from one call per second to one call per several minutes. This nominal rate is controlled by a midbrain neuronal oscillator (Schmidt 1992). When two calling frogs can hear each other, however, very particular changes in the timing of calls may occur such that there is synchronization or alternation of calls. There have been numerous studies of these temporal interactions addressing the three basic issues: the extent to which frogs alternate or synchronize, the reason this might be important, and the mechanism of alternation or synchrony.

In at least some species, the timing of this oscillator is rapidly adjusted based on nearby acoustic events, such as the calls of neighbors (Zelick and Narins 1985; Schwartz 1991). The adjustments allow entrainment, which yields two observed outcomes: synchronization in which calls partially overlap, or alternation causing the perception by a human observer of dueting or antiphonal calling. In some cases the synchronization is fast enough that the calls of one individual almost completely overlap with those of a neighbor (Tuttle and Ryan 1982; Ryan 1986).

It is likely that both alternation and synchrony are two extremes of a single continuum and both share the same mechanism. Evoked calling is common in frogs, and the finer ability to rapidly adjust the neural call oscillator, allowing tracking of even randomly placed acoustic events (Zelick and Narins 1985), could grow out of this simple behavior. A particular delay relative to the acoustic trigger will yield the appearance of either synchrony or alternation. Interestingly, the original selective pressure to allow rapid adjustments in call oscillator timing may have been a female's inability to localize or judge overlapping or massed calls (Greenfield 1994). If the female prefers the leading call, simply on grounds of ease of localization or feature detection, then every frog would want to lead. One strategy to be a leader is to abort a call that is scheduled to coincide with your neighbor, and try again after an interval. This sort of communication strategy is not unlike modern packet transmission protocols of computer networks.

Call alternation or at least avoidance of temporal overlap has been seen in many frog species (Lemon 1971; Loftus-Hills 1974; Rosen and Lemon 1974; Awbrey 1978; Lemon and Struger 1980; Narins 1982; Schwartz and Wells 1983; Given 1993b), and one might assume a common obvious significance to the behavior. A reasonable presumption is that alternation avoids the deleterious alternative, namely jamming your neighbor's calls. Is there evidence that female frogs have difficulty localizing male frogs when their calls overlap? While this is widely assumed to be true, in at least one case females were found to accurately localize calls played from speakers even if there was complete overlap (Passmore and Telford 1981). Rather, it may be that call alternation is more important for (1) detection of fine-temporal structure within the call and (2) male-male spacing and territory maintenance. Calls that overlap in time and have particular pulses or trills within them will have those temporal patterns obscured by overlap (Schwartz 1987; Sullivan and Leek 1987; Given 1993b). Relative to acoustic territory maintenance, if a male frog is less able to hear another male's calls during, and for a short period after he vocalizes, it would be advantageous for neighbors to call following this refractory period. In this way the nearest neighbors maximize their own detectability (Schwartz 1987; Given 1993b).

Call alternation also occurs in mixed-species choruses. The Central American frog *Hyla ebraccata* calls in dense choruses with other *H. ebraccata* males but also with *Hyla microcephala* males. Female *H. ebraccata* are less attracted to male *H. ebraccata* calls when the male *H. ebraccata* calls overlap *H. microcephala* calls (Schwartz and Wells 1984, 1985). *H. ebraccata* males normally alternate calls with *H. microcephala* and presumably increase their chance of mating.

The issue of male-male spacing in a chorus is an interesting one. Male frogs are often aggressive toward conspecific males; thus, a balance must be struck between the tendency to be aggressive and the need to participate in the chorus. This is particularly true when the density of calling males and their vocal activity is high. As the number of frogs in a chorus increases, one would expect that if male-spacing and aggression are regulated by acoustic cues, then the chorus boundary should grow so that chorus density would stay the same. In fact, the density of the chorus increases (Fellers 1979; Gerhardt et al. 1989; Dyson and Passmore 1992). What happens in choruses when the density of males increases? If the chorus has males which *alternate* calls, then only the nearest neighbors will alternate and further neighbors are ignored (Brush and Narins 1989; Schwartz 1993, 1994). In addition, the tolerance threshold for aggressive behavior increases. This is a compromise because time spent in aggressive interactions lessens the time available for advertisement calling (Wells 1988). In such choruses density can only increase to a point (Narins 1982). In the case of *H. marmorata* choruses, in which males do not alternate calls, the regular spacing of calling frogs at low densities becomes random at high densities (Dyson and Passmore 1992). In these densest of choruses females can still find males because they seem to attend to a local set of males. While females prefer spaced males, they can still localize and evaluate more crowded calling frogs (Telford 1985; Gerhard and Klump 1988; Brush and Narins 1989).

It is clear that there are advantages to alternation of calls, but what is the advantage of synchrony? There is one case where it appears to be adaptive to synchronize calls: The tropical frog *Smilisca sila* produces advertisement calls spaced widely in time (about one call every minute) and there is no obvious rhythm in the calling. Due to the ability of *Smilisca* to rapidly track acoustic signals (with an evoked call delay as short as 55 ms!), the calls of neighbors sometimes overlap almost completely. This behavior makes a given frog more difficult to localize by a predatory bat, which uses the frog's call as a homing signal (Tuttle and Ryan 1982; Ryan 1986). Antiphonal calling can also depend on very short latencies of around 60 ms (Walkowiak 1992), and more work should be done on the neuronal bases of these fast non-reflex behaviors.

Finally, there is the case of *H. microcephala* males, which permit gross temporal overlap of calls with neighbors but in a very specific way (Schwartz and Wells 1985). Each call is composed of a series of pulses and the overlapping calls are temporally positioned so that individual pulses of one call and those of the neighbor's call interdigitate. Again such behavior requires remarkably precise triggered oscillator timing.

3.10 Costs of Communication

3. 10.1 Energetic Costs

Displays used for mating advertisement can be essentially free. For example, the colorful plumage of birds and other animals does not represent a significant energetic cost to make or maintain. Acoustic displays are another matter, however. The long-term acoustic output of a frog is proportional to the power in each note of the call, the duration of the notes, and the rate at which notes are given. In most cases frogs do not modulate the intensity of their calls, and from a theoretical standpoint they should call at the highest rate sustainable if calling effort is correlated with mating success (Ryan 1988).

Interestingly, the first study to examine the energetic cost of calling in frogs was done on the same species for which another very interesting cost of display has been studied. Bucher and colleagues (1982) placed calling male *Physalaemus pustulosus*, a small (2 g) Central American frog, in respirometer chambers and measured the oxygen consumption during resting and calling periods. They found that the mean energy expenditure of calling males is twice the expenditure during resting. In other words, advertisement calling is costly! Interestingly, at higher call rates, the energetic cost per call goes down, possibly because air is shuttled more efficiently between the lungs and vocal sac.

Frogs that give more intense calls spend even more energy on advertisement. The gray treefrog (*Hyla crucifer*) produces extraordinarily intense calls of around 110 dB (re: 20 μPa) at 50 cm, and makes over 1200 such calls per hour. This is sustained for 2 to 3 hours each night (Gerhardt 1973; Rosen and Lemon 1974). A 10-g gray treefrog may, just by calling, increase its energy expenditure from a resting value of 13 joules/hour to near 300 joules/hour (Taigen and Wells 1985; Wells and Taigen 1986). Thus it is not surprising that male frogs, but not female frogs, lose considerable body mass over a breeding season (Grafe et al. 1992) and that the number of hours during which a particular frog chorus is active declines throughout the breeding season (Runkle et al. 1994).

The high metabolic cost of sound production implies further that calling is not very efficient, and indeed this seems to be true. Prestwich (1994) estimates an efficiency (acoustic power/net metabolic power) of between 0.05% and 6.0%, considerably less than the efficiency of locomotion, which has an efficiency between 10% and 20%, and for which the frog uses less total energy. Finally, the number of calls given per unit time is also temperature dependent and varies linearly with oxygen consumption (Wells et al. 1996). Thus the metabolic cost of calling increases with ambient temperature.

3.10.2 Predation Costs

Predators use all sensory means available to them to locate potential prey items. Thus in many cases organisms have become extremely stealthful, using such techniques as cryptic coloration, to avoid becoming a meal. This poses a problem for *intra*-specific communication, because it is at the same time impossible to be entirely cryptic yet broadcast information about yourself. One of the most elegant studies of the evolutionary consequences of incidental communication to a predator involves the same Central American frog described above, *Physalaemus pustulosus* (Ryan 1985). This frog gives an advertisement call of variable complexity. The first part is a frequency modulated "whine" and is always produced. The second part is one to six harmonically rich "chucks." Females are attracted to the whine, but are more attracted to the whine plus chucks. Thus natural selection should favor males who append lots of chucks onto their whines. Unfortunately for *Physalaemus*, the bat *Trachops cirrosus* is a predator who finds the calling males by listening to their advertisement calls. Presumably because of the broad frequency spectrum of the chucks, the bats more easily localize frogs producing chucks compared with those that produce only whines. Thus there is competing selection pressure on the frogs to produce no chucks. The compromise is that when males do not detect conspecifics calling, they produce only whines. Without competition from other

Physalaemus, they will be attractive enough to females. In the presence of other calling males, however, the males begin adding chucks, balancing the need to be more attractive than their neighbors with the increased likelihood that they will be eaten by a bat.

3.11 Physical Environmental Factors

Amphibians find themselves in a special situation with regard to their environmental physiology. Amphibians are ectothermic, thus they have only behavioral means to regulate their body temperature. In practice the body temperature of most terrestrial frogs is not well controlled and this has consequences for acoustic communication. In addition, while nearly all species of frogs and toads are terrestrial as adults, they have no skin barriers to water loss and indeed lose water at the same rate as a free surface of water of the same surface area (Shoemaker et al. 1992). Terrestrial amphibians must thus contend with both temperature and serious dehydrational stress.

3.11.1 Temperature Effects

Reptiles are well known for behaviors such as basking and making postural changes to regulate their body temperature (Avery 1972). On theoretical grounds it should be less advantageous for amphibians to bask because they, unlike reptiles, cannot use basking to maintain a body temperature much higher than the ambient air temperature: as the amphibian warms up, evaporative cooling compensates for the thermal radiation. As a consequence frogs are best considered eurythermal, that is, operating as well as possible over a wide range of temperatures (Putnam and Bennett 1981; Renaud and Stevens 1983). This strategy can work because, compared with most reptiles, frogs tend not to be active foraging predators, instead using a sit-and-wait strategy for obtaining food.

   Although not as useful as for reptiles, there are nevertheless some frogs that do seem to behaviorally thermoregulate by basking in the sun. Members of three families of anurans (Ranidae, Hylidae, and Bufonidae) have been observed to bask in the wild and in some cases shuttle between a sunlit bank and the water during the day to maintain a high temperature (Lillywhite 1970; Valdiviesio and Tamsitt 1974; Carey 1978; Bradford 1984). Compared to the small number of field observations, there are numerous lab studies showing that both larval (tadpoles) and adult amphibians select preferred body temperatures (for review see Hutchison and Dupre 1992). The difference between the field and laboratory data may be explained by a cost-benefit analysis. In the wild, behavioral means of body temperature regulation only seems to occur when it is not very costly. Thus a lizard might move from shade to sun if this is a relatively short distance, but the lizard will not climb a 10-m tree to find sun (Withers and Campbell 1985).

For acoustic communication the expected consequence of these temperature effects is for one or more call features to vary with temperature, and sometimes in the range of temperatures within which the frog must defend a territory or advertise for a mate. Clearly such temperature variation in call parameters could create problems for female frogs attempting to find a male of the correct species when the localization is based on that male's call!

As a rule the spectral parameters of frog calls do not change greatly with temperature (Blair 1958; Lorcher 1969; Heinzmann 1970; Schneider and Eichelberg 1974). The temperature Q_{10} for the spectral features of a number of frogs and toads is approximately 0.2, which suggests remarkably good stability. For comparison, a typical value for the change in vertebrate nerve conduction velocity with temperature is 1.8, or nine times larger (Schmidt-Nielsen 1993).

Despite the small changes in vocalization carrier frequency with temperature, several studies have examined this effect in relation to the frequency tuning of the peripheral auditory system. Ideally, any temperature-dependent change in the male's call carrier frequency should be matched by an equal change in female preference, given two important assumptions. First, it must be the case that both females and the males they are trying to find are at the same temperature. If there are microenvironmental differences between male calling sites and the paths females take to find the males, the usefulness of temperature matching is questionable. Second, we assume that a given temperature-dependent feature of the male's call is important for the female in terms of localization and/or judgment of quality. Perhaps the most labile call parameter will respect to temperature is pulse or trill modulation rate. For this parameter the female's preference for the species-specific advertisement call shows a matched temperature dependence: Colder female gray tree frogs (*Hyla versicolor*), for example, prefer male calls that have a modulation pulse rate expected from a cold male, and warmer females prefer the pulse rate associated with a warmer male (Gerhardt 1978; Gerhardt and Mudry 1980).

3.12 Genetic Coupling of Sound Production and Generation

The temperature coupling experiments described above suggest a close association between the vocal control system and the sound detection systems of anurans. Such an association has been more directly demonstrated in experiments where hybrids of two different frog species (*Hyla chrysoscelis* and *H. femoralis*) were made by artificial crossing (Doherty and Gerhardt 1984). Such hybrid males produce a call that is intermediate in temporal modulation properties, and it is this hybrid call that female hybrids find more attractive, when given the choice between it and either of the parental-type calls. Figure 9.8 shows a similar situation from one of the first documented records of hybridization effects on vocalization. Thus there is a genetic linkage between the sound generating portions of the frogs brain, and those involved in recognizing important features of another frog's vocalizations. This is similar to the situation described for insects with acoustic communication such as crickets (Hoy et al. 1977). Interestingly, the fact that hybrids reveal themselves as acoustic intermediates has also been used as a tool to verify the occurrence of natural hybrids (Gerhardt et al. 1994b).

It is interesting that some of the coupling between the sender and receiver comes about because of mechanical morphometric factors. Wilczynski and colleagues (1993) found that in three related species of hylids both the temporal and spectral features of the advertisement call change with head and laryngeal muscle size, the latter being characteristically different in the three species. Head size, as it relates to outer and middle ear structure size, also defines the preferred transmission of sound to the inner ear. Thus a female of a larger species will prefer sound generated by a male of the larger species, etc.

3.13 Conclusions

There is now quite a large amount of behavioral data on frog acoustic communication. Studies of the underlying neural substrates is lagging. For example, frogs would be ideal for examination of the neuronal processes involved in making context-dependent judgments, as is done when call types are switched following detection of a particular acoustic cue. In the area of ecology and evolution of signaling and mating systems, again we know much but there are interesting issues to be resolved. For instance, sometimes variation in call parameters is linked to female choice and mating success, but in other cases there is no link. Why should this be so? Furthermore, in some cases female frogs are interested in extreme variants of the advertisement call, but in other cases females are both behaviorally and neurophysiologically "tuned" to a particular value of a parameter (such as trill rate). Finally, more work would be welcome providing field data on hormonal variation during such social activities as aggression, mating, and advertisement calling with the notion of learning the interplay between endocrine modulation and the operation of a frog's acoustic communication paradigm. Recent work on a voiceless frog with male parental care (Emerson et al. 1992) is a good example as are the very nice studies of Walkowiak (e.g., Walkowiak and Luksch 1994).

4. Summary

This chapter has presented overviews of what is known about sound communication by two of the major vertebrate groups, bony fishes (there are no indications that cartilaginous or jawless fishes produce sounds) and frogs. It is apparent from this overview that considerably more is known about sound communication in frogs than in fishes, and much of the explanation for these differences result from the problems associated with studying acoustic behavior underwater. Moreover, the investigators working on frogs can benefit not only from the relative ease of studying their species, but also from the methodology developed for parallel studies in birds.

For the most part, investigators interested in fish communication would like to be able to ask many of the same questions that have already been asked about frogs. Important questions concerning intraspecific and interspecific variation in calls, effects of selective pressure on sound content, and detailed analyzes of male-female interactions and the use of sounds that have been so elegantly answered for frogs need to be investigated for fishes. While it is improbable that the sounds produced by fishes are learned, there have been no hybridization studies on fishes, as there have been for frogs. Finally, while much is known about the energetics of frog sound production, nothing is known about fishes. How much energy does it take for a midshipman to contract its sonic muscles and produce sound for hours?

Perhaps the one area in which fish acoustics investigations lead those for frogs is in the understanding of sound detection (see Fay and Megala Simmons, Chapter 7). Since fishes are far more amenable to conditioned behavioral investigations of hearing capabilities than frogs, we have a good sense of the kinds of sounds fish can detect and discriminate. Although we still need more data for sound-producing fishes, far fewer data are available for frogs due to the inherent difficulty of training frogs to respond behaviorally in the presence of sounds. This does not mean, of course, that we don't know a good deal about frog hearing. Indeed, as described by Fay and Megala Simmons (Chapter 7) and Lewis and Narins (Chapter 4), we know a good deal about what frogs hear, and both peripheral and central (McCormick, Chapter 5) mechanisms and structures associated with sound detection. Research on the neurophysiological bases of alternative mating tactics in fishes is one area in which studies on fish communication can provide a guide for future studies in frog communication.

At this point, investigations of fish acoustic communication lags behind that of frog acoustic communication. With the advent of new techniques, one would hope that more extensive and sophisticated data will become available for fishes. Indeed, with fishes being by far the largest of all vertebrate groups (e.g., 25,000–30,000 extant species), it would be a wonder if fishes did not only parallel many of the behaviors seen in frogs, but also demonstrate a range of behaviors and uses of sound that are vastly different.

Acknowledgments. R.Z. expresses thanks to C. Cookus for library assistance. The authors thank Drs. Richard R. Fay and William N. Tavolga for reading and commenting on a draft of the manuscript. Preparation of this chapter was supported in part by National Institutes of Health (NIH) training grant DC-00046-02 from the National Institute of Deafness and Other Communicative Disorders to D.M.

References

Aitkin PG, Capranica RR (1984) Auditory input to a vocal nucleus in the frog *Rana pipiens*: hormonal and seasonal effects. Exp Brain Res 57:33–39.

Amiet JL (1989) Photographs of amphibians from Cameroon Island burying behavior and open mouth phonation in *Conraua crassiper* (Buchholz and Peters 1875). Alytes (Paris) 8:99–104.

Avery RA (1972) Field studies of body temperatures and thermoregulation. In: Gans C, Pough FH (eds) Biology of the Reptilia. London: Academic Press, pp. 93–166.

Awbrey FT (1978) Social interaction among chorusing Pacific Tree Frogs, *Hyla regilla*. Copeia 1978:208–214.

Bass AH (1985) Sonic motor pathways in teleost fishes: a comparative HRP study. Brain Behav Evol 27:115–131.

Bass AH (1990) Sounds from the intertidal zone: vocalizing fish. Am Sci 40:249–258.

Bass AH (1992) Dimorphic male brains and alternative reproductive tactics in a vocalizing fish. Trends Neurosci 15:139–145.

Bass AH (1993) From brains to behaviour: hormonal cascades and alternative mating tactics in teleost fishes. Rev Fish Biol Fish 3:181–186.

Bass AH, Baker R (1990) Sexual dimorphisms in the vocal control system of a teleost fish: morphology of physiologically identified neurons. J Neurobiol 21:1155–1168.

Bass AH, Baker R (1991) Evolution of homologous vocal control traits. Brain Behav Evol 38:240–254.

Bass AH, Marchaterre MA (1989) Sound-generating (sonic) motor system in a teleost fish (*Porichthys notatus*): sexual polymorphisms and general synaptology of sonic motor nucleus. J Comp Neurol 286:154–169.

Blair WF (1958) Mating call in the speciation of anuran amphibians. Am Nat 92:27–51.

Bogert Cochlear Microphonics (1960) The influence of sound on the behavior of amphibians and reptiles. In: Lanyon WE, Tavolga WN WE (eds) Animal Sounds and Communication. Washington, DC: American Institute of Biological Sciences, pp. 137–320.

Bourne GR (1992) Lekking behavior in the neotropical frog *Ololygon rubra*. Behav Ecol Sociobiol 31:173–180.

Bourne GR (1993) Proximate costs and benefits of mate acquisition at leks of the frog *Ololygon rubra*. Anim Behav 45:1051–1059.

Boyd SK (1992) Sexual differences in hormonal control of release calls in bullfrogs. Horm Behav 26:522–535.

Bradbury JW (1981) The evolution of leks. In: Alexander RD, Tinkle D (eds) Natural Selection and Social Behavior. New York: Chiron Press, pp. 138–169.

Bradford DF (1984) Temperature modulation in a high elevation amphibian, *Rana muscosa*. Copeia 1984:966–976.

Brawn VM (1961) Sound production by the cod (*Gadus callarias* L.). Behav 18:239–255.

Breder CM Jr (1968) Seasonal and diurnal occurrences of fish sounds in a small Florida bay. Bull Am Mus Nat Hist 138:327–378.

Brush JS, Narins PM (1989) Chorus dynamics of a neotropical amphibian assemblage: comparison of computer simulation and natural behavior. Anim Behav 37:33–44.

Bucher TL, Ryan MJ, Bartholomew A (1982) Oxygen consumption during resting, calling, and nest building in the frog *Physalaemus pustulosus*. Physiol Zool 55:10–22.

Burkenroad MD (1930) Sound production in the Haemulidae. Copeia 17–18.

Capranica RR (1968) The vocal repertoire of the bullfrog (*Rana catesbeiana*). Behavior 31:302–325.

Carey C (1978) Factors affecting body temperatures of toads. Oecologica (Berlin) 35:179–219.

Catz DS, Fischer LM, Moschella MC, Tobias ML, Kelley DB (1992) Sexually dimorphic expression of a laryngeal-specific, androgen-regulated myosin heavy chain gene during *Xenopus laevis* development. Dev Biol 154:366–376.

Chen K-C, Mok H-K (1988) Sound production in the anemonefishes, *Amphiprion clarkii* and *A. frenatus* (Pomacentridae), in captivity. Jpn J Ichthyol 35:90–97.

Cherry MI (1993) Sexual selection in the raucous toad *Bufo rangeri*. Anim Behav 45:359–373.

Clay CS, Medwin H (1977) Acoustical Oceanography: Principles and Applications. New York: Wiley.

Cocroft RB, Ryan MJ (1995) Patterns of advertisement call evolution in toads and chorus frogs. Anim Behav 49:283–303.

Connaughton MA, Taylor MH (1996) Drumming, courtship, and spawning behavior in captive weakfish, *Cynoscion regalis*. Copeia 195–199.

Crawford JD, Hagedorn M, Hopkins CD (1986) Acoustic communication in an electric fish, *Pollimyrus isidori* (Mormyridae). J Comp Physiol [A] 159:297–310.

Crawford JD, Cook AP, Heberlein AS (1997a) Bioacoustic behavior of African fishes (Mormyridae): potential cues for species and individual recognition in *Pollimyrus*. J Acoust Soc Am 102:1–13.

Crawford JD, Jacob P, Bénch V (1997b) Sound production and reproductive ecology of strongly acoustic fish in Africa: *Pollimyrus isidori*, Mormyridae. Behavior 134:1–49.

Darwin C (1874) The Descent of Man, 2nd ed. New York: H.H. Caldwell.

Daugherty J, Marshall JA (1976) The sound producing mechanism of the croaking gourami, *Trichopsis vittatus*, (Pisces, Belontiidae). Am Zool 11:227–244.

Davies NB, Halliday TR (1979) Competitive mate searching in male common toads *Bufo bufo*. Anim Behav 27:1253–1267.

Diakow C (1978) A hormonal basis for breeding behavior in female frogs: vasotocin inhibits the release call of *Rana pipiens*. Science 199:1456–1457.

Diakow C, Nemiroff A (1981) Vasotocin, prostaglandin, and female reproductive behavior in the frog, *Rana pipiens*. Horm Behav 15:86–93.

Doherty JA, Gerhardt HC (1984) Acoustic communication in hybrid treefrogs: sound production by males and selective phonotaxis by females. J Comp Physiol 154:319–330.

Dôtu Y (1951) On the sound producing mechanism of a scorpaenoid fish, *Sebasticus marmoratus*. Kyushu Imp Univ Dept Agri Bull Sci 13:286–288.

Duellman WE (1967) Courtship isolating mechanisms in Costa Rican hylid frogs. Herpetologica 23:169.

DuFossé M (1874) Recherches sur les bruits et les sons expressifs que font entendre les poissons d'Europe. Ann Sci Nat Ser 5 19:1–53; 20:1–134.

Dyson ML, Passmore NI (1992) Inter-male spacing and aggression in African painted reed frogs, *Hyperolius marmoratus*. Ethology 91:237–247.

Dyson ML, Henzi SP, Passmore NI (1994) The effect of changes in the relative timing of signals during female phonotaxis in the reed frog, *Hyperolius marmaoratus*. Anim Behav 48:679–685.

Emerson SB, Rowsemitt CN, Hess DL (1992) Androgen levels in a Bornean voiceless frog, *Rana blythi*. Can J Zool 71:196–203.

Fay RR (1970) Auditory frequency discrimination in the goldfish (*Carassius auratus*). J Comp Physiol Psychol 73:175–180.

Fay RR (1972) Perception of amplitude-modulated auditory signals by the goldfish. J Acoust Soc Am 52:660–666.

Fay RR (1980) Psychophysics and neurophysiology of temporal factors in hearing by the goldfish: amplitude modulation detection. J Neurophysiol 44:312–332.

Fay RR (1982) Neural mechanisms of an auditory temporal discrimination by the goldfish. J Comp Physiol 147:201–216.

Fay RR (1988) Hearing in Vertebrates: A Psychophysics Databook. Winnetka, IL: Hill-Fay Associates.

Fay RR, Passow B (1982) Temporal discrimination in the goldfish. J Acoust Soc Am 72:753–760.

Fellers G (1979) Aggression, territoriality, and mating behaviour in North American treefrogs. Anim Behav 27:107–119.

Fine ML (1978) Seasonal and geographical variation of the mating call of the oyster toadfish *Opsanus tau* L. Oecologica 36:45–57.

Fine ML, Lenhardt ML (1983) Shallow-water propagation of the toadfish mating call. Comp Biochem Physiol 76A:225–231.

Fine ML, Mosca PJ (1989) Anatomical study of the innervation pattern of the sonic muscle of the oyster toadfish. Brain Behav Evol 34:265–272.

Fine ML, Mosca PJ (1995) A Golgi and horseradish peroxidase study of the sonic motor nucleus of the oyster toadfish. Brain Behav Evol 45:123–137.

Fine ML, Economos D, Radtke R, McClung JR (1984) Ontogeny and sexual dimorphism of the sonic motor nucleus in the oyster toadfish. J Comp Neurol 225:105–110.

Fish JF, Cummings WC (1972) A 50-dB increase in sustained ambient noise from fish (*Cynoscion xanthulus*). J Acoust Soc Am 52:1266–1270.

Fish MP, Mowbray WH (1970) Sounds of Western North Atlantic Fishes. Baltimore: Johns Hopkins Press.

Forester DC, Lykens DV (1986) Significance of satellite males in a population of spring peepers (*Hyla crucifer*). Copeia 719–724.

Forrest TG (1994) From sender to receiver: propagation and environmental effects on acoustic signals. Am Zool 34:644–654.

Forrest TG, Miller GL, Zagar JR (1993) Sound propagation in shallow water: implications for acoustic communication by aquatic animals. Bioacoustics 4:259–270.

Gainer H, Kusano K, Mathewson RF (1965) Electrophysiological and mechanical properties of squirrelfish sound-producing muscle. Comp Biochem Physiol 14:661–671.

Gans C (1973) Sound production in the Salientia: mechanism and evolution of the emitter. Am Zool 13:1179–1194.

Gerhardt HC (1973) Reproductive interactions between *Hyla crucifer* and *Pseudacris ornata* (Anura: Hylidae). Am Mid Nat 89:81–88.

Gerhardt HC (1974) Behavioral isolation of the treefrogs, *Hyla cinerea* and *Hyla andersonii*. Am Mid Nat 91:424–433.

Gerhardt HC (1975) Sound pressure levels and radiation patterns of the vocalizations of some North American frogs and toads. J Comp Physiol 102:1–12.

Gerhardt HC (1978) Temperature coupling in the vocal communication system of the gray treefrog, *Hyla versicolor*. Science 199:992.

Gerhardt HC, Klump GM (1988) Masking of acoustic signals by the chorus background noise in the green tree frog: a limitation on mate choice. Anim Behav 36:1247–1249.

Gerhardt HC, Mudry KM (1980) Temperature effects on frequency preferences and mating call frequencies in the green treefrog, *Hyla cinerea*. J Comp Physiol 137:1–6.

Gerhardt HC, Klump GM, Diekamp B, Ptacek M (1989) Inter-male spacing in choruses of the spring peeper, *Pseudacris (Hyla) crucifer*. Anim Behav 38:1012–1024.

Gerhardt HC, Dyson ML, Tanner SD, Murphy CG (1994a) Female treefrogs do not avoid heterospecific calls as they approach conspecific calls: implications for mechanisms of mate choice. Anim Behav 47:1323–1332.

Gerhardt HC, Ptacek MB, Barnett L, Torke KG (1994b) Hybridization in the diploid-tetraploid treefrogs: *Hyla chrysoscelis* and *Hyla versicolor*. Copeia 51–59.

Given MF (1993a) Male response to female vocalizations in the carpenter frog, *Rana virgatipes*. Anim Behav 46:1139–1149.

Given MF (1993b) Vocal interactions in *Bufo woodhousii fowleri*. J Herpet 27:447–452.

Graft TU, Schmuck R, Linsenmair KE (1992) Reproductive energetics of the African reed frogs *Hyperolius viridiflavus* and *Hyperolius marmoratus*. Physiol Zool 65:153–171.

Gray G-A, Winn HE (1961) Reproductive ecology and sound production of the toadfish, *Opsanus tau*. Ecology 42:274–282.

Greene CW (1924) Physiological reactions and structure of the vocal apparatus of the California singing fish *Porichthyes notatus*. Am J Physiol 70:496–499.

Greenfield MD (1994) Synchronous and alternating choruses in insects and anurans: common mechanisms and diverse functions. Am Zool 34:605–615.

Guest WC (1978) A note on courtship behavior and sound production of red drum. Copeia 337–338.

Hannigan P, Kelley DB (1986) Androgen-induced alterations in vocalizations of female *Xenopus laevis*: modifiability and constraints. J Comp Physiol [A] 158:517–527.

Hawkins AD, Rasmussen KJ (1978) The calls of gadoid fish. J Mar Biol Assoc UK 58:891–911.

Hawkins AD, Myrberg AA Jr (1983) Hearing and sound communication under water. In: Lewis B (ed) Bioacoustics, A Comparative Approach. London: Academic Press, pp. 347–405.

Heinzmann U (1970) Untersuchungen zur Bio-Akustic und Okologie der Geburtshelferkrote, *Alytes o obstretricans* (Laur). Oecologica 19–55.

Henwood K, Fabrick A (1979) A quantitative analysis of the dawn chorus: temporal selection for communicatory optimization. Am Nat 114:260–274.

Horch K, Salmon M (1973) Adaptations to the acoustic environment by the squirrelfishes *Myripristis violaceus* and *M. pralinius*. Mar Behav Physiol 2:121–139.

Howard RD (1978) The evolution of mating strategies in bullfrogs, *Rana catesbeiana*. Evolution 32:850–871.

Hoy RR, Hahn J, Paul RC (1977) Hybrid cricket auditory behavior: evidence for genetic coupling in animal communication. Science 195:82–83.

Hutchison VH, Dupre RK (1992) Thermoregulation. In: Feder ME, Burggren WW (eds) Environmental Physiology of the Amphibians. Chicago: University of Chicago Press, pp. 206–249.

Ibara RM, Penny LT, Ebeling AW, Dykhuizen GV, Cailliet G (1983) The mating call of the plainfin midshipman fish, *Porichthys notatus*. In: Noakes DLG (ed) Predators and Prey in Fishes. Netherlands: Dr. W. Junk, pp. 205–212.

Jacobs DW, Tavolga WN (1968) Acoustic frequency discrimination in the goldfish. Anim Behav 16:67–71.

Kacelnik A, Krebs JR (1983) The dawn chorus in the great tit (*Parus major*): proximate and ultimate causes. Behavior 83:287–309.

Karino K (1995) Male-male competition and female mate choice through courtship display in the territorial damselfish *Stegates nigricans*. Ethology 100:126–138.

Kelley DB, Morrell JJ, Pfaff DW (1975) Autoradiographic localization of hormone-concentrating cells in the brain of an amphibian, *Xenopus laevis*. I. Testosterone. J Comp Neurol 164:47–62.

Kelley DB, Fenstemaker S, Hannigan P, Shih S (1988) Sex differences in the motor nucleus of cranial nerve IX–X in *Xenopus laevis*: a quantitative Golgi study. J Neurobiol 19:413–429.

Knapp RA (1995) Influence of energy reserves on the expression of a secondary sexual trait in male bicolor damselfish, *Stegastes partitius*. Bull Mar Sci 57:672–681.

Knapp RA, Kovach JT (1991) Courtship as an honest indicator of male parental quality in the bicolor damselfish, *Stegastes partitus*. Behav Ecol 2:295–300.

Knudsen VO, Alford RS, Emling JW (1948) Underwater ambient noise. J Mar Res 7:410–429.

Ladich F, Fine ML (1992) Localization of pectoral fin motoneurons (sonic and hovering) in the croaking gourami *Trichopsis vittatus*. Brain Behav Evol 39:1–7.

Ladich F, Fine ML (1994) Localization of swimbladder and pectoral motoneurons involved in sound production in Pimelodid catfish. Brain Behav Evol 44:86–100.

Lance SL, Wells KD (1993) Are spring peeper satellite males physiologically inferior to calling males? Copeia 1162–1166.

Lemon RE, Struger J (1980) Acoustic entrainment to randomly generated calls by the frog, *Hyla crucifer*. J Acoust Soc Am 67:2090–2095.

Lewis ER, Narins PM (1985) Do frogs communicate with seismic signals? Science 227:187–189.

Licht LE (1976) Sexual selection in toads (*Bufo americanus*). Can J Zool 54:1277–1284.

Lillywhite HB (1970) Behavioral temperature regulation in the bullfrog. Copeia 1970:158–168.

Liu CC (1935) Types of vocal sac in the Salientia. Proc Boston Soc Nat Hist 41:19–40.

Lobel PS (1991) Sounds produced by spawning fishes. Environ Biol Fish 33:351–358.

Lobel PS, Mann DA (1995) Spawning sounds of the domino damselfish, *Dascyllus albisella* (Pomacentridae), and the relationship to male size. Bioacoustics 6:187–198.

Loftus-Hills JJ (1974) Analysis of an acoustic pacemaker in Streker's chorus frog, *Pseudacris streckeri* (Anura, Hylidae). J Comp Physiol 90:75–97.

Loftus-Hills JJ, Littlejohn MJ (1971) Mating call sound intensities of anuran amphibians. J Acoust Soc Am 49:1327–1329.

Lopez PT, Narins PM, Lewis ER, Moore SW (1988) Acoustically induced call modification in the white-lipped frog, *Leptodactylus albilabris*. Anim Behav 36:1295–1308.

Lorcher K (1969) Vergleichende bio-akustische Untersuchugen and der Rot-und Gelbbauchunke, *Bombina bombina* (L) und *Bombina v variegata* (L). Oecologica 3:84–124.

Lucas JR, Howard RD, Palmer JG (1996) Callers and satellites: chorus behavior in anurans as a stochastic dynamic game. Anim Behav 51:501–518.

Lugli M, Pavan G, Torricelli P, Bobbio L (1995) Spawning vocalizations in male freshwater gobiids (Pisces, Gobiidae). Environ Biol Fish 43:219–231.

Luh HK, Mok HK (1986) Sound production in the domino damselfish *Dascyllus trimaculatus* (Pomacentridae) under laboratory conditions. Jpn J Ichthyol 33:70–74.

Mann DA (1995) Bioacoustics and reproductive ecology of the damselfish *Dascyllus albisella*. Ph.D. Thesis, MIT/Woods Hole Oceanographic Institution.

Mann DA, Lobel PS (1995) Passive acoustic detection of sounds produced by the damselfish, *Dascyllus albisella* (Pomacentridae). Bioacoustics 6:199–213.

Mann DA, Lobel PS (1997) Propagation of damselfish (Pomacentridae) courtship sounds. J Acoust Soc Am 101:3783–3791.

Mann DA, Bowers-Altman J, Rountree RA (1997) Sounds produced by the striped cusk-eel *Ophidion marginatum* (Ophidiidae) during courtship and spawning. Copeia ••.

Marshall NB (1967) Sound-producing mechanisms and the biology of deep-sea fishes. In: Tavolga, WN (ed) Marine Bio-Acoustics II. Oxford: Pergamon Press, pp. 123–133.

Marten K, Marler P (1977) Sound transmission and its significance for animal vocalization. Behav Ecol Sociobiol 2:271–290.

Martin WF (1971) Mechanics of sound production in toads of the genus *Bufo*: passive elements. J Exp Zool 176:273–294.

Martin WF (1972) Evolution of vocalization in the Genus *Bufo*. In: Blair WF (ed) Evolution in the Genus *Bufo*. Austin, TX: University of Texas Press, pp. 279–309.

Martin WF, Gans C (1972) Muscular control of the vocal tract during release signaling in the toad, *Bufo valliceps*. J Morphol 137:1–28.

Matsui M, Wu GF, Yong HS (1993) Acoustic characteristics of three species of the genus *Amolops* (Amphibia, Anura, Ranidae). Zool Sci (Tokyo) 10:691–695.

Miyamoto MM, Cane JH (1980) Behavioral observations of non-calling males in Costa Rican *Hyla ebraccata*. Biotropica 12:225–227.

Morris MR (1991) Female choice of large males in the tree frog *Hyla ebraccata*. J Zool (Lond) 223:371–378.

Moulton JM (1960) Swimming sounds and the schooling of fishes. Biol Bull 119:210–223.

Murphy CG (1994a) Chorus tenure of male barking treefrogs, *Hyla gratiosa*. Anim Behav 48:763–777.

Murphy CG (1994b) Determinants of chorus tenure in barking treefrogs (*Hyla gratiosa*). Behav Ecol Sociobiol 34:285–294.

Myrberg AA Jr (1972) Ethology of the bicolor damselfish, *Eupomacentrus partitus* (Pisces: Pomacentridae); A comparative analyiss of laboratory and field behavior. Anim Behav Monogr 5:197–283.

Myrberg AA Jr (1980) Fish bio-acoustics: its relevance to the "not so silent world". Environ Biol Fish 5:297–304.

Myrberg AA Jr (1981) Sound communication and interception in fishes. In: Tavolga WN, Popper AN, Fay RR (eds) Hearing and Sound Communication in Fishes. New York: Springer-Verlag, pp. 395–425.

Myrberg AA Jr, Riggio RJ (1985) Acoustically mediated individual recognition by a coral reef fish (*Pomacentrus partitus*). Anim Behav 33:411–416.

Myrberg AA Jr, Kramer E, Heinecke P (1965) Sound production by cichlid fishes. Science 149:555–558.

Myrberg AA Jr, Spanier E, Ha SJ (1978) Temporal patterning in acoustical communication. In: Reese E, Lighter F (eds) Contrasts in Behavior. New York: Wiley, pp. 138–179.

Myrberg AA Jr, Mohler M, Catala JD (1986) Sound production by males of a coral reef fish (*Pomacentrus partitus*): its significance to females. Anim Behav 34:913–923.

Myrberg AA Jr, Ha SJ, Shamblott MJ (1993) The sounds of bicolor damselfish (*Pomacentrus partitus*): predictors of body size and a spectral basis for individual recognition and assessment. J Acoust Soc Am 94:3067–3070.

Narins PM (1982) Effects of masking noise on evoked calling in the Puerto Rican coqui (Anura: Leptodactylidae) J Comp Physiol 147:439–446.

Narins PM, Capranica RR (1978) Communicative significance of the two-note call of the treefrog, *Eleutherodactylus coqui*. J Comp Physiol 127:1–9.

Nelson K (1964) The evolution of a pattern of sound production associated with courtship in the characid fish, *Glandulocauda inequalis*. Evolution 18:526–540.

Ovaska K, Hunte W (1992) Male mating behavior of the frog *Eleutherodactylus johnstonei* (Leptodactylidae) in Barbados West Indies. Herpetologica 48:40–49.

Packard A (1960) Electrophysiological observations on a sound-producing fish. Nature 187:63–64.

Passmore NI, Telford SR (1981) The effect of chorus organization on mate localization in the painted reed frog (*Hyperolius marmoratus*). Behav Ecol Sociobiol 9:291–293.

Passmore NI, Capranica RR, Harned GD (1985) Discrimination of frequency-modulated sounds by the frog *Kassina senegalensis*. HAA Newsletter 6:10.

Passmore NI, Bishop PJ, Caithness N (1992) Calling behaviour influences mating success in male painted reed frogs *Hyperolius marmoratus*. Ethology 92:227–241.

Paulsen K (1967) Das Prinzip der Stimmbildung in der Wirbeltierreihe und beim Menschen. Frankfurt: Akad Verl Ges.

Perrill SA, Gerhardt HC, Daniel RE (1982) Mating strategy shifts in male green treefrogs (*Hyla cinerea*)—an experimental study. Anim Behav 30:43–48.

Petersen CW (1995) Male mating success and female choice in permanently territorial damselfishes. Bull Mar Sci 57:690–704.

Platz JE (1993) *Rana subaquavocalis*, a remarkable new species of leopard frog (*Rana pipiens* complex) from southeastern Arizona that calls under water. J Herpet 27:154–162.

Prestwich KN (1994) The energetics of acoustic signaling in anurans and insects. Am Zool 34:625–643.

Putnam R, Bennett A (1981) Thermal dependence of behavioral performance of anuran amphibians. Anim Behav 29:502–509.

Rand AS, Dudley R (1993) Frogs in helium: the anuran vocal sac is not a cavity resonator. Physiol Zool 66:793–806.

Rauther M (1945) Über die Schwimmblase und die zu ihr in Beziehung tretenden somatischen Muskein be den Triglidae und anderens Scleroparei. Zool Jb Abt Anat 69:159–250.

Recchia CA, Read AJ (1989) Stomach contents of harbour porpoises, *Phocoena phocoena* (L.), from the Bay of Fundy. Can J Zool 67:2140–2146.

Renaud JM, Stevens ED (1983) A comparison between field habits and contractile performance of frog and toad sartorius muscle. J Comp Physiol 151:127–131.

Ritke ME, Semlitsch RD (1991) Mating behavior and determinants of male mating success in the gray treefrog *Hyla chrysoscelis*. Can J Zool 69:246–250.

Rogers PH, Cox M (1988) Underwater sound as a biological stimulus. In: Atema J, Fay RR, Popper AN, Tavolga WN (eds) Sensory Biology of Aquatic Animals. New York: Springer-Verlag, pp. 131–149.

Rosen M, Lemon RE (1974) The vocal behavior of spring peepers *Hyla crucifer*. Copeia 940–950.

Runkle LS, Wells KD, Robb CC, Lance SL (1994) Individual, nightly, and seasonal variation in calling behavior of the gray tree frog, *Hyla versicolor*: implications for energy expenditure. Behav Ecol 5:318–325.

Ryan MJ (1983a) Frequency modulated calls and species recognition in a neotropical frog. J Comp Physiol 150:217–221.

Ryan MJ (1983b) Sexual selection and communication in a neotropical frog, *Physalaemus pustulosus*. Evolution 37:261–272.

Ryan MJ (1985) The Tungara Frog. Chicago: University of Chicago Press.

Ryan MJ (1986) Synchronized calling in a treefrog (*Smilisca sila*). Brain Behav Evol 29:196–206.

Ryan MJ (1988) Energy, calling and selection. Am Zool 28:885–898.

Ryan MJ (1991) Sexual selection and communication in frogs. Trends Ecol Evol 6:351–355.

Ryan MJ, Keddy-Hector A (1992) Directional patterns of female mate choice and the role of sensory biases. Am Nat 139:S4–S35.

Ryan MJ, Rand AS (1993) Species recognition and sexual selection as a unitary problem in animal communication. Evolution 47:647–657.

Ryan MJ, Sullivan BK (1989) Transmission effects on temporal structure in the advertisement calls of two toads, Bufo woodhousii and *Bufo valliceps*. Ethology 80:182–189.

Ryan MJ, Wilczynski W (1991) Evolution of intraspecific variation in the advertisement call of a cricket frog (*Acris crepitans* Hylidae). Biol J Linnean Soc 44:249–271.

Ryan MJ, Cocroft RC, Wilczynski W (1991) The role of environmental selection in intraspecific divergence of mate recognition signals in the cricket frog *Acris crepitans*. Evolution 44:1869–1872.

Ryan MJ, Perrill SA, Wilczynski W (1992) Auditory tuning and call frequency predict population-based mating preferences in the cricket frog *Acris crepitans*. Am Nat 139:1370–1383.

Salmon M, Winn HE, Sorgente N (1968) Sound production and associated behavior in triggerfishes. Pac Sci 22:11–20.

Sassoon D, Kelley DB (1986) The sexually dimorphic larynx of *Xenopus laevis*: development and androgen regulation. Am J Anat 177:457–472.

Saucier MH, Baltz DM (1993) Spawning site selection by spotted seatrout, *Cynoscion nebulosus*, and black drum, *Pogonias cromis*, in Louisiana. Environ Biol Fish 36:257–272.

Schlefer EK, Romano MA, Guttman SI, Ruth SB (1986) Effects of twenty years of hybridization in a disturbed habitat on *Hyla cinerea* and *Hyla gratiosa*. J Herpetol 20:210–221.

Schmidt RS (1974) Neural correlates of frog calling: trigeminal tegmentum. J Comp Physiol 92:229–254.

Schmidt RS (1982) Sexual dimorphism in succinic dehydrogenase staining of toad pretrigeminal nucleus. Exp Brain Res 45:449–450.

Schmidt RS (1992) Neural correlates of frog calling production by two semi-independent generators. Behav Brain Res 50:17–30.

Schmidt RS, Kemnitz CP (1989) Anuran mating calling circuits: inhibition by prostaglandin. Horm Behav 23:361–367.

Schmidt-Nielsen K (1993) Animal Physiology. Cambridge, England: Cambridge University Press.

Schneider H (1964) Bioakustische untersuchungen an anemonenfischen der gattung Amphiprion (Pisces). Z Morph Okol Tiere 53:453–474.

Schneider H (1967) Morphology and physiology of sound production in teleost fishes. In: Tavolga WN (ed) Marine Bioacoustics, vol. 2. Oxford: Pergamon Press, pp. 135–158.

Schneider H (1988) Peripheral and central mechanisms of vocalization. In: Fritzsch B, Ryan MJ, Wilczynski W, Heatherington TE, Walkowiak W (eds) The Evolution of the Amphibian Auditory System. New York: Wiley, pp. 537–558.

Schneider K, Eichelberg H (1974) The mating call of hybrids of the fire-bellied toad and the yellow-bellied toad (*Bombina bombina* (L), *Bombina v variegata* (L) Discoglossidae, Anura). Oecologica 16:61–71.

Schwartz JJ (1987) The function of call alternation in anuran amphibians: a test of three hypotheses. Evolution 41:461–470.

Schwartz JJ (1991) Why stop calling? A study of unison bout singing in a neotropical treefrog. Anim Behav 42:565–578.

Schwartz JJ (1993) Male calling behavior, female discrimination and acoustic interference in the Neotropical treefrog *Hyla microcephala* under realistic acoustic conditions. Behav Ecol Sociobiol 32:401–414.

Schwartz JJ (1994) Male advertisement and female choice in frogs: recent findings and new approaches to the study of communication in a dynamic acoustic environment. Am Zool 34:616–624.

Schwartz JJ, Wells KD (1983) An experimental study of acoustic interference between two species of neotropical treefrogs. Anim Behav 31:181–190.

Schwartz JJ, Wells KD (1984) Interspecific acoustic interactions of the neotropical treefrog *Hyla ebraccata*. Behav Ecol Sociobiol 14:211–224.

Shoemaker VH, Hillman SS, Hillyard SD, et al. (1992) Exchange of water, ions, and respiratory gases in terrestrial amphibians. In: Feder ME, Burggren WW (eds) Environmental Physiology of the Amphibians. Chicago: University of Chicago Press, pp. 125–151.

Shpigel M, Fishelson L (1986) Behavior and physiology of coexistence in two species of *Dascyllus* (Pomacentridae, Teleostei). Environ Biol Fish 17:253–265.

Skoglund, CR (1961) Functional analysis of swimbladder muscles engaged in sound production of the toadfish. J Biophys Biochem Cytol Suppl 10:187–199.

Smith HM (1905) The drumming of the drum-fishes (Sciaenidae). Science 22:376–378.

Spanier E (1979) Aspects of species recognition by sound in four species of damselfishes, genus Eupomacentrus (Pisces: Pomacentridae). Z Tierpsychol 51:301–316.

Steinberg JC, Cummings WC, Brahy BD, Spires JYM (1965) Further bio-acoustic studies off the west coast of North Bimini, Bahamas. Bull Mar Sci 15:942–963.

Sullivan BK (1982) Significance of size, temperature and call attributes to sexual selection in *Bufo woodhousei australis*. J Herpetol 16:103–106.

Sullivan BK, Leek MR (1987) Acoustic communication in Woodhouse's toad (*Bufo woodhousii*). II. Response of females to variation in spectral and temporal components of advertisement cells. Behavior 103:16–26.

Sullivan MS (1990) Assessing female choice for males when the male's characters vary during the sampling period. Anim Behav 40:780–782.

Taigen TL, Wells KD (1985) Energetics of vocalization by an anuran amphibian (*Hyla versicolor*). J Comp Physiol 155:163–170.

Tavolga WN (1956) Visual, chemical and sound stimuli as cues in the sex discriminatory behavior of the gobiid fish, *Bathygobius soporator*. Zoologica 41:49–64.

Tavolga WN (1958a) The significance of underwater sounds produced by males of the gobiid fish, *Bathygobius soporator*. Phys Zool 31:259–271.

Tavolga WN (1958b) Underwater sounds produced by two species of toadfish, *Opsanus tau* and *Opsanus beta*. Bull Mar Sci 8:278–284.

Tavolga WN (1962) Mechanisms of sound production in the ariid catfishes *Galeichthys* and *Bagre*. Bull Am Mus Nat Hist 124:1–30.

Tavolga WN (1971) Acoustic orientation in the sea catfish, *Galeichthys felis*. Ann NY Acad Sci 188:80–97.

Tavolga WN (1976) Acoustic obstacle detection in the sea catfish (Arius felis). In: Schuijf A, Hawkins AD (eds) Sound Reception in Fish. Amsterdam: Elsevier, pp. 185–204.

Tavolga WN (ed) (1977) Sound Production in Fishes. Benchmark Papers in Animal Behavior, vol. 9. Stroudsburg, PA: Dowden, Hutchinson, and Ross.

Telford SR (1985) Mechanisms and evolution of intermale spacing in the painted reed frog. Anim Behav 33:1353–1361.

Tower R (1908) The production of sound in the drumfishes, the sea-robins, an the toadfish. Ann NY Acad Sci 18:149–180.

Townsend DS, Stewart MM (1994) Reproductive ecology of the Puerto Rican frog *Eleutherodactylus coqui*. J Herpet 28:34–40.

Trewavas E (1933) The hyoid and larynx of the Anura. Philas Trans R Soc Lond [B] 222:401–527.

Tuttle, MD, Ryan MJ (1982) The roles of synchronized calling, ambient noise and ambient light in the anti-bat-predator behavior of a treefrog. Behav Ecol Sociobiol 11:125–131.

Urick RJ (1983) Principles of Underwater Sound. New York: McGraw-Hill.

Valdiviesio D, Tamsitt JR (1974) Thermal relations of the neotropical frog *Hyla labialis* (Anura: Hylidae). R Ontario Mus Life Sci Occas Papers 26:1–10.

Wagner WE Jr (1992) Deceptive or honest signaling of fighting ability? A test of alternative hypotheses for the function of changes in call dominant frequency by male cricket frogs. Anim Behav 44:449–462.

Walkowiak W (1992) Acoustic communication in the fire-bellied toad: an integrative neurobiological approach. Ethol Ecol Evol 4:63–74.

Walkowiak W, Luksch H (1994) Sensory motor interfacing in acoustic behavior of anurans. Am Zool 34:685–695.

Watson JT, Kelley DB (1992) Testicular masculinization of vocal behavior in juvenile female *Xenopus laevis* reveals sensitive periods for song duration, rate, and frequency spectra. J Comp Physiol [A] 171:343–350.

Wells KD (1977) The social behavior of anuran amphibians. Anim Behav 25:666–693.

Wells KD (1988) The effect of social interaction on anuran vocal behavior. In: Fritzsch B, Ryan MJ, Wilczynski W, Hetherington TE, Walkowiak W (eds) The Evolution of the Amphibian Auditory System. New York: Wiley, pp. 433–454.

Wells KD, Bard KM (1987) Vocal communication in a neotropical treefrog, *Hyla ebraccata*: responses of females to advertisement and aggressive calls. Behavior 101:200–210.

Wells KD, Schwartz JJ (1982) The effect of vegetation on the propagation of calls in the neotropical frog *Centrolenella fleischmanni*. Herpetologica 38:449–455.

Wells KD, Schwartz JJ (1984) Interspecific acoustic interactions of the neotropical treefrog *Hyla ebraccata*. Behav Ecol Sociobiol 14:211–224.

Wells KD, Taigen TL (1986) The effect of social interactions on calling energetics in the gray treefrog (*Hyla versicolor*). Behav Ecol Sociobiol 19:9–18.

Wells KD, Taigen TL, O'Brien JA (1996) The effect of temperature on calling energetics of the spring peeper (*Pseudacris crucifer*). Amphibia-Reptilia 17:149–158.

Wetzel DM, Haerter UL, Kelley DB (1985) A proposed neural pathway of vocalization in South African clawed toads, *Xenopus laevis*. J Comp Physiol 157:749–761.

Whitney CL, Krebs JR (1975) Mate selection in Pacific treefrogs. Nature 255:325–326.

Wilczynski W, McCleland BE, Rand AS (1993) Acoustic, auditory, and morphological divergence in three species of neotropical frog. J Comp Physiol [A] 172:425–438.

Wiley RH, Richards DG (1978) Physical constraints on acoustic communication in the atmosphere: implications for the evolution of animal vocalizations. Behav Ecol Sociobiol 3:69–94.

Winn HE (1964) The biological significance of fish sounds. In: Tavolga W (ed) Marine Bio-Acoustics. New York: Pergamon Press, pp. 213–231.

Winn HE, Marshall JA, Hazlett B (1964) Behavior, diel activities, and stimuli that elicit sound production and reactions to sounds in the longspine squirrelfish. Copeia 413–425.

Withers PC, Campbell JD (1985) Effects of environmental cost on thermoregulation in the desert iguana. Physiol Zool 58:329–339.

Yager DD (1992) A unique sound production mechanism in the pipid anuran *Xenopus borealis*. Zool J Linnean Soc 104:351–375.

Zelick R, Narins PM (1985) Characterization of the advertisement call oscillator in the frog, *Eleutherodactylus coqui*. J Comp Physiol [A] 156:223–229.

Index

Since this volume is highly comparative, the index can get complex in terms of finding specific species or groups of species. Accordingly, terms have generally been indexed under scientific names of animals. Where possible, we have included common names of species to point to the scientific name. Since there is information germane to portions of larger taxa (e.g., Anurans rather than all amphibians, teleosts rather than all bony fishes), readers should look under these broader terms in addition to information about specific species. Broad terms, such as Ear, Fish, Amphibian, are only used as main indexing terms in a limited way in order for these words not to include most every item in this book.

DATE DUE

JUN 1 1 2007			